AUTOCAD® DRAFTING

Don H. Grout

Paul J. Resetarits

Jody A. James

GLENCOE

McGraw-Hill

New York, New York   Columbus, Ohio   Mission Hills, California   Peoria, Illinois

AutoCAD Drafting is a work-text for those who wish to learn how to use AutoCAD in their drafting endeavors. AutoCAD is a computer-aided drafting and design package produced by Autodesk, Inc. For information on how to obtain the AutoCAD software, contact Autodesk at 1-800-964-6432 or 415-332-2344.

AutoCAD and AutoLISP are registered trademarks of Autodesk, Inc.
AME is a trademark of Autodesk, Inc.

Send all inquiries to:
Glencoe/McGraw-Hill
3008 W. Willow Knolls Drive
Peoria, IL 61614-1083

ISBN 0-02-677135-7 (Work-text)
ISBN 0-02-677136-5 (Instructor's Guide)
ISBN 0-02-677137-3 (Disk)

Printed in the United States of America

1 2 3 4 5 6 7 8 9 10 RRW 98 97 96 95 94

Acknowledgments

The publisher and authors gratefully acknowledge the cooperation and assistance received from many persons and companies during the development of *AutoCAD Drafting*. Individuals and corporations who provided illustrations are listed in the credits in the back of the book. Special recognition is given to the following:

Autodesk, Inc.
Sausalito, California

Reviewers

Ron Shea
Quality Corporation
Rochester, Washington

Helen Everts
Peoria, Illinois

Contributing Writer

James J. Kirkwood
Department of Industry and Technology
Ball State University
Muncie, Indiana

Table of Contents

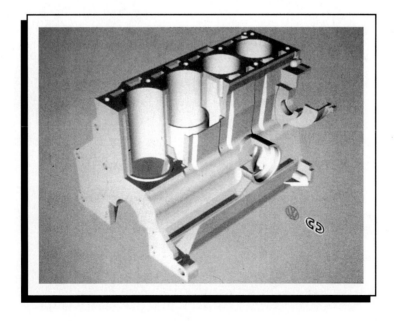

"Using CAD" Features

Introduction

AutoCAD Drafting is a work-text that combines drafting instruction with instruction on how to use AutoCAD to perform basic drafting procedures. The work-text is written primarily for AutoCAD Release 12 for DOS, but it also can be used with Release 10, Release 11, and Release 12 for Windows. No knowledge of drafting or AutoCAD is assumed, although some prior computer knowledge may be helpful. The purpose of the work-text is to provide a single source for people who wish to learn CAD drafting techniques along with basic drafting principles.

Format

AutoCAD Drafting provides a unique format developed especially for teaching CAD drafting techniques. For example, after the introductory chapter, two chapters are devoted to manual drafting techniques. These chapters build a framework that will help students understand the computer-assisted techniques taught later in the book. They also expose students to techniques and skills they may need on the job but that are not necessarily used in CAD-based drafting.

Chapters 4 through 7 introduce basic AutoCAD commands and explain their application in drafting. By working through these chapters, students become familiar with the AutoCAD software and begin to understand how it can be used effectively in drafting. Beginning with Chapter 8, the contents of the work-text parallel the contents of standard drafting textbooks.

The important difference is that this text explains the drafting procedures from a CAD perspective. CAD procedures for creating views, adding text and dimensions to drawings, and developing working drawings are included in these chapters. Mechanical and structural engineering applications are discussed, as well as architectural drafting and three-dimensional modeling.

Each major section in the work-text includes at least one hands-on activity to help students understand the concepts taught in the section. Some of the more complex sections include five or more activities. "Your Turn" activities present step-by-step instruction whenever new commands and techniques are required to accomplish the drafting task.

Features

Other helpful features in *AutoCAD Drafting* include:
- Notes and "Hot Tip" pointers are located throughout the work-text to provide students with additional help. Notes give further information about a topic that may be useful under certain circumstances. "Hot Tips" describe how to perform tasks in the most efficient manner. These tips are useful because AutoCAD is an extensive software program in which there is often more than one way to accomplish the same task. Some methods are easier and faster, and sometimes even more accurate, than others.
- Icons appear within the text whenever a procedure is different in Release 10 or Release 11. Cross-references in Appendixes B and C describe the differences.

For each icon in the work-text, an entry is present in the appropriate appendix. This icon system allows users of Releases 10 and 11 to use the work-text easily.

- Questions and problems at the end of each chapter help ensure that students have met the chapter objectives.
- Feature stories at the end of each chapter demonstrate various uses of AutoCAD in the commercial and industrial sectors. The stories are designed to expose students to a wide range of applications that may be beyond the scope of a single drafting course.
- A Pictorial Glossary at the end of the work-text clearly explains new terms using graphics as well as text.
- Appendixes provide useful tables and ANSI drafting standards as well as cross-references for AutoCAD Release 10 and Release 11.

Disk

A disk is available for use with *AutoCAD Drafting.* The disk comes in two sizes — 5¼" and 3½". (Both sizes are included in the package.) The disk includes symbol libraries, prototype drawings, and drawing practice problems. Extra practice is provided for all fifteen CAD chapters. To order the disk, call 1-800-334-7344.

Instructor's Resource Guide

An Instructor's Resource Guide is also available to accompany *AutoCAD Drafting.* The guide includes information on the AutoCAD certification examination, AutoCAD configuration, hardware options, an introduction to DOS, teaching strategies, answers to chapter questions, chapter tests and answers, visual masters, activity handouts, and diskette drawings. Instructors may order the guide at 1-800-334-7344.

To the Student

This book is designed to provide you with a comprehensive explanation of the use of the AutoCAD software in drafting applications. By reading each chapter and doing the related activities, you will learn both standard drafting techniques and AutoCAD drafting procedures.

Certain features and conventions have been used in this work-text to make it easier to use. Understanding these features and conventions will help you get the most benefit from the work-text.

Features

- If you are using AutoCAD Release 10 or Release 11, look for the R10 and R11 icons in the margins of the page. These icons appear whenever you must follow a different procedure in those releases to accomplish the task. When you see an icon that applies to your release of AutoCAD, turn to Appendix B (for Release 10) or C (for Release 11). In the appendix, find the chapter and page number that corresponds to the icon in the text. The appendix entry explains the difference and tells how to perform the procedure.

- When you encounter a term that you do not understand, turn to the Pictorial Glossary at the end of the work-text. The Pictorial Glossary contains many drafting terms and illustrations to help you understand the terminology used in the work-text.
- Be sure to read the Notes and "Hot Tip" pointers included in each chapter. These features will provide you with additional information that you may need to perform the procedure correctly.
- Questions and problems are provided at the end of each chapter to help you check your understanding of the chapter material.

Conventions

- Text that appears at the command line in AutoCAD appears in *italic type.*
- Text that you must enter at the keyboard appears in **bold type**.
- All AutoCAD command names and variable names appear in uppercase letters. For example, to create a line in AutoCAD, you would use the LINE command.

AutoCAD Reference Manuals

When you (or your school) purchase the AutoCAD software from Autodesk, Inc., you also will receive several reference manuals. These reference manuals are a valuable resource. Refer to them whenever you want information about commands or procedures that are beyond the scope of this text.

Chapter

1

Key Terms

drafting

communication

graphic
    communication

pictograms

ideograms

hieroglyphics

Computer-Aided
    Drafting (CAD)

cartographers

technical illustrators

architects

Introduction to Drafting

Objectives

When you have completed this chapter, you will be able to:
- describe graphic communication and drafting.
- understand some of the careers available to people with a drafting background.
- discuss drafting careers and the education and experience required for these careers.
- apply a seven-step decision-making process to career planning.

"What will you do when you graduate?"
"Get a job."
"Doing what?"
"Drafting."
"What's drafting?"

Drafting is the use of pictures to describe an object precisely so that it can be constructed. Drafting forms a visual bridge between an idea and its creation.

Graphic Communication

We communicate with people every day, and we frequently complain because others don't communicate with us. In its most general form, **communication** means having a relationship with others. We communicate mentally, and we communicate physically. We communicate by talking, and we communicate with pictures. Communication that relies on pictures is known as **graphic communication**. But what about communicating with written words, such as the words in this text? Written words are a form of pictures, too. Books, magazines, and newspapers are types of graphic communication — even if they don't contain illustrations!

Background

Drawing as a form of communication goes back into early history. Cave dwellers drew on their walls, and the Egyptians recorded events in pictures, or **pictograms**, on their temples *(Figs. 1-1 and 1-2)*. Pictograms developed into more abstract **ideograms**, variations of which appear as modern highway signs *(Fig. 1-3)*. The pictogram for water progressed through the Egyptian hieroglyph to become the Roman letter M *(Fig. 1-4)*.

Early languages such as **hieroglyphics** also came from pictograms. Fig. 1-5 shows the hieroglyphic text for the name of an Egyptian king, Ptolemy. Japanese, Chinese, and other languages currently use pictures to convey ideas *(Fig. 1-6)*. The American Sign Language is a modern example of pictures used to communicate with people who are deaf *(Fig. 1-7)*.

Fig. 1-1 Drawings on the walls of caves were an early example of communication using pictures.

Fig. 1-2 Early Egyptians recorded events as pictograms on temple walls.

Fig. 1-3 Abstract ideograms evolved from pictograms. Highway signs are examples of ideograms in modern use.

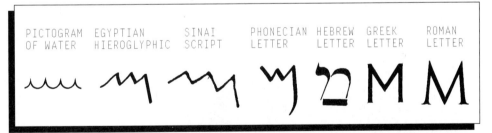

Fig. 1-4 The early pictogram for water developed through several languages and societies to become the Roman letter M.

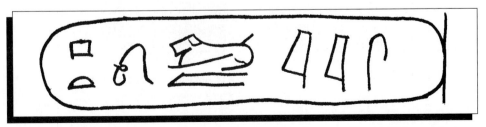

Fig. 1-5 Hieroglyphics were used to communicate in early Egypt. This symbol stands for the Egyptian King Ptolemy.

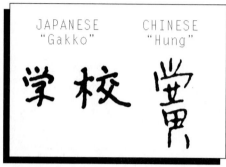

Fig. 1-6 The Japanese and Chinese written languages are based on ideograms. Above are the Japanese and Chinese characters for the word *school.*

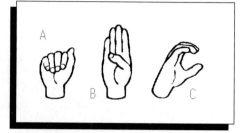

Fig. 1-7 The American Sign Language uses pictorial language, too. These characters represent the letters A, B, and C.

Drafting

Sketches by Leonardo da Vinci and later by Thomas Jefferson and others led to the use of drawings for developing designs and for construction *(Figs. 1-8 and 1-9).* The precise drawings that took the place of sketches for many applications were said to be *drafted.* Soon drawing instruments and drafting machines became common, and organizations were formed to standardize drafting. Two such organizations are the American National Standards Institute (ANSI) and the International Standards Organization (ISO). ANSI Y14 is the set of standards most often followed by drafters. Individual companies also design formats to meet their own needs.

 Refer to Appendix G, "Reference Tables and Standards," for more information about the ANSI Y14 standards.

Drafting has become an international language of graphic communication used in industry. Drawings created in the United States can be understood in France, Japan, Egypt, Sweden, Mexico, and most other countries *(Fig. 1-10).* The text may need an interpreter, but the lines, layout, style, and general techniques have been standardized. It is important for you, the drafter, to know and follow the accepted techniques and standard practices of drafting.

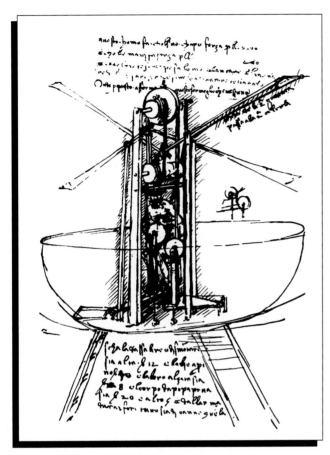

Fig. 1-8 This sketch of a helicopter was made by Leonardo da Vinci (1452-1519). Note the detail and the accuracy of the sketch. The person in the center powers the rotating wings.

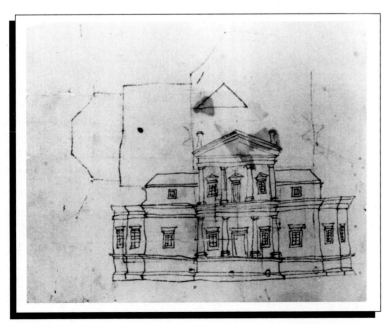

Fig. 1-9 Thomas Jefferson drew this sketch for Monticello, his home, in 1770.

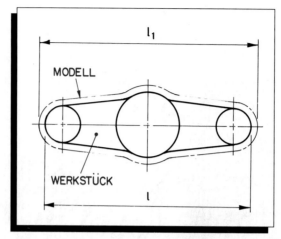

Fig. 1-10 Even though an American drafter may not understand the German on this drawing, he or she can understand the drawing because the drafter used internationally recognized standards.

Computer-Aided Drafting

Computerized drafting began with the mainframes of the 1950s. However, it took the microcomputer revolution of the late 1970s and early 1980s to develop **Computer-Aided Drafting (CAD)** as a practical tool for industry. CAD is an extremely powerful drawing tool which offers a tremendous advantage over manual drafting in many respects. For example, it allows drafters to create and modify precise drawings quickly and easily.

Drafting in Transition

Drafting is now in a state of transition. Although many companies have adopted CAD exclusively, others use manual drafting or manual/CAD combinations *(Figs. 1-11 and 1-12)*. Ultimately CAD will replace manual drafting; CAD skills are necessary for your drafting future. During the transition, however, you will need to be familiar with manual drafting techniques and procedures.

This text teaches the fundamentals of computer-aided drafting. It also introduces you to manual drafting. Study and practice both methods, because in today's job market you will need every edge available to you.

Fig. 1-11 Many companies now use CAD exclusively for drafting tasks. The drafter in this room is working on a detailed drawing of a machine part.

Fig. 1-12 Some companies find it more convenient to use all manual or a combination of CAD and manual drafting.

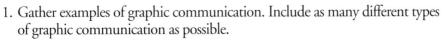

ACTIVITY

To familiarize yourself further with the ideas presented in this section, do the following activities.

1. Gather examples of graphic communication. Include as many different types of graphic communication as possible.

2. Use your examples to support an oral report on the topic.

Drafting as a Career

When you consider drafting as a career, you should consider the following:

- the personal traits required for a drafter
- the education you will need to begin
- the education and experience necessary for advancement

Drafting requires neatness, accuracy, and speed. It involves a love of detail and a knowledge of computer operation. Drafters also need a basic knowledge of mathematics, science, and English. They must know both the materials and the processes used in their chosen industry to manufacture the drafted parts.

Drafters usually enter the field as beginners or trainees. Depending on their interests, skills, and education, they may follow one of several paths of professional growth. The American Design Drafting Association (ADDA) recognizes three professional groups: drafters, designers, and checkers. The hierarchy rises from Drafter I through designers to Checker II. Each level requires more education and/or experience. At the top of the profession is the Principal Designer. In addition to these professional positions, the ADDA lists two levels of supervision: Design/Drafting Supervisor and Design/Drafting Manager.

Drafters

Professional positions for drafters include Drafter I, Drafter II, and Drafter III. Drafter I positions are beginning or entry level positions that require little experience. This position may also be posted as Junior Drafter or Drafter Trainee. People working in this position work under direct supervision. They provide drafting support and revise drawings as directed. Their duties and responsibilities include the following:

- preparing detailed drawings
- revising and updating drawings
- repairing or redrawing old or damaged drawings
- learning company and industrial drafting standards
- developing drafting skills

Drafter I positions require two years of post-secondary education. High-school graduates with one year of algebra and one year of geometry may also be accepted.

The next professional drafting level is Drafter II, also called Drafter or General Drafter. This position forms the middle level of the drafter group. Its duties are more and its responsibilities are greater than those of the Drafter I position. Under normal supervision, drafters in this position are expected to develop drawings of varying levels of complexity. They must comply with company and industrial standards and practices. They should also be able to read and interpret reference materials, specifications, and technical publications and standards. People in this position also perform preliminary layout. They may sometimes be called upon to make freehand sketches of designs and ideas.

To be a Drafter II, you must have a general understanding of established drafting standards and procedures. Most positions require a minimum of two years of post-secondary education and two years of drafting experience. High-school graduates with six years of drafting experience also meet the general criteria for Drafter II.

Drafter III, Senior Drafter, and Design Drafter are titles given to people at the upper level of the drafting group. These people work with minimum supervision. Their primary purpose is to provide drafting and limited design support in the development of drawings and sketches.

People in Drafter III positions must be able to do all of the tasks required for Drafter I and Drafter II positions. They also assist in the preliminary layout of complex assignments. They are often called upon to make freehand sketches of designs and ideas. They may also prepare design modifications from sketches and data. In addition, these drafters are qualified to review and check the work of others.

To become a Drafter III, you must have a demonstrated knowledge of established drafting standards and procedures, as well as two years of post-secondary education and five years of drafting experience. High-school graduates with nine years of drafting experience are also accepted at this level.

Designers

Professional designer positions include Designer I and Designer II. Designer I employees (also known as Designers or Senior Design Drafters) develop designs and drawings and provide technical design support. They work independently with general supervision. They develop designs of varying technical complexity and develop preliminary layouts. They also prepare design modifications and make freehand sketches of designs and ideas. They may also assist in the development of drawing specifications for the manufacture of parts.

Designer I job candidates must have a demonstrated knowledge of drafting procedures. They must also have a demonstrated ability to communicate ideas, both verbal and written. This position requires four years of post-secondary education and four years of drafting experience or two years of post-secondary education and eight years of drafting experience. High school graduates with 12 years of drafting experience may also fill this position.

The Designer II classification is the next rung on the career ladder. In some companies, this position may also be called Senior Designer or Senior Engineering Designer. The responsibilities for Designer II are basically the same as those for Designer I. However, people in Designer II positions are expected to be more experienced, and they work with minimal supervision.

Candidates for the Designer II position must have a sound knowledge of drafting standards and procedures. They should have a demonstrated ability to communicate ideas, both verbal and written. They must also have four years of post-secondary education and six years of design/drafting experience or two years of post-secondary education and ten years of design/drafting experience. A high-school graduate with 14 years of drafting experience may also fulfill the requirements for this position.

Checkers and Principal Designer

The Checker I, Checker II, and Principal Designer positions are the top three professional design/drafting positions. People in these positions require less supervision than drafters or designers. Their responsibilities are greater, and their duties are more complex. In most cases, people who fill these positions also have more education and/or experience than people in other drafting jobs.

Supervision

Design/Drafting Supervisors and Managers oversee and run design or drafting departments and organizations. Although they need the knowledge and skills associated with design and drafting, their primary duties and responsibilities are supervisory. In fact, many professionals prefer to stay in drafting or design positions rather than to move into management because of the different duties, responsibilities, and skills required.

Drafting tasks are important for many reasons. One of the biggest responsibilities of drafters is to prevent manufacturing errors. For example, what would the consequences be if the Jarvis heart valve were manufactured incorrectly? Drafting plays an important part in quality control.

Your Turn

ACTIVITY

To find out more about the ideas presented in this section, do the following activities.

1. Read the classified ads in your local newspaper for one week or contact firms that do drafting work in your area. Note available drafting positions of any type.
2. Create a chart of the drafting/design positions and the qualifications needed. What other information do the ads or contact people give about the positions?

Drafting-Related Fields

Although drafting is a career in itself, many people use their drafting skills in other related fields. They may use their skills to read and interpret drawings, make sketches, inspect parts, and build products. Examples of people whose careers require drafting knowledge include engineers, technicians, assemblers, electricians, architects, inspectors, instructors, welders, contractors, and many more. These people study drafting to learn the standards and techniques which will enable them to communicate graphically.

Drafting does not exist in a vacuum. It always relates to an industry or trade. An architectural drafter may not be a contractor, but he or she knows "how the sticks go together" *(Fig. 1-13)*. A mechanical drafter understands the difference between fusion welding and resistance welding *(Fig. 1-14)*. New materials, processes, and procedures continually challenge drafters and change their methods. Plastic pipes for building and integrated circuits for electronics are examples of innovations which have changed the drafting world.

Fig. 1-13 An architectural drafter understands building construction as well as drafting techniques.

Fig. 1-14 Mechanical drafters understand the application of proper mechanical materials and techniques. This drafter is placing the appropriate welding symbols on a drawing.

In other occupations, people use their drafting skills to create special kinds of drafted documents. Some examples are given in this section. Talk with your drafting instructor or guidance counselor about opportunities in these and other related fields which may be of interest to you.

Cartographers are specialized drafters who prepare maps. They frequently work from aerial photographs, depth charts, and surveys. Their job is to transform these materials into road maps, aeronautical charts, topographical maps, and specialized area maps and charts.

Technical illustrators create drawings to illustrate ideas or products. Illustrations done by technical illustrators differ from other illustrations in precision and accuracy. Technical illustrations are much more precise and accurate than other types of illustrations. Many of the drawings in this text were done by technical illustrators.

Drafting instructors come from a variety of backgrounds, but most have both drafting experience and a college degree. Specific skills and educational requirements depend upon the level of drafting taught. Drafting is usually taught in junior/senior high schools, vocational/technical schools, post-secondary schools, and colleges.

Architects design buildings, construction projects, and landscapes. They also prepare sketches for architectural drafters. Some architects, especially in small firms, actually draft their designs.

ACTIVITY

To learn more about the ideas in this section, do the following activities.

1. With a group of five classmates, brainstorm to create a list of non-drafting careers in which people need drafting skills. Include as many careers as possible that are not mentioned in this text.

2. Choose one of these drafting-related careers to find out more about: cartographers, technical illustrators, drafting instructors, and architects. Write a short summary about the career.

Career Planning and Research

When you meet a new subject and it "clicks," your first reaction may be: "This is fun! I'd like to be a ...!" Then reality sets in: "This is tough!" But if it's still "fun," then maybe the new subject warrants serious career consideration. The trouble is, there may be several "fun" subjects, and there are many places in the world of work where you may fit in. How do you decide what to do? One way to begin is to do some formal career planning and research.

Careers and Your Life-Style

Your future life-style will be determined to a large extent by the work you do. Your job will probably determine your standard of living, your location, your friends, and the amount of time you spend with your family. When other people know what you do, they may be able to guess a great deal about you. They may be able to guess how and where you live, how much you earn, how much education you have, and so on.

Career Planning

For some students, employment is by chance. It is just something to do when school is over. For others, employment comes as a result of careful planning and well-thought-out decisions. Since your work will continue for most of your adult life, influence your life-style, and provide your identity, it is important that you make your decision carefully. The decision-making process described by Kimbrell and Vineyard in their book *Succeeding in the World of Work* can help you make the correct choice *(Fig. 1-15)*. Following these steps can help you make sure you don't overlook any possibilities. These steps can also help prevent you from making a career decision that is not right for you.

SEVEN-STEP DECISION-MAKING PROCESS

STEP 1: Define your needs or wants. What kind of life-style do you want 5, 10, or 20 years from now?

STEP 2: Analyze your resources. Your resources are what make you who you are: your personality, values, skills, interests.

STEP 3: Identify your choices. Which careers will help you achieve your needs or wants?

STEP 4: Gather information. Research the careers you identified in Step 3. Investigate other careers and areas of interest.

STEP 5: Evaluate your choices. Check your career information against your resources. Look for careers that suit your proposed life-style.

STEP 6: Make a decision. Choose your career.

STEP 7: Plan the steps necessary to reach your career goal.

Fig. 1-15 The seven-step decision-making process, as described by Kimbrell and Vineyard in *Succeeding in the World of Work*.

Career Research

Traditional methods of career research such as job fairs, literature reviews, and personal interviews may not provide enough information to help you reach an intelligent decision. Changing technology rapidly changes employment. Not only are there more choices, but there are more variations of the choices. There are new careers and new specialties within existing careers. To help students investigate possible careers, many schools have established internships, apprenticeships, job shadowing, cooperative educational programs, and other paid and unpaid work experiences. In addition, guidance counselors, cooperative education coordinators, and teachers often have special arrangements with local employers for visits, seminars, and other forms of career research and investigation. Don't be afraid to ask; after all, it's your future!

Traits of a Good Employee

A positive work attitude plays as important a role in career development as specific job skills. Some motivational experts have gone so far as to say that "attitude determines aptitude." Look at the people you know: your family, friends, other students, teachers. Are they successful in what they do? Are they happy in their jobs? Are they well liked by others in their workplace? If your answers are "yes," these people probably have a positive work attitude.

Initiative and responsibility go hand-in-hand with attitude as traits that will help you keep your job once you have landed it. When giving raises and promotions, employers look for those who "go the extra mile" and do more than is required. They want people who accept responsibility and ask for more. They want people who get along well with others, people who are ambitious and take pride in their work. They want people who are on time and don't miss work. In short, they want people who possess all the traits of a good employee.

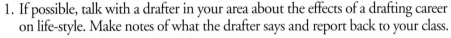

ACTIVITY

To learn more about the ideas presented in this section, do the following activities.

1. If possible, talk with a drafter in your area about the effects of a drafting career on life-style. Make notes of what the drafter says and report back to your class.

2. Use the seven-step decision-making process to investigate career opportunities that interest you. Make notes on what you discover at each step.

Chapter 1 Review

1. Describe the relationship between graphic communication and drafting.
2. What two organizations standardize drafting procedures?
3. Name at least one advantage of CAD over manual drafting.
4. Describe the personal traits a drafter should have.
5. What three professional drafting groups are recognized by the American Design Drafting Association (ADDA)?
6. Explain the basic difference between a drafter and a designer.
7. Name at least three non-drafting careers in which people must use drafting skills.
8. Describe the effects a career can have on your life-style.
9. Briefly describe the seven-step decision-making process explained in this book.

Chapter 1 Problems

1. Visit your school's media center or a public library in your area to find out more about ancient languages and the pictures upon which many of them were based. Collect examples of pictograms, ideograms, and hieroglyphics.
2. Find an example of an architectural drawing done by Thomas Jefferson. Compare that drawing with a modern architectural plan done by a drafter. What differences do you see? In what ways are the two plans the same?
3. Locate companies in your area that employ drafters. Find out how many of them use manual drafting and how many use CAD. Do any companies currently use both? Make a chart of your findings.
4. Contact high schools and community colleges in your area to find out what types of drafting courses they offer. Make a list of available courses.
5. Visit your school's media center or a public library in your area. Find out what resources are available to you for investigating career options.

Using CAD
Drafting Student Discovers Career

Some people plan their careers; other people discover theirs. As an architect who now owns his own firm, Rob Axton discovered his. His high school experiences influenced him. He fondly remembers his high school drafting class: "I had to take this drafting class, and I absolutely loved it," he said. But Axton didn't see drafting in his future. Instead, he was deeply interested in astronomy and fine arts.

In college he studied astrophysics, far removed from art and drafting. To support himself, Axton worked part-time for architects and civil engineering firms. That's how he improved his high school drafting skills. "I watched other people," said Axton. "For example, from older, skilled drafters I learned how to use a compass to draw inked lines on linen."

Still, Axton, hadn't discovered his future career. Not until he studied architecture as a graduate student did Axton see how his skill in drafting could help him. His creative mind envisioned homes and other building. "Knowing drafting was necessary to get my designs on paper so someone could build from them," he said.

Axton recorded his designs on paper and then later on a computer. For Axton, drafting on the board or on the computer is a way of communicating an idea that he or somebody else can bring to life. He found a "nice transition" from board drafting to CAD drawing. "Quite frankly, it's just substituting one tool for another," he said.

Axton's architectural firm is not large — he employs 30 to 40 professional people. Almost all of them do their own drafting on the computer. In larger firms with specialized workers, drafters without architectural training are part of the design team, working under the supervision of project designers. The individual drafter does not have to be an architect or a designer. As one of the first architectural firms to use AutoCAD, Axton wanted all his architects to be their own drafters. In addition, Axton expects all his employees to have CAD skills. He explains, "My philosophy is, one machine, one architect." All architects at Axton's firm quickly learn to use AutoCAD competently because they apply AutoCAD to real problems every day.

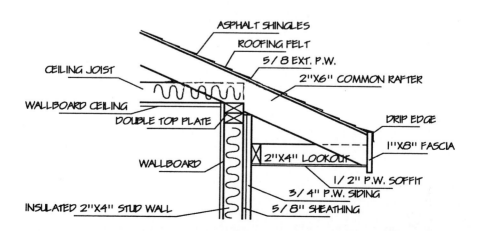

OVERHANG DETAIL

Chapter 2

Key Terms

sketching	proportions
inclined lines	apparent dimensions
axis	actual dimensions
origin	view
arc	pictorial
ellipse	oblique sketch
major axis	isometric sketch
minor axis	perspective sketches

Sketching

Objectives

When you have completed this chapter, you will be able to:

- sketch lines used in technical drawing.
- sketch rectangles, squares, circles, arcs, and ellipses.
- estimate the proportions of an object.
- determine how many views are required to describe the shape of an object completely.
- identify oblique, isometric, and perspective sketches.

Y ou find it difficult to describe an idea, concept, or modification adequately with words. You need a drawing, but a computer is not available. What to do? Make a sketch!

Sketching is simply freehand drawing; it is a quick way to create a visual bridge between conception and reality. Some sketches are crudely made, while others are highly sophisticated. With enough practice, your sketches may take on the characteristics of an engineering drawing *(Fig. 2-1)*.

Understanding sketching is the first step to understanding drafting. Ideas become sketches, and sketches form the basis of designs. Sketching, like all drawing, is a form of visual communication. The message must be transmitted accurately and efficiently.

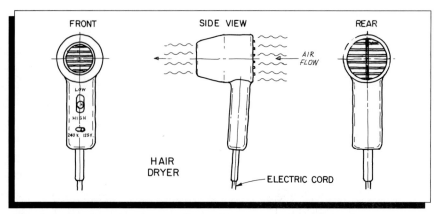

Fig. 2-1 Precise sketches such as this one are the result of a great deal of practice.

Materials

To sketch, you need something to draw with and something to draw on. In its broadest sense, sketching takes many forms. An architect may sketch an idea in the dirt for a contractor *(Fig. 2-2)*. On a larger scale, a farmer creates an intricate pattern of furrows on a field *(Fig. 2-3)*. For drafting purposes, however, you need to be more formal and more accurate. Rather than the earth, you will use paper; and rather than a stick or a tractor, you will use a pencil. Add to them an eraser for making corrections, and you have the complete set of sketching tools.

Fig. 2-2 Sometimes you may need to make a sketch when even paper and pencil are not available. This architect is sketching an idea in the dirt for his client, a contractor.

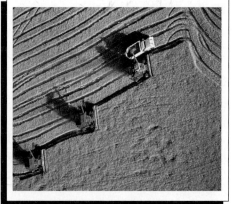

Fig. 2-3 Even the intricate pattern formed by this combine can be considered a "sketch."

Paper

Although any type of paper can be used, from an envelope to a lunch bag, formal sketching is usually done with plain, unlined paper. You can use graph paper for greater ease and accuracy. However, it is not always available, and as a beginning drafter, you should learn not to rely on it.

Pencils

As with paper, almost any kind of pencil can be used for sketching. Formal sketching, however, requires the use of high-quality pencils or leads. Lead is graded from 9H, the hardest, to 7B, the softest.

Use a pencil or lead holder with medium soft F or HB lead for sketching. The sharpness of the point determines the type of line drawn *(Fig. 2-4)*.

- Sharp points make thin lines for hidden lines, center lines, dimension lines, extension lines, and construction lines.
- Medium points make medium lines for object lines.
- Dull points make thick lines for borders and cutting planes.

The pencils used for various drafting applications are discussed in more detail in Chapter 3, "Mechanical Drafting."

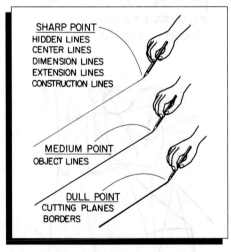

SHARP POINT
HIDDEN LINES
CENTER LINES
DIMENSION LINES
EXTENSION LINES
CONSTRUCTION LINES

MEDIUM POINT
OBJECT LINES

DULL POINT
CUTTING PLANES
BORDERS

Fig. 2-4 The sharpness of the pencil point determines the type of line you draw.

Erasers

Avoid the erasers on the ends of pencils. They are likely to harden and smear your work. Many drafters prefer the pink block erasers with tapered ends. Art gum erasers also work well.

Visit a store in your area that sells drafting supplies or look through a drafting supply catalog. Survey the sketching materials and talk with the proprietor or a salesperson if necessary to complete the following activities.

1. Make a list of the types of pencils available. For what task is each type of pencil recommended? Which pencils are recommended for sketching?
2. Make a list of the types of papers available. Which papers are recommended for sketching? Why?
3. Describe the erasers recommended for sketching. What characteristics make an eraser a good choice for sketching?

Sketching Lines

Freehand lines are never perfectly straight, but you should make every attempt to follow the correct path. Your lines should look sharp and crisp. Sketched lines should not have the rigid look of a mechanical line, but neither should they look inaccurate, wavy, or sloppy. Sketched lines should look sketched *(Fig. 2-5)*.

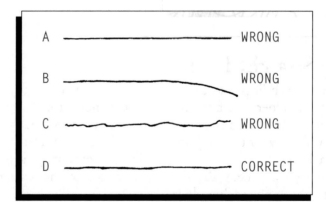

Fig. 2-5 A well-sketched line does not look like a mechanically drawn line (A), but it is not wavy (B) or crooked (C), either. A well-sketched line should look like what it is: a straight line drawn without mechanical aids (D).

Holding the Pencil

There are almost as many ways to hold a pencil and draw a line as there are people holding pencils and drawing lines. Hold the pencil in a way that is comfortable for you, but follow these guidelines:

- Use a relaxed, comfortable grip.
- Maintain the pencil at about a 50- to 60-degree angle *(Fig. 2-6)*.

Drafters use the pencil's point to control line weights and thicknesses. If you hold the pencil too flat, you may lose control of the pencil point and of your line.

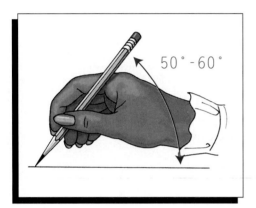

Fig. 2-6 Hold the pencil at a 50- to 60-degree angle for the best results while you are sketching.

Making Straight Lines

When you sketch lines, move your arm from the shoulder, not from the wrist or elbow. Rotate your pencil slightly so it will not wear unevenly and create a tapered line.

You can make short lines with a single pencil stroke. However, long lines may require a series of short lines with or without small spaces between them. Do not overlap the short lines, or you will create fuzzy, sloppy-looking lines *(Fig. 2-7)*.

Another method frequently used for sketching long lines is the point-to-point method *(Fig. 2-8)*. Mark the starting point and the ending point of the line; then move your pencil from the starting point to the ending point. It is like throwing a ball. Keep your eyes on the receiver (ending point), not the ball (pencil tip). Before you actually draw, make a few passes with your pencil off the paper. This will establish the direction of your line and give you a feel for the smooth movement of your pencil over the desired path.

As you become comfortable with sketching, you will learn to use your wrist and extend your fingers to keep the line straight. You will learn with practice to shift your pencil to compensate for the type of line you are drawing. You will learn other techniques that will help you sketch better. Remember your objective, however; your lines must continue in the correct path, and they must be sharp and crisp.

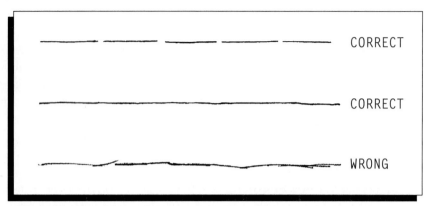

Fig. 2-7 Long lines, when properly drawn, have a neat appearance. Both of the first two line styles are acceptable.

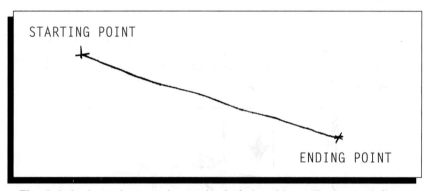

Fig. 2-8 In the point-to-point method of sketching a line, you define the starting and ending points and draw from the starting point to the ending point. For horizontal lines, remember to have your starting points on the left if you are right-handed, or on the right if you are left-handed. The starting point for this line is shown for a right-handed drafter.

▶ Vertical and Horizontal Lines

Draw vertical lines down and horizontal lines from left to right *(Fig. 2-9)*.

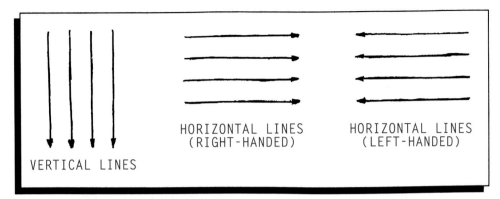

Fig. 2-9 Drawing patterns for vertical and horizontal lines.

If you are left-handed, the opposite is true. You should draw horizontal lines from right to left.

You can draw lines parallel to the edge of your paper by placing a rigid finger against the paper edge as a guide *(Fig. 2-10).*

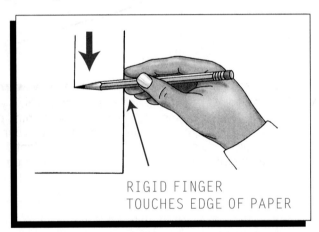

Fig. 2-10 Use your finger as a guide for drawing lines parallel to edge of paper.

▶ Inclined Lines

All lines that are neither horizontal nor vertical are **inclined lines**; that is, they have a nonzero slope. You can draw inclined lines with a shallow slope in the same way that you draw horizontal or vertical lines, depending on the angle of incline. For sharply inclined lines, turn your paper so that the line becomes horizontal or vertical, whichever you prefer *(Fig. 2-11).*

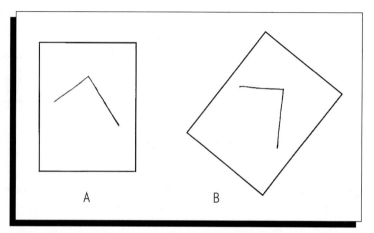

Fig. 2-11 The easiest way to draw the inclined lines shown in A is to turn your paper to the angle shown in B, so that you can draw the lines horizontally or vertically.

ACTIVITY

Practice drawing straight lines by following these steps.

1. Fold a sheet of sketching paper into two halves lengthwise.

2. On the right side of the drawing sheet, practice drawing vertical lines by making several short lines that touch or have small spaces between them, as described on pages 37-38.

3. On the left side of the drawing paper, practice drawing vertical lines using the point-to-point method. First, establish a starting point at the top of the page and an ending point at the bottom of the page. Then draw the line as described on pages 37-38.

4. On either side of the same paper, make at least 10 horizontal lines. Remember to start at the left side of the page if you are right-handed, and on the right side if you are left-handed.

5. On a clean sheet of paper, or on the back of the one you just used, sketch the simple diagram shown in Fig. 2-12. Remember to turn your paper to sketch the inclined lines, as described on pages 39-40.

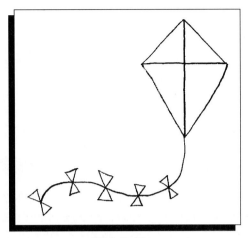

Fig. 2-12

Sketching Geometric Shapes

To sketch geometric shapes such as squares, rectangles, and circles with a fair degree of accuracy, drafters use the construction techniques discussed in this section. To understand these techniques, you must first understand a few basic terms.

An **axis**, for example, is an imaginary line that shows the direction or size of the object. Drafters usually describe two-dimensional objects using two axes *(Fig. 2-13)*. The horizontal axis, or X axis, shows the width of an object. The vertical axis, or Y axis, shows the height of the object. For three-dimensional objects, a third axis, called the Z axis, is added to show the depth of the object. All the axes are at right (90°) angles to each other. The **origin** is the point at which all the axes meet, or intersect.

Additional terms are explained as they are needed. If you forget the meaning of a term, you can always look it up in the Pictorial Glossary near the end of this text.

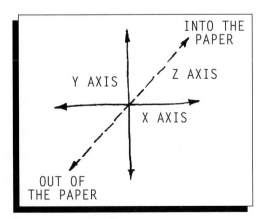

Fig. 2-13 Many sketches are made by imagining (and lightly sketching) axes through the objects to be sketched. The Z axis is used only for three-dimensional sketches. To understand the Z axis, think of it as running from a point somewhere in back of the paper (away from you) to a point in front of the paper (toward you). The Z axis is always placed at a 90-degree angle from the X and Y axes.

Squares and Rectangles

Squares and rectangles may seem to be the easiest geometric shapes to sketch. As you probably know, a square is a four-sided figure in which all four sides are the same length. A rectangle is a four-sided figure in which opposite sides are the same length. There are a couple of ways to make sure your sketched squares and rectangles are approximately the correct size and shape.

ACTIVITY

The first method of drawing a square allows you to locate the middle of the square precisely. Follow these steps to draw a square using this method.

1. Decide where on your sketch paper you want the center of the square to be. With your pencil, make a light mark at that location.

2. Draw horizontal and vertical axes at right angles to each other so that they cross at the mark you made on the paper *(Fig. 2-14A)*.

3. Using a pencil as a measure, make a mark on the X axis to the left of the origin. The distance between the mark and the origin should be half the width of your intended square. Then make a corresponding mark on the right side of the origin. Again, the distance from the origin to the mark should be half the width of your intended square.

4. Repeat the procedure in step three to make two marks on the Y axis. One mark should be above the origin and the other below it. The distance from the origin to each mark should be half the distance of the intended square.

5. Sketch light lines through each point parallel to the axes *(Fig. 2-14B)*.

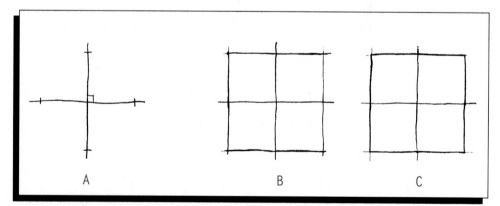

Fig. 2-14 A) Draw the X and Y axes and mark half the width of the square in each direction from the center point. B) Sketch light lines through the points you marked. C) Darken in the square.

Two lines are parallel if every point on the first line is exactly the same distance from the corresponding point on the second line.

6. Darken in the square (*Fig. 2-14C*).

This completes your sketch of a square. Since opposite sides of a rectangle are the same length, you can use the same process to create a rectangle.

There may be times when you need to place a square or rectangle by placing a corner of the object in a precise location. You can do that using the method shown below. Follow these steps to draw a rectangle using this method (*Fig. 2-15*).

1. Decide where on your sketch paper you want the lower left corner of the rectangle to be. With your pencil, make a light mark on the paper.

2. Draw horizontal and vertical axes as you did before. They should intersect, or cross, at the mark you made.

3. From the intersection of the lines, mark the horizontal and vertical distances (width and length) for the sides of the rectangle. Make the rectangle roughly twice as wide as it is high.

4. Draw parallel lines through the marks for the other two sides.

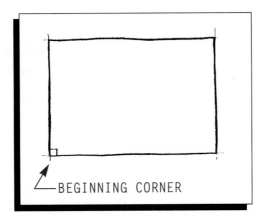

BEGINNING CORNER

Fig. 2-15 In this method, the square or rectangle shares two sides with the X and Y axes. This method is useful when you need to place a corner of a square in a certain location. The axes can form any two sides of the square or rectangle.

Circles and Arcs

Circles and arcs are more difficult to draw than squares and rectangles. (An **arc** is a segment, or part, of a circle.) Most circles and arcs made without any guides end up flat, oblong, or generally distorted. It should come as no surprise that drafters have devised various techniques to help avoid distorted circles and arcs.

One way to draw a circle is to use a piece of paper marked with the circle's radius. Set off the radius at various points from the center of the circle and connect the points *(Fig. 2-16)*. Use as many points as you need to create a smooth circle. Since an arc is part of a circle, you can use this technique to create arcs, too.

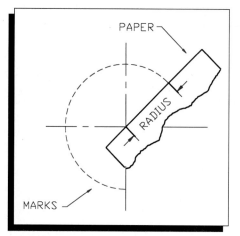

Fig. 2-16 One quick way to create a circle is to mark off the radius of the circle using a strip of paper.

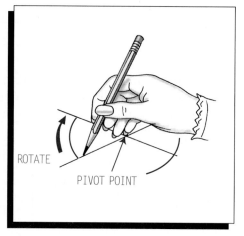

Fig. 2-17 Another way to create circles is to use your hand.

For larger circles, you can use your fingers or the heel of your hand as a pivot point for the circle's center. When you use your hand as a compass, it is usually easier to turn the paper than to turn your hand. Other aids, such as a piece of string, cardboard, or another pencil, can also help you make smooth, well-rounded circles *(Fig. 2-17)*.

Two very reliable methods for sketching circles are to draw the circle within a square or to locate the center of the circle and use "spokes" to define the circle. Since arcs are parts of circles, you can use the same methods to draw them. Just draw the part of the circle you need. You will practice using these methods in the following "Your Turn."

Follow these steps to sketch a circle by placing it within a square. The finished circle should have a diameter of about 2 inches.

1. Sketch a square about 2 inches on each side.

2. Mark the midpoint of each side of the square *(Fig. 2-18A)*.

3. Draw diagonals to locate the center of the circle. Diagonals are lines that go from the top left to the bottom right and from the bottom left to the top right of the square. The center of the circle will be where the two diagonals cross.

4. Mark the radius of the circle (half of the diameter, or 1 inch) on each of the diagonals. To do this, begin at the point where the diagonals cross, and make a light mark about 1 inch from that point on each side of each diagonal *(Fig. 2-18B)*.

5. Connect each of the points with a smooth circle *(Fig. 2-18C)*.

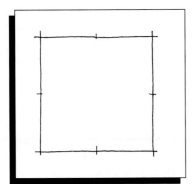

Fig. 2-18(A) To use a square to make a circle: Draw the square so that each side of the square is equal to the diameter of the circle. Mark the midpoint of each side of the square.

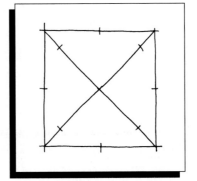

Fig. 2-18(B) Draw the diagonals inside the square and mark the circle's radius on each side of each diagonal.

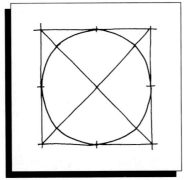

Fig. 2-18(C) Connect the points with a smooth circle.

This is a good method for drawing arcs, as well. For arcs of 90 degrees or less, use a quarter of a square. For arcs between 90 and 180 degrees, use half a square *(Fig. 2-19)*.

Now try the other method of making a circle. In this method, you start with the center point and use radial spokes as a guide.

1. Determine where on the paper the center point of your circle will be. Make a light mark there.

2. Draw axes at right angles through the center point.

3. Draw two 45-degree diagonals through the center point *(Fig. 2-20A)*.

4. Mark the circle's radius (again, approximately 1 inch) on each of the eight spokes *(Fig. 2-20B)*.

5. Sketch the circle through each of the eight marks *(Fig. 2-20C)*.

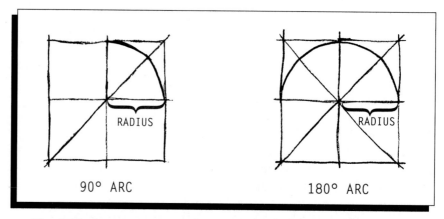

Fig. 2-19 Squares can also be used to create arcs. Just fill in the diagonal or diagonals you need and proceed as for a whole circle.

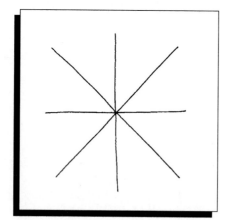

Fig. 2-20(A) Draw X and Y axes through the center point of the circle, and then draw 45-degree diagonals through the center point.

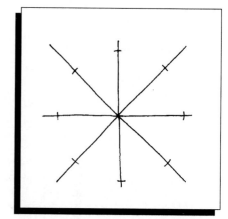

Fig. 2-20(B) Mark the radius of the circle on each "spoke" coming out from the center point.

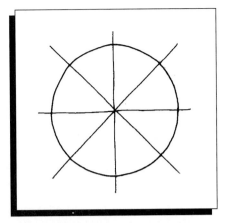

Fig. 2-20(C) Sketch the circle
through the marks on the spokes.

Ellipses

Hold a coin, a roll of tape, or a screw-top jar lid directly in front of your eyes and look at it. It is round; it can be drawn as a circle *(Fig. 2-21A)*. Now tip it so that the left side moves away from you *(Fig. 2-21B)*. It no longer looks round. Keep tipping it until the left side disappears. Now it looks flat *(Fig. 2-21C)*. That shape between round and flat, while you can still see a curved surface, is called an **ellipse**. A circle appears as an ellipse when you view it at an angle.

Since an ellipse is not round, it has two diameters at right angles to each other. The long diameter is the **major axis**, and the short diameter is the **minor axis** *(Fig. 2-22)*.

You can draw an ellipse using the same method you used for drawing a circle. The only difference is that you sketch an ellipse within a rectangle rather than within a square.

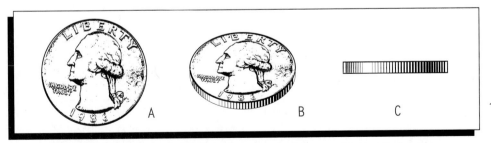

Fig. 2-21 A) A round object looks round when you look at it directly.
B) A round object seen at an angle no longer appears round. C) A round object seen from the side appears to be flat.

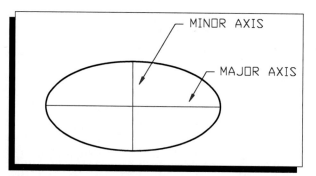

Fig. 2-22 The major axis and minor axis of an ellipse.

ACTIVITY

1. To draw an ellipse, lay out the major axis and the minor axis and sketch a rectangle around them *(Fig. 2-23A)*.
2. Draw the diagonals inside the rectangle.
3. Estimate a distance of one-third of each half of the diagonal measured from the corner of the rectangle and mark the locations *(Fig. 2-23B)*.
4. Sketch light arcs at the midpoints of the sides of the rectangle *(Fig. 2-23C)*.
5. Complete the ellipse and darken it in *(Fig. 2-23D)*.

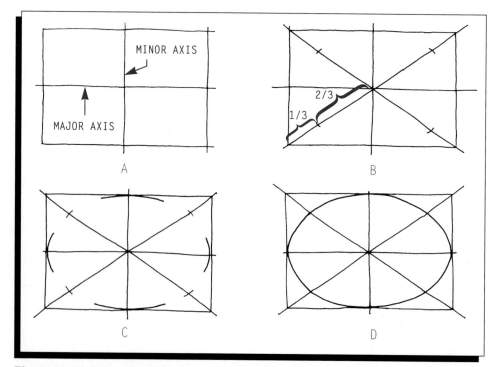

Fig. 2-23 A) Draw the major and minor axes and sketch a rectangle around them. B) Draw the diagonals and mark the ⅓ distances from the corners. C) Sketch arcs at the midpoints of the sides of the rectangle. D) Complete the ellipse and darken it.

Estimating Proportion

Any sketch is only as good as its **proportions,** or the relative size of each part as compared with the size of all the other parts. You must maintain the proportions between the **apparent** **dimensions.** Apparent dimensions are the distances between points on an object as you see them. They are important in sketching because the apparent dimensions are what you sketch.

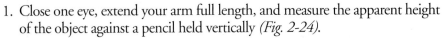

ACTIVITY

Your teacher will assign an object in your classroom for you to sketch. Determine the proportions of the object using the pencil measurement procedure outlined below. Then sketch the object, remembering the sketching techniques you've studied in this chapter.

1. Close one eye, extend your arm full length, and measure the apparent height of the object against a pencil held vertically *(Fig. 2-24)*.

2. Now measure the apparent width of the object with the pencil held horizontally.

3. Compare the two measurements. The apparent dimensions will be in the same proportion as the **actual dimensions**, or measured dimensions *(Fig. 2-25)*.

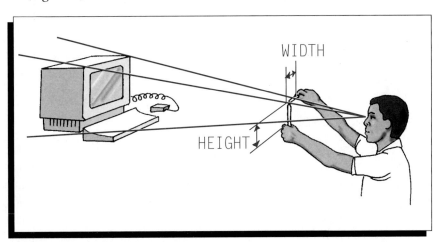

Fig. 2-24 Estimate the dimensions of an object by holding a pencil at arm's length and comparing the apparent length and width of an object.

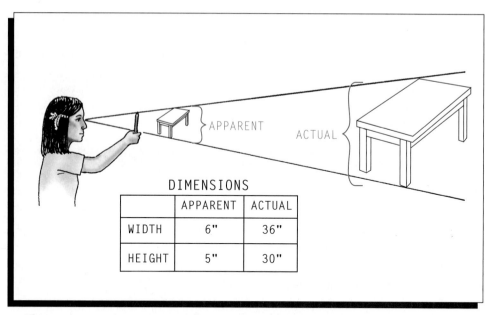

DIMENSIONS

	APPARENT	ACTUAL
WIDTH	6"	36"
HEIGHT	5"	30"

Fig. 2-25 The apparent size of the object, as measured by your pencil, is proportional to the actual size of the object.

Views

A **view** of an object shows a single side of an object or shows the object from a single angle. Some sketches may only need one view to get the point across. Others may need two or more views.

One-View Sketches

Drafters use one-view sketches for objects that can be described adequately in two dimensions. This is true for many objects that have a uniform thickness. A simple example is a door key. The key can be shown in two dimensions. The drafter adds a note to show the thickness of the key *(Fig. 2-26)*.

One-view sketches may also be made of parts of complex objects. If a change is being made to an object or to a drawing of an object, only the part that changes needs to be sketched. If the change affects only one view of the object, then a one-view drawing is all that is necessary *(Fig. 2-27)*.

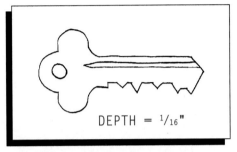

Fig. 2-26 A simple key can be shown in one view. Since the depth is constant, it can be shown in a note that accompanies the sketch.

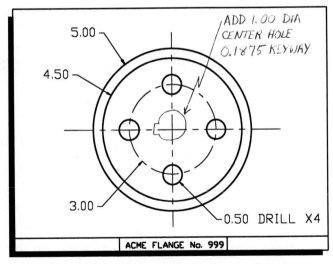

Fig. 2-27 Sketching a change on a drawing can help drafters understand precisely what changes are required.

Multiview Sketches

More than one view may be necessary to describe complex features. Multiview sketches may also be necessary if an entire object is to be sketched. If more than one view is used, the views should be placed as shown in Fig. 2-28. You will learn more about multiview drawing in Chapter 8, "Views and Techniques of Drawing."

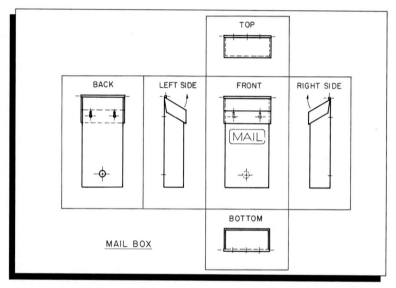

Fig. 2-28 Locations of views for a multiview sketch.

ACTIVITY

One of the most important skills in sketching—and in drafting—is understanding what views to draw to describe the shape of an object entirely. To practice deciding what views are necessary, sketch the following objects. Use only as many views as you think are necessary to describe the shape of the object entirely. (Do not worry about text at this time.)

1. Sketch a sharpened pencil.
2. Sketch a 1-quart milk carton.
3. Sketch a credit card.

How many views did you need to sketch each of the three objects? In most cases, you should need two views for the pencil, three views for the milk carton, and one view for the credit card. If your sketches had a different number of views, go back and look at them more carefully. Did you describe the shape adequately? Could you have done it in fewer views? Ask your instructor if you need help.

Pictorial Sketches

If an entire object needs to be drawn, it is frequently drawn as a **pictorial**, or picture-like, sketch. Pictorials are usually done as oblique or isometric sketches.

Oblique Sketches

An **oblique sketch** is one that shows the front view of an object as though you were looking at it directly in front of you. The top and side views are drawn at an angle other than 90 degrees *(Fig. 2-29)*. Oblique sketches are easy to draw because the front view, which contains the width and height of the object, represents the true size of the object.

Since no distortion occurs in the front view, circles appear round, not elliptical. The size of the angle for the top and side views determines whether the emphasis is placed on the top or the side of the object. A steep angle, greater than 45 degrees, emphasizes the top. A shallow angle, less than 45 degrees, emphasizes the side *(Fig. 2-30)*. Two special types of oblique drawings are the "cabinet oblique," which has a half-size depth, and the "cavalier oblique," which has a full-size depth *(Fig. 2-31)*.

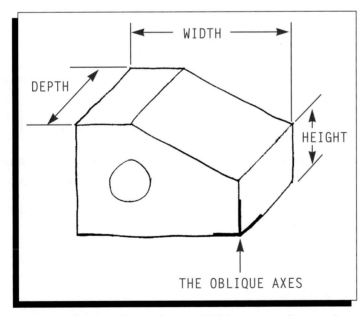

Fig. 2-29 An oblique sketch is fairly easy to draw and shows all three of the overall dimensions—height, width, and depth—of an object.

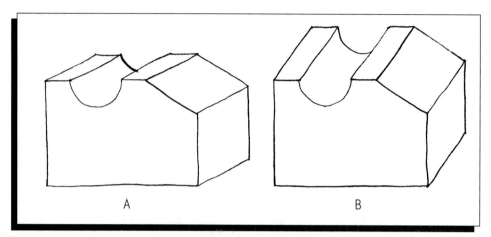

Fig. 2-30 The angle of the third axis determines the angle of the view. A) Angle is less than 45 degrees, so the side is emphasized. B) Angle is greater than 45 degrees, so the top is emphasized.

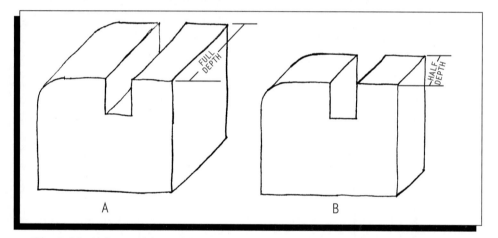

Fig. 2-31 A) In a cabinet oblique sketch, the depth is drawn to full scale. B) In a cavalier oblique sketch, the depth is drawn to half-scale. Although cabinet oblique is the more accurate of the two, it may make the object look distorted. Cavalier oblique provides a more natural-looking sketch.

Isometric Sketches

An **isometric sketch** has three axes spaced 120 degrees apart. The width and depth axes are 30 degrees from horizontal, and the height is vertical *(Fig. 2-32)*. Lines parallel to the axes are called "isometric lines." True size distances are measured on the isometric lines. Lines that do not lie on the isometric axes are called "nonisometric lines." Noniso-metric lines connect points located by isometric lines *(Fig. 2-33)*.

Isometric grid paper is available to make isometric sketching easier *(Fig. 2-34)*. However, since isometric paper may not always be available, it is better to learn how to sketch accurately without the aid of a paper grid.

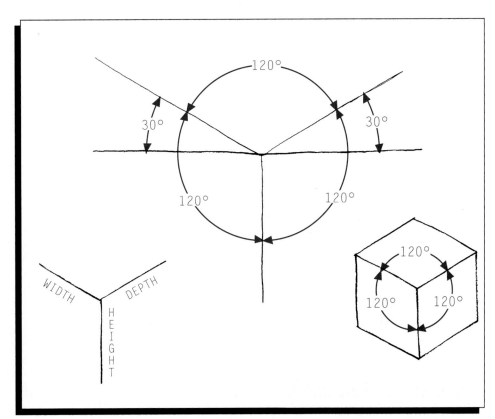

Fig. 2-32 The isometric axes are spaced 120 degrees apart, as shown. The axes that show depth and width are 30 degrees above the horizontal line that passes through the *origin*, or point where all the axes meet.

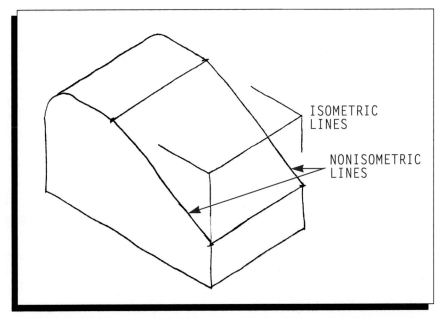

Fig. 2-33 Nonisometric lines are lines that are not parallel to any of the three isometric axes.

55

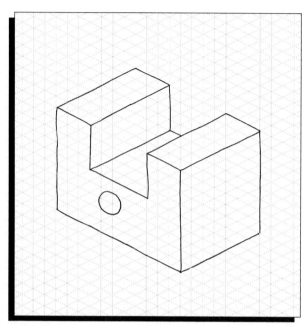

Fig. 2-34 The lines on isometric grid paper run along the three isometric axes.

ACTIVITY

To practice creating isometric sketches, imagine that you are an architect. Your client wants to build an office in a nearby city. The front of the building is to be 30 feet wide, and the building will be a perfect bar shape; in other words, all of its surfaces are rectangular. The building is to be 50 feet long and 20 feet high *(Fig. 2-35)*. For this practice example, do not worry about the roof, doors, and windows.

1. Make a mark near the center of a clean sheet of sketching paper. This mark will be the upper left corner of the office building.

2. Draw the axes around which you will sketch the building *(Fig. 2-36A)*. First, sketch a line that goes straight down from the mark on your paper. This will be the vertical axis, which defines the front corner of the building.

3. From the mark on your paper, draw another line about 120 degrees to the right of the first line. This line should point toward the top right corner of your paper.

4. Finally, from the mark on your paper, draw a third line about 120 degrees to the right of the line you drew in step 3. This line should point to the upper left corner of your paper.

You have completed the axes around which you will draw the office building. Now you can begin drawing the building.

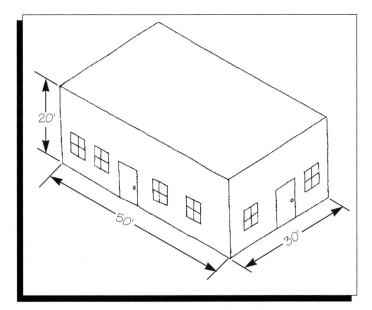

Fig. 2-35 A client's idea for an office building.

5. Using arbitrary units, mark the height of the building (20 feet) on the vertical axis. This locates the lower left corner of the building *(Fig. 2-36B)*.

6. Using the same arbitrary units, mark the front of the building on the axis that points to the upper right corner of your paper. Recall that the front of the building is 30 feet wide *(Fig. 2-36C)*.

7. Using the same units, mark the depth of the building (50 feet) on the third axis *(Fig. 2-36D)*.

8. Sketch light lines from the points you made on the two nonvertical axes down toward the bottom of the page. The lines should be parallel to the vertical axis *(Fig. 2-36E)*.

9. Measuring from the points you made on the two nonvertical axes, make a mark 20 feet down on each of the lines you sketched in step 8.

10. Sketch lines from the marks you made in step 9 through the mark you made on the vertical axis *(Fig. 2-36F)*.

This completes the front and left side of the building.

11. Sketch a line that goes from the top right corner of the front of the building, parallel to the axis that points to the top left corner of your paper *(Fig. 2-36G)*.

12. Using the same units you used before, measure 50 feet along that line and make a mark. This will be the top right corner of the back of the building *(Fig. 2-36H)*.

13. Sketch a line that goes from the top left corner of the back of the building to the top right corner. If you have sketched accurately, this line should be about 3 units long *(Fig. 2-36I)*.

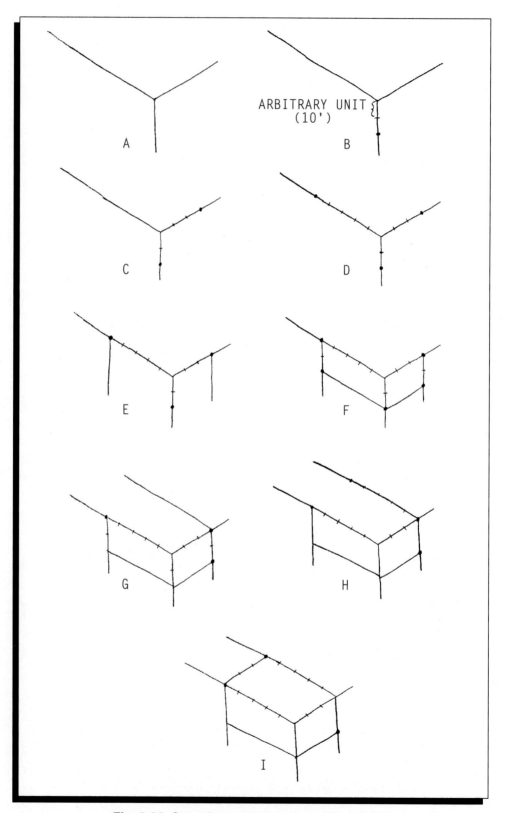

ARBITRARY UNIT
(10')

A

B

C

D

E

F

G

H

I

Fig. 2-36 Steps for making an isometric drawing.

Perspective Sketches

Another common way to sketch three-dimensional objects is to use perspective. **Perspective sketches** do not necessarily contain accurate proportions; instead, they picture objects as the eye sees them. When you look down a street, for example, objects that are farther away seem to be smaller, even though they are not. Perspective sketches look more natural than oblique or isometric sketches, but they are not as accurate.

Stand at a window and look at a building across the street; it appears, as it is, in three dimensions. When you reach out to touch the peak of the roof, however, your hand is stopped by the glass in the window. You touch the window glass at the point where the roof peak appears to be. Each point of the house appears to be at a fixed point on the glass. If you imagine that the image of the house is projected onto the window glass, this projected image is actually in perspective form *(Fig. 2-37)*. The apparent distances on the house, which give the house its depth (third dimension), are projected onto the two-dimensional window glass.

Fig. 2-37 When you look through a glass window at an object, you can imagine that the object is projected onto the glass.

■ Chapter 2 Review

1. What materials do you need to make a sketch of an object?
2. When you are sketching, at what angle from the paper should you hold your pencil?
3. Describe the correct procedure for sketching a vertical line.
4. What is the easiest method for sketching an inclined line?
5. Explain the role axes play in sketching.
6. What is a view? Why do some sketches need to show more than one view of an object?
7. Briefly describe the three types of pictorial sketches mentioned in this chapter.

■ Chapter 2 Problems

Using the principles you have learned in this chapter, sketch each of the following objects. Do not include the dimensions; they are there for your reference only. Remember that these are sketches! Do not use a ruler or any other instrument except your sketch paper, a pencil or lead holder, and possibly an eraser.

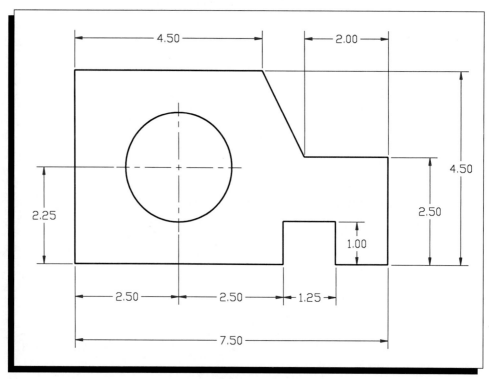

Problem 2-1

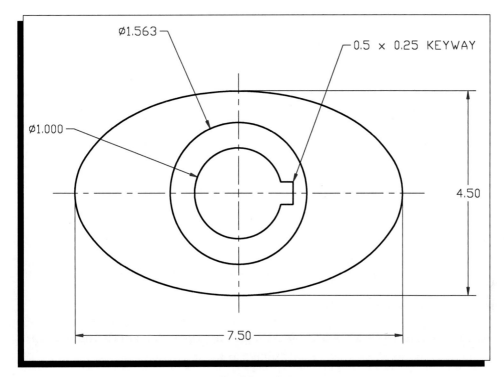

Ø1.563

0.5 × 0.25 KEYWAY

Ø1.000

4.50

7.50

Problem 2-2

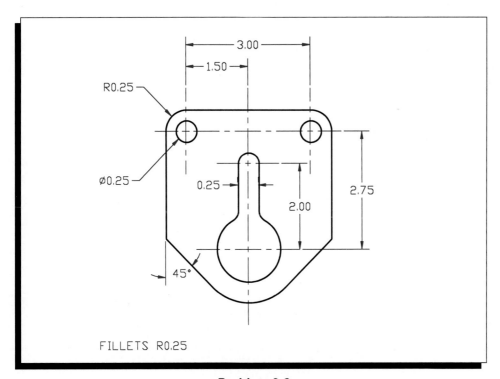

3.00

1.50

R0.25

Ø0.25

0.25

2.00

2.75

45°

FILLETS R0.25

Problem 2-3

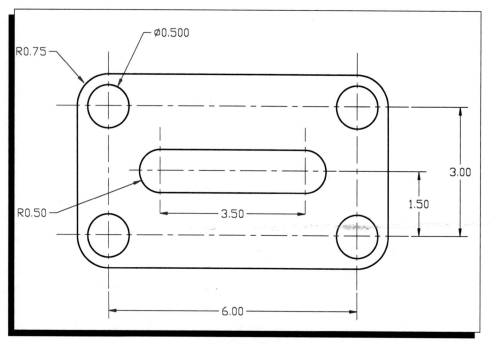

Problem 2-4

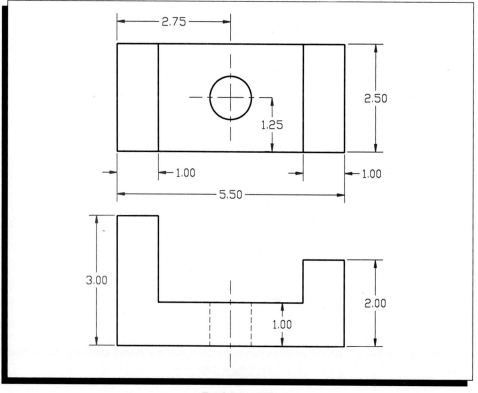

Problem 2-5

Using CAD
Architect Packs Up Drawing Boards

Joe MacRae is an architect who began using AutoCAD in 1984 after drawing on a board professionally for 32 years. Since packing up his boards, Joe has done one finished drawing by hand. That was done with markers on sketch paper at the kitchen table because someone needed a quick landscape plan for a zoning meeting.

On the other hand, Joe often uses sketches to get a point across. He says that people see and remember best with pictures, so it's the most effective way to communicate with them. It might be a sketch of the floor plan or of the elevation of a wall so they can visualize it. Joe says, "I've found that the sketch doesn't have to be a Picasso. Sometimes just a few lines will get some point across. But in the long run, the better the sketch, the better the reception of the idea and the longer they will remember it."

Professionally, Joe has used everything from pencils and water color brushes on paper to ruling pens on cloth. He did one lettering job with a turkey quill, shaping and splitting the point himself. The sketch shown here was done while talking to carpenters about building a new kitchen.

Joe says, "I guess you could say that I also sketch at the computer. But with a machine, I let the program worry about keeping the lines straight. The output doesn't look like a sketch in the traditional sense; but when you start with a blank screen and wind up with a design, there has to be some sketching in there." It's like using an electronic pencil rather than a lead one. He even uses a stylus with a digitizer because it's shaped like a pencil.

In the early days of CAD, people like Joe who were the first to begin using CAD programs had no one to turn to except themselves for help. One way they found each other, and continue to find each other, is through the Auto-CAD forums on CompuServe. CompuServe is an information service that can be reached with a modem, which is connected to the telephone lines. Joe is now a System Operator for the AutoCAD forums, which means he will probably be the first to answer if you ask a question about AutoCAD.

Joe started out doing house plans on a computer with an 8088 processor, 640K of RAM, and 40M of hard disk space—a system that is very limited by today's standards. However, he says the use of that machine with AutoCAD Release 2.05 provided a good income and the opportunity to upgrade his equipment over the years. The biggest surprise was that it also brought him friends from all over the world.

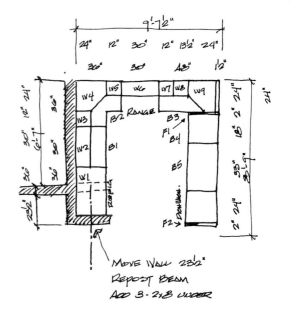

Chapter 3

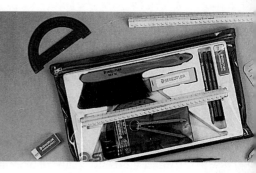

▶ Key Terms

T-square
triangle
drafting machines
alphabet of lines
scales
irregular curves
compasses
dividers

bisect
tangent
regular hexagons
inscribed
circumscribed
lettering
orthographic projection

Mechanical Drafting

Objectives

When you have completed this chapter, you will be able to:

- identify the tools of mechanical (manual) drafting.
- demonstrate an understanding of geometric construction with mechanical drafting.
- letter a drawing.
- use the drawing tools and techniques to lay out an A-size drawing sheet.
- create one-view and multiview drawings.

Traditional drafting has changed little over the past decades. The drawing equipment, techniques, and procedures of the 1950s are not only recognizable but are still used in the 1990s. Drafting texts have also remained consistent. New editions reflect new materials and techniques, but the mechanical drafting process remains fairly constant.

Then came the computer—and CAD!

Computer-aided drafting is often referred to by its acronym: CAD. The term *CAD* will be used throughout this textbook.

However, computers are not always available. Both instructors and students may often feel the need for knowledge of the traditional drafting methods. Into the world of CAD, this chapter introduces manual drafting.

Drawing Equipment and Techniques

Fig. 3-1 shows some of the primary equipment used by mechanical (manual) drafters. This equipment includes the T-square, a 30-60-degree triangle, a 45-degree triangle, scales, pencils, and various instruments. In addition, the drafter needs a drawing board or table.

Fig. 3-1 Tools for mechanical drafting.

Drawing Boards and Tables

Drawing boards and tables are made from wood or wood-composition materials and are frequently covered with vinyl. The height and angle of the tables are adjustable. One side edge of the board or table must be *true*, or perfectly straight. Use this as the working edge for the T-square.

T-Square

The **T-square** is an instrument used to draw horizontal lines. It also forms a base for the triangles used to draw vertical and inclined lines *(Fig. 3-2)*. The T-square has a head and a blade. The working edge of the T-square is the upper edge of the blade, which must be straight and at right angles to the head. For general drawing, keep the head of your T-square firmly against the working edge of your drafting table or board. If you are right-handed, this should be the left side, and if you are left-handed, it should be the right side.

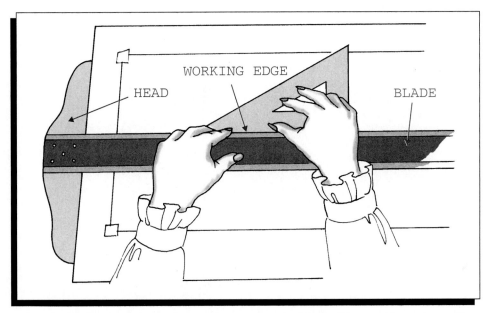

Fig. 3-2 A T-square is made up of a head and a blade. The working edges of the blade are usually made of a clear plastic.

Drawing Paper

Use the paper supplied by your instructor or your employer. If you are on your own, use tracing paper, velum, or any good grade of drawing paper. Refer to Chapter 4 for both U.S. and metric drawing sheet standard sizes.

Pencils

Always use high-quality drawing pencils or leads for manual drafting, never ordinary writing pencils. Most drafters prefer mechanical lead holders to wooden pencils. Lead holders need only to be pointed, not sharpened, and they never shrink in size. Fine line lead holders do not even need to be pointed.

If you are using one of the other types of pencils, always remember to keep your pencil sharp. A dull pencil makes fuzzy lines—the sign of an inexperienced drafter.

Lead comes in eighteen grades from 9H, the hardest, to 7B, the softest. Drafting is usually done with the medium grades of 3H to B *(Fig. 3-3)*. As a general rule, use 2H and H for finished lines, 3H to 5H for construction lines, and F and HB for sketching.

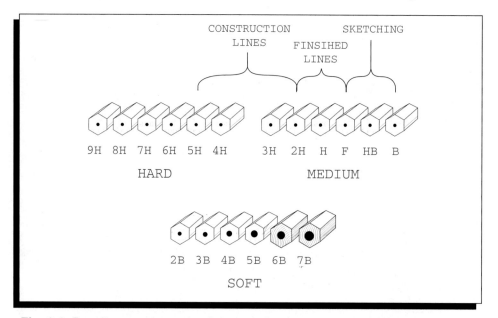

Fig. 3-3 Pencils come in grades from very hard to very soft. In general, harder leads make finer, lighter lines, and softer leads make heavier, darker lines.

Triangles

A **triangle** is a drafting tool that contains three sides at known angles. Drafters use the 30-60-degree triangle and the 45-degree triangle to draw all vertical and most sloping lines. When used in combination with the T-square, the two triangles can make inclined lines every 15 degrees from horizontal *(Fig. 3-4)*. You can also use the T-square and one triangle to create parallel lines by sliding the triangle along the blade of the T-square *(Fig. 3-5)*.

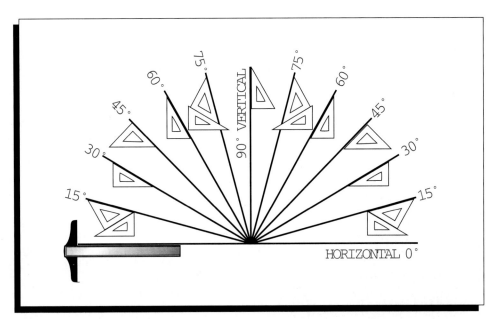

Fig. 3-4 Working with the T-square and the two triangles, you can draw lines accurately at 15-degree intervals.

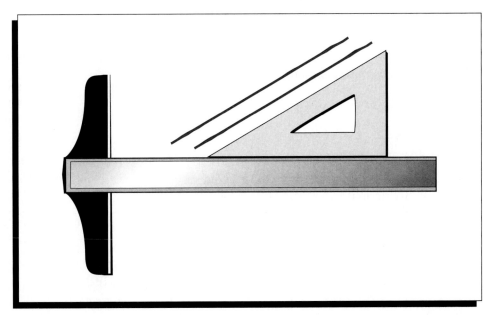

Fig. 3-5 Parallel lines are easy to draw using the T-square and a triangle. Draw the first line, then slide the triangle along the T-square and draw the second line.

Drafting Machines

Track-type **drafting machines** have replaced the T-square and triangles in the mechanical drafting rooms of most industries *(Fig. 3-6)*. Two scales at 90 degrees to each other act as straight edges. They move on tracks horizontally and vertically and can be rotated to any angle to create inclined lines.

Fig. 3-6 Track-type drafting machines have taken the place of the traditional T-square, triangles, scales, and protractors in many places.

Alphabet of Lines

The **alphabet of lines** is the standard library of line symbols used in drafting. This standard is set by the American National Standards Institute (ANSI) *(Fig. 3-7)*. You will learn more about the use of various types of lines in Chapter 4.

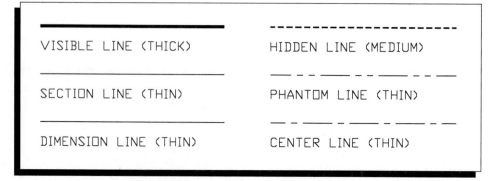

VISIBLE LINE (THICK)

HIDDEN LINE (MEDIUM)

SECTION LINE (THIN)

PHANTOM LINE (THIN)

DIMENSION LINE (THIN)

CENTER LINE (THIN)

Fig. 3-7 The alphabet of lines provides a standard for drafters so that the same types of lines are used consistently for the same applications.

Scales

Scales are instruments that let you measure distances accurately on drawings. You can use a scale to lay out full-size distances or distances that are longer or shorter than full size. You should draw objects at their full size whenever possible. However, some objects cannot reasonably be drawn at full scale. For example, a house must be scaled down to fit on a drawing, and a watch part must be scaled up so that it is easier to see.

Most scales are marked in several types of units to make it easier to mea-sure distances on drawings that have been scaled up or down. In fact, the proportional dimensions for both the house and the watch part can be read directly from a drafter's scale just as easily as the full-size dimensions of a box.

Scales are usually flat or triangular and can be obtained in many configurations. Common drafting scales include the architect's scale, mechanical engineer's scale, civil engineer's scale, and metric scale *(Fig. 3-8)*.

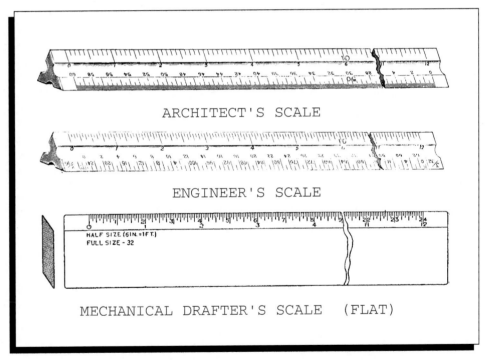

Fig. 3-8 A variety of drafting scales make it possible for drafters to use an appropriate scale for the object to be drawn—however large or small the object may be.

Irregular Curves

Irregular curves, also called French curves, are used to draw non-circular curves *(Fig. 3-9)*. As the name implies, irregular curves are *not* regular curves, which are established by center points. Irregular curves are made to fit. They do not establish original curves; rather, they blend points and form a smooth transition among points, lines, and/or arcs. Many different types of irregular curve tools are available for special purposes in drafting.

In addition to the standard irregular curves, "flexible curves" are now available that allow you to follow or create any curved surface. You just bend the flexible curve to the correct shape. It holds that shape so that you can use it on the drafting board as a template *(Fig. 3-10)*.

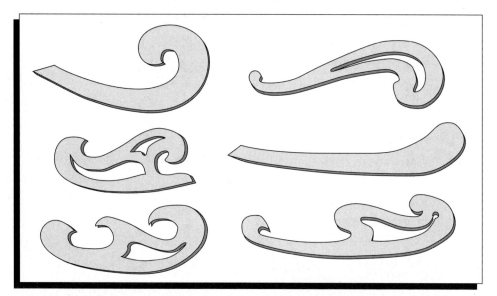

Fig. 3-9 Irregular curves are available to create smooth curves for just about any drafting purpose.

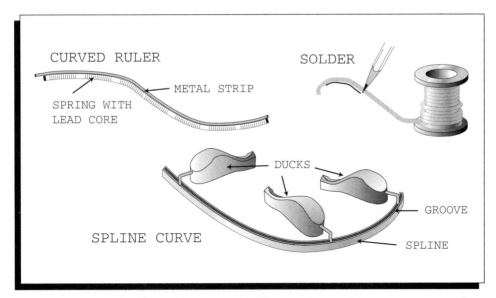

Fig. 3-10 Flexible curves allow you to define the shape of the curve you need.

Drafting Instrument Sets

Drafting instrument sets come in different sizes and configurations. Most basic sets contain a small bow compass, a large bow compass, and friction dividers *(Fig. 3-11)*.

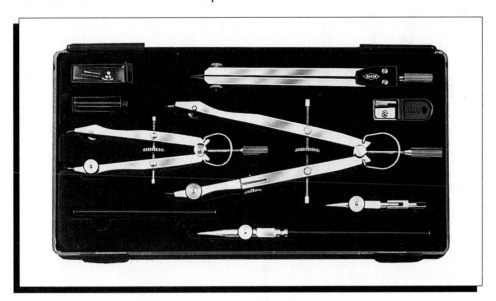

Fig. 3-11 A basic drafting instrument set.

▶ Compasses

Compasses are drawing instruments used to draw circles. Small compasses draw small circles, and large compasses draw large circles. Brilliant comment! But you might be surprised at how many beginning drafters, and even some not-so-beginning drafters, try to draw a small circle with a giant bow or a large circle by stretching a small bow. The size of the circle determines the type of compass you should use. A drop-bow compass draws a very small circle, and a beam compass draws a very large circle *(Fig. 3-12)*.

To draw a circle, set the point in the center, adjust the radius, and swing the compass by holding the handle between the thumb and forefinger. Tip the compass slightly in the direction of rotation *(Fig. 3-13)*. If you are right-handed, draw the circle in the clockwise direction; if you are left-handed, draw it in the counterclockwise direction.

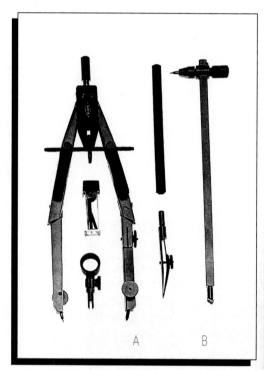

Fig. 3-12 A) Drop bow compass with attachment. B) Beam compass. Attachment can go on either end.

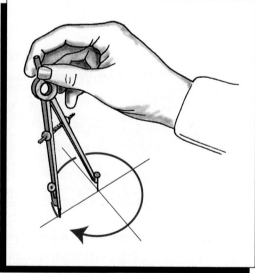

Fig. 3-13 Tip the compass in the direction of rotation and swing in a clockwise (right-handed) or counterclockwise (left-handed) direction.

▶ Dividers

Dividers are used to divide lines and to mark off equal spaces along a line or circle *(Fig. 3-14)*. They may also be used to transfer lines or distances from one place to another. Dividers may be friction or center screw instruments.

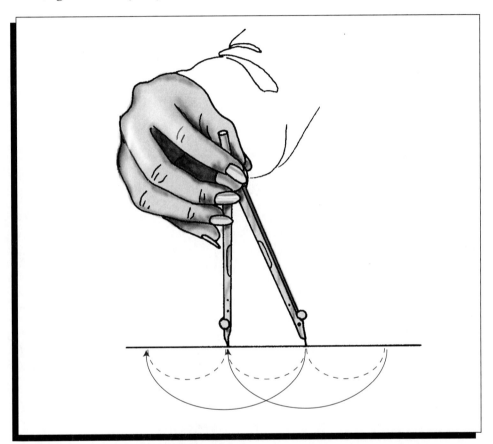

Fig. 3-14 To set off equal distances with dividers, set the legs of the dividers at the distance you want to mark off. Put one of the legs on the starting point on the line or circle. Swing the dividers as shown until the other leg of the divider touches the line or circle, and make a mark there. Holding that leg down on the paper, swing the first leg around to touch the line or circle at the next division point, and make a mark. Continue until you have marked the required number of equal distances.

Drafting machines have made the drafting process easier in many ways, yet some drafters prefer the traditional tools. To find out more about the alternatives, do the following activities.

1. Look up at least two types of drafting machines and compare them with the use of a table with separate tools such as the T-square. What are the advantages and disadvantages of each?

2. Use a catalog of drafting supplies or go to a drafting supply store to find out how the cost of a drafting machine compares with the combined cost of all the tools it can replace. Be sure to include the cost of a drafting table, T-square, triangles, protractor, and compass in your calculations.

Geometric Construction

Drafters make extensive use of geometric shapes and constructions to create drawings and solve problems. Chapter 6, "Geometric Drawing," discusses geometric constructions from the CAD viewpoint. However, mechanical drafting constructions differ considerably. The following examples are typical of the procedures and tools used for mechanical constructions.

Construction lines are light lines a drafter may draw to use for construction purposes. These lines are either not present in the final version, or they are so light that they do not reproduce. Refer to the Pictorial Glossary at the back of this book for definitions of other terms you may not understand.

Constructing and Dividing Lines

To construct a line manually, always use a straight edge. The straight edge may be the upper edge of the T-square, any edge of a triangle, or the blade of a drafting machine. Draw with your hand over your straight edge. Tip your pencil about 50 to 60 degrees in the direction of the line you are drawing *(Fig. 3-15)*. Move the pencil away from you: right-handers left to right and left-handers right to left for horizontal lines, and bottom to top for vertical lines.

Recall that this is different from sketching, in which you make vertical lines from top to bottom.

Right-handers should draw inclined lines as in Figs. 3-16 and 3-17. Left-handers should draw in the opposite direction. *Never* push your pencil! Rotate your pencil as you move it to keep a conical point.

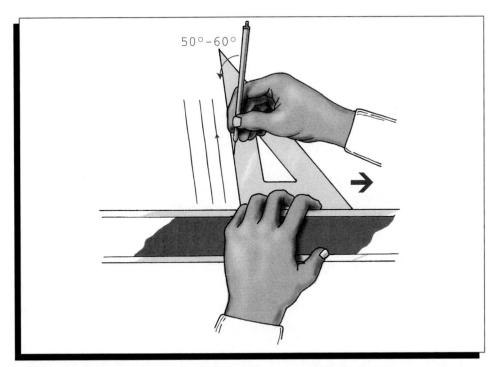

Fig. 3-15 Correct hand position for drawing a vertical line. (Arrows indicate direction.)

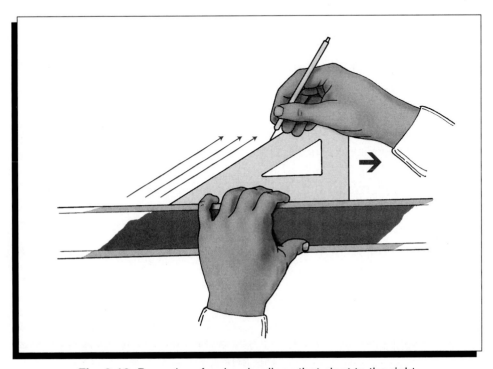

Fig. 3-16 Procedure for drawing lines that slant to the right.

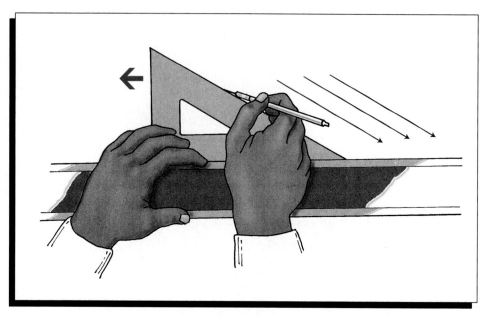

Fig. 3-17 Procedure for drawing lines that slant to the left.

ACTIVITY

You may recall from geometry courses that lines are often referred to by named points through which the lines run. For example, a line that runs through points A and B may be called "line AB." This naming system is used in the following procedure, in which you will construct a line and divide it into equal parts. Fig. 3-18C shows the finished exercise.

1. Construct a horizontal line AB. Points A and B should be 6 inches apart.
2. Draw line AC at any length and at any angle from AB.
3. Draw a line to connect point C with point B *(Fig. 3-18A)*.
4. Mark off 4 equal spaces on line AC *(Fig. 3-18B)*.
5. Draw lines parallel to line BC from the other points you marked on line AC so that they intersect line AB *(Fig. 3-18C)*.

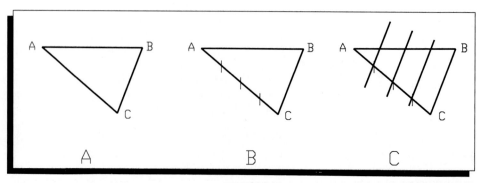

Fig. 3-18

Bisecting an Angle

There may be times when you need to **bisect** an angle, or divide it into two equal parts. The following procedure shows you how to do this accurately and efficiently. If you need help drawing the angle, refer again to Fig. 3-4.

ACTIVITY

Your Turn

In this procedure you will draw an angle and then bisect it. The angles in this procedure are named according to standard convention: angle ABC is an angle formed by two lines, AB and BC, where the common point, or *vertex*, is point B. With this in mind, follow these steps.

1. Construct an angle similar to the one in Fig. 3-19A.

2. Set your compass to any radius that will intersect the two sides of the angle when the pin of the compass is at the vertex of the angle (point A).

3. With the pin on point A, swing an arc to intersect the sides of the angle at points B and C *(Fig. 3-19B)*.

4. Set your compass to a radius greater than half the distance between points B and C.

5. Swing arcs from points B and C to locate point D *(Fig. 3-19C)*.

6. Draw a bisecting line through point D to point A *(Fig. 3-19D)*.

   The new line creates two triangles, CAD and BAD. Since the line bisects angle BAC, angle CAD equals angle BAD.

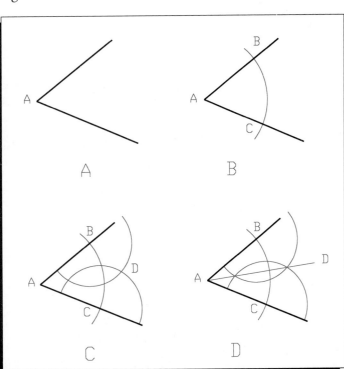

Fig. 3-19 A) Construct the angle to be bisected with vertex at point A. B) Swing arcs to locate points B and C. C) With the pin of the compass on point B and the radius adjusted to more than half the width of BC, swing an arc inside the angle. Then place the pin on point C and repeat this process. The two arcs intersect at point D. D) Draw a line from point D to point A to bisect the angle.

Constructing Tangent Lines and Arcs

A line, arc, or circle is **tangent** to another arc or circle if the two touch, or intersect, at exactly one point *(Fig. 3-20)*.

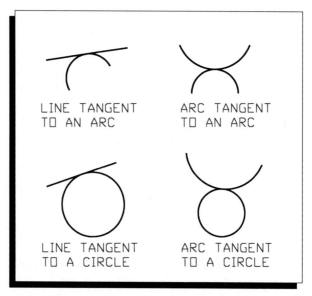

Fig. 3-20 Tangency occurs when a line, arc, or circle touches another arc or circle at exactly one point.

Two lines cannot be tangent to one another.

Tangent lines are important in mechanical drafting because they are often used to construct geometric shapes and objects.

ACTIVITY

The following procedure will show you how to construct an arc that is tangent to both a line and another arc. Refer to Fig. 3-21 as you follow the steps. Fig. 3-21D shows the completed exercise.

1. Draw a horizontal line and an arc with a radius (R) of 2 inches as shown in Fig. 3-21A. These will be the "given line" and the "given arc" *(Fig. 3-21A)*.

2. Draw a line 1 inch above the given line and parallel to it. The distance between the two lines establishes the radius (R1) of the third arc—the one that will be tangent to the given arc and the given line. (For convenience, this discussion will refer to this arc as the "tangent arc.") *(Fig. 3-21B.)*

3. Draw an arc parallel to the original arc with a radius R+R1 *(Fig. 3-21C)*.

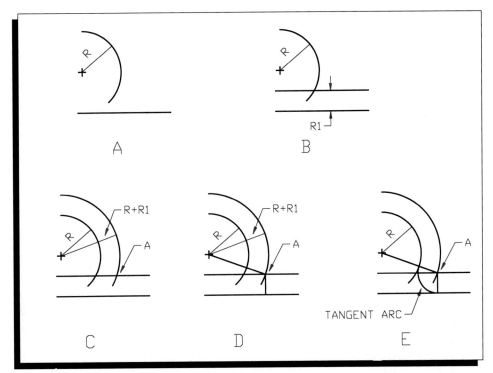

Fig. 3-21 A) Draw the "given" arc and line. B) Draw a line parallel to the given line. C) Draw an arc parallel to the given arc, with a radius of R + R1. D) Draw a perpendicular line from point A to the original line and from point A to the center of the original arc. E) Construct the tangent arc from the tangent point on the line to the tangent point on the arc.

Since R equals 2 inches and R1 equals 1 inch, R + R1 equals 3 inches. Remember that arcs are parallel when they have the same center.

The intersection of the parallel line and the parallel arc is point A. Point A will be the center of the tangent arc.

4. Establish the tangent point on the given line by drawing a perpendicular line from point A to the line, as shown in Fig. 3-21D. The point at which this line intersects the given line is the tangent point.

5. Establish the tangent point on the given arc by drawing a line from point A to the center of the given arc, as shown in Fig. 3-21D. The point at which this line intersects the given arc is the tangent point.

6. With the point of the compass at point A, swing an arc from tangent point to tangent point *(Fig. 3-21E)*.

Creating Regular Hexagons

Regular hexagons are six-sided figures in which all the sides are the same length. They can be measured across the corners or across the flat sides *(Fig. 3-22)*. If the hexagon is measured across the corners, it is said to be **inscribed** within a circle. If it is measured across the flat sides, it has been **circumscribed** around a circle.

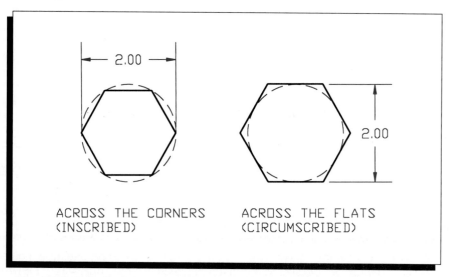

Fig. 3-22 Regular hexagons can be measured across the flat sides or across the corners.

ACTIVITY

The following step-by-step procedures show you how to create hexagons by inscribing them and by circumscribing them. In both sequences, you will use a circle that is 3 inches in diameter. Note the difference in the sizes of the two hexagons.

1. To inscribe a hexagon in a circle, first draw a circle with a diameter of 3 inches. This diameter will be equal to the distance across the corners of the finished hexagon.

2. Draw line AB as a horizontal diameter *(Fig. 3-23A)*.

By using arcs equal to the radius of the circle, you can divide the circle into six equal parts. This forms the basis for the inscribed hexagon.

3. Swing arcs equal to the radius of the circle (1½ inches) from points A and B *(Fig. 3-23B)*.

4. Connect each of the six points to form the hexagon *(Fig. 3-23C)*.

Now follow these steps to circumscribe a hexagon around a circle.

1. Draw a circle with a diameter of 3 inches. In this case, the diameter of the circle is equal to the distance across the flat sides of the hexagon.

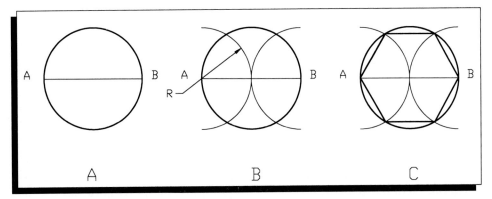

Fig. 3-23 A) Draw line AB as a horizontal diameter of the circle. **B)** Swing arcs equal to the radius of the circle from points A and B. **C)** Connect the six points you marked on the circle to form the hexagon.

2. Draw vertical and horizontal center lines (lines that run through the center point of the circle). *(Fig. 3-24A.)*

3. Draw horizontal lines tangent to the circle at the top and bottom. (Remember that a tangent line is one that intersects, or touches, a circle or arc at only one point.) *(Fig. 3-24B.)* These lines make up the top and bottom segments of the hexagon.

4. Draw tangent lines at 60 degrees from the existing horizontal lines to finish the circumscribed hexagon *(Fig. 3-24C)*.

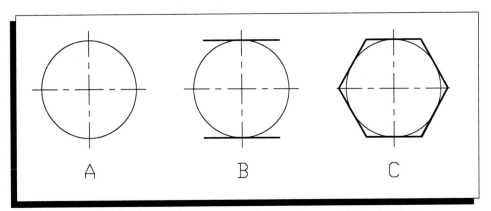

Fig. 3-24 A) Draw vertical and horizontal center lines. **B)** Draw horizontal lines tangent to the circle at the top and bottom. **C)** Draw tangent lines at 60 degrees from the existing horizontal lines.

Lettering

Drafters convey information with both pictures and text. Mechanical drafters refer to text as **lettering**. The lettering style most commonly used on engineering drawings is uppercase, single-stroke Gothic. The text can be either vertical or slanted.

Uniformity

One of the most important criteria for lettering is uniformity. You should never mix upper and lower case, and you should never mix vertical and slanted letters. Use guidelines for uniform height. Strokes that make letters should be uniformly thick. In overall appearance, the lettering should look neat and precise, because lettering can "make or break" an otherwise good drawing.

Spacing

The spacing of letters and words relative to each other plays an important role in the visual impact of the lettering on a drawing. The spacing between words in your lettering should be equal. Spacing between letters should *appear* equal. Study Fig. 3-25 and notice that the letters are not equally spaced. It is the *background* areas of the letters that must appear equal.

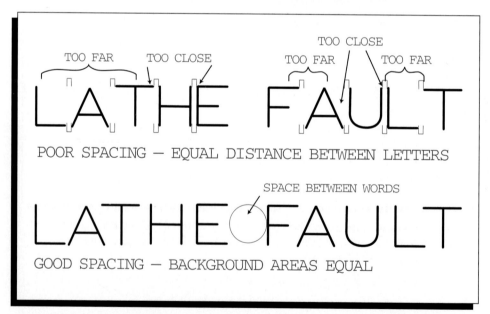

Fig. 3-25 Spacing of letters and words: Top) Equal distances between letters provides poor spacing because letters appear to be unevenly spaced.
Bottom) Good spacing has unequal spaces between letters to make the background areas approximately equal; the letters then appear evenly spaced, even though they are not.

Lettering Guides

Numerous devices have been developed to aid with lettering. Popular lettering devices include the Ames Lettering Instrument and a triangle made by Braddock-Rowe and other companies. Both are used for drawing guidelines. Lettering stencils are also used to aid with lettering *(Fig. 3-26)*.

An effective device for use with tracing paper is a plastic sheet with various horizontal and vertical guidelines. You place the plastic sheet under the tracing paper and position it as necessary. Then create the text using the guidelines that show through the tracing paper. When you are finished, you remove the plastic sheet and no guidelines remain.

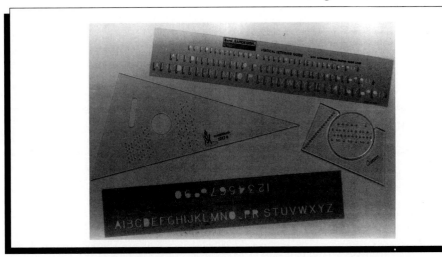
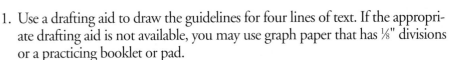

Fig. 3-26 Common lettering aids are the Vanderburg or Braddock-Rowe triangle, the Ames Lettering Instrument, and stencils.

ACTIVITY

Your Turn

The alphabet with vertical Gothic upper-case letters and numbers is shown in Fig. 3-27. Study the proportions of each letter, and learn the order of strokes that creates each one. Lettering is a form of freehand sketching. The only way to become a good letterer is to practice. Follow these steps to practice forming uppercase Gothic letters.

1. Use a drafting aid to draw the guidelines for four lines of text. If the appropriate drafting aid is not available, you may use graph paper that has ⅛" divisions or a practicing booklet or pad.

2. Following the examples of letters and letter strokes given in Fig. 3-27, form the letters to write your name, your full address, and the date.

3. Evaluate your efforts using Fig. 3-27. Are your letters straight and well formed? How could you improve them?

4. Continue to practice making the various letters and numbers. You may wish to choose material from other texts that will give you practice using letters and numbers that you did not use in step 2.

THESE LETTERS ARE 5/6 AS WIDE AS THEY ARE HIGH.

J D R F N Z C G U P B L E H

THESE LETTERS ARE AS WIDE AS THEY ARE HIGH.

O Q S T A K V M X Y &   W I

NUMERALS ARE 5/6 AS WIDE AS THEY ARE HIGH.

1 2 3 4 5 6 7 8 9 0

THE W IS
WIDER THAN
IT IS HIGH.

Fig. 3-27 Vertical Gothic uppercase letters and numbers.

Sheet Layout

Sheet layout includes preparing the paper by fastening it to the drawing board and constructing the borders and title block. The actual layout depends on the size of paper you are using. The problems at the end of this chapter are for A-size paper, so the following procedure shows you how to lay out an A-size sheet. An A-size sheet of drawing paper measures 8½" × 11" or 9" × 12".

ACTIVITY

Follow these steps to lay out an A-size drawing sheet on your drafting board. These steps are based on an 8½" × 11" sheet.

1. Tape an A-size sheet approximately 2 inches from the working edge of your drawing board and a comfortable distance from the bottom.
2. Lay out the sheet by drawing ¼" borders on all sides and a ⅜" title block across the bottom.
3. Divide the title block into four sections measuring 3¼", 4¼", and 1½" from left to right. If you have measured correctly, the fourth section will be 1½".
4. Draw two *very* light horizontal construction lines ⅛" apart, equally spaced in the title block *(Fig. 3-28)*.

Use this layout for all of the drawing problems at the end of this chapter. Your working space is 7⅜" × 10½" for an 8½" × 11" sheet.

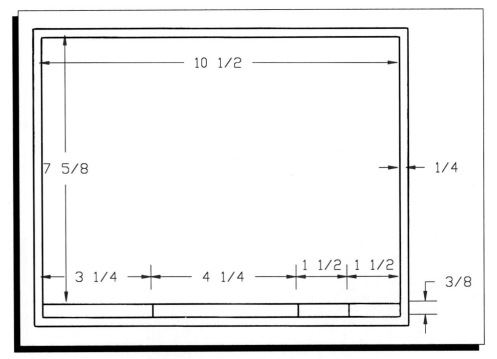

Fig. 3-28 A sheet layout for A-size drawing paper.

Views of Objects

Pick up a closed book and look squarely at the front cover *(Fig. 3-29A)*. All you see is the front cover with the binding on the left. You see the width of the book and the height of the book. This is the *front* view of the book.

Now turn the book so that you look down at it from the top *(Fig. 3-29B)*. You still see the binding on the left, but from a different perspective, and you no longer see the front of the cover. Instead, you see the top edge of the front cover, the pages, and the top edge of the back cover. You no longer see the height of the book, either. You still see its width, but now you see how thick it is: its depth. This is the *top* view of the book.

Turn the book back to its front cover, and then turn it so that you can see its right side *(Fig. 3-29C)*. Now what do you see? You see the right edges of both covers and the pages. This time you see the height and thickness of the book, but you don't see its width. You don't see the front of the cover, and you don't see the binding except at the top and bottom of the pages. This is the *right-side* view of the book.

You have just cycled the book through the three views most commonly used in drafting. The other views are the left-side, back, and bottom views. However, most objects can be adequately described using only the front, top, and right-side views.

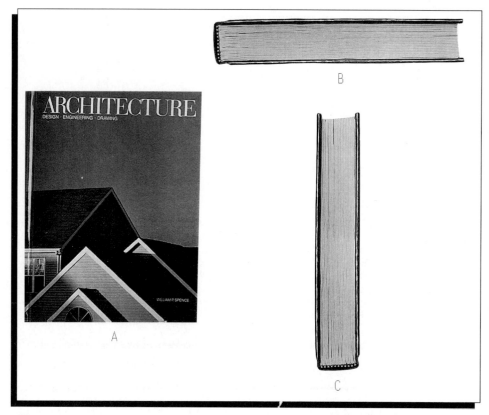

Fig. 3-29 The three views most commonly used in multiview drawings: A) front view; B) top view; C) right-side view.

Orthographic Projections

When you look at an object as a series of individual views, in two dimensions, you are seeing it as a drafter sees it: as an **orthographic projection.** Objects are drawn as orthographic projections so that each surface can be seen in its true size and shape. When you draw an object, show only the surfaces necessary to describe the object fully. Usually the front, top, and right-side views are shown, as mentioned above. However, sometimes one or two views are enough, or more than three may be necessary.

When you draw orthographic projections, always locate your views as shown in Fig. 3-30. Never draw views out of place! Make sure that all the views are the correct size, that all features are shown, and that the object lines up in the proper views. Refer to Chapter 8, "Views and Techniques of Drawing," for more details.

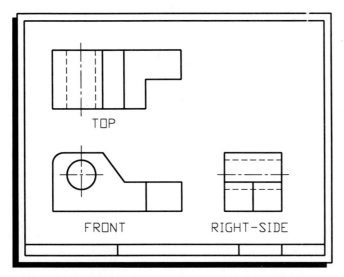

Fig. 3-30 Arrange the views of a multiview drawing as shown here.

Drawing Layout

Numerous factors, such as views, dimensions, and visual impact, determine drawing layouts. As a general rule, center your drawings and leave enough space between them for dimensioning. Figs. 3-31, 3-32, and 3-33 illustrate layouts for one-, two-, and three-view drawings without dimensions or notes.

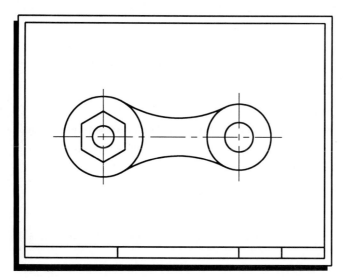

Fig. 3-31 A representative layout for a one-view drawing.

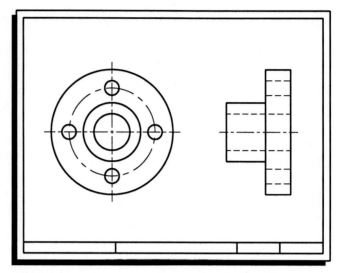

Fig. 3-32 A representative layout for a two-view drawing.

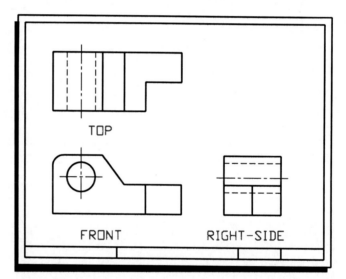

Fig. 3-33 A representative layout for a three-view drawing.

ACTIVITY

Follow the procedure below to make a drawing of a letter E stencil *(Fig. 3-34)*. This is an example of a drawing that requires only one view.

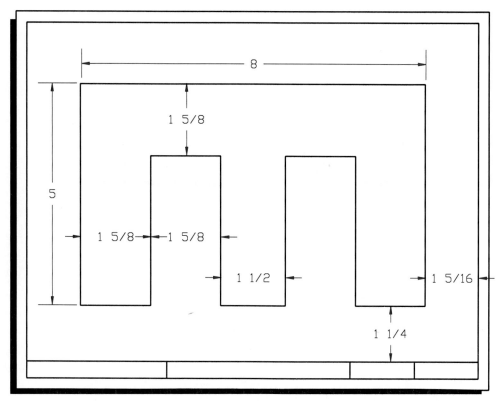

Fig. 3-34 A stencil of the letter E.

1. Lay out an A-size sheet as described in the "Sheet Layout" section earlier in this chapter.

2. Determine the placement of the E stencil on the sheet. The overall dimensions of the stencil measure 5" × 8", and your working space is 7⅝" × 10½". To center the stencil, subtract 5" from 7⅝" and subtract 8" from 10½". Divide each answer by 2 to find out how much space to leave on each side of the E stencil. For the top and bottom of the paper, the number comes out to 1¼". For the sides, the answer is 1⁵⁄₁₆". Therefore, leave a bottom space of 1¼" and a right-side space of 1⁵⁄₁₆" *(Fig. 3-35A)*.

3. Mark off the horizontal and vertical dimensions *(Fig. 3-35B)*.

4. Lightly block in the stencil with construction lines *(Fig. 3-35C)*.

5. Darken and finish the stencil *(Fig. 3-35D)*.

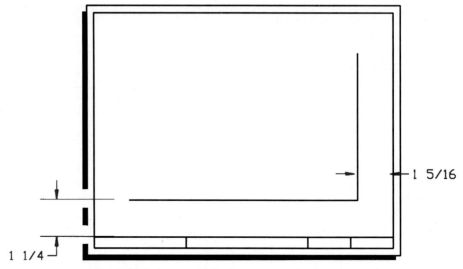

Fig. 3-35A Leave 1¼" of space on the bottom and on the right side.

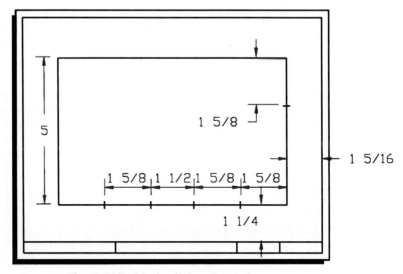

Fig. 3-35B Mark off the dimensions.

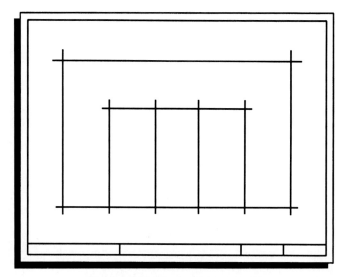

Fig. 3-35C Block in the stencil.

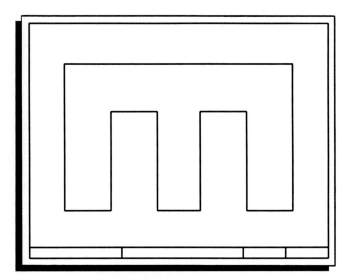

Fig. 3-35D Darken and finish.

To create multiview drawings, you follow the same general procedure as for one-view drawings. Follow these steps to create the multiview drawing shown in Fig. 3-36E.

1. Lay out an A-size sheet of drawing paper *(Fig. 3-36A)*.
2. Determine the size of each view and center the views in the available space. If possible, leave slightly more space between the views than between views and borders *(Fig. 3-36B)*.
3. Mark off the horizontal and vertical dimensions *(Fig. 3-36B)*.
4. Lightly block in each view with construction lines *(Fig. 3-36C)*.
5. Lightly add features *(Fig. 3-36D)*.
6. Darken and finish the views *(Fig. 3-36E)*.

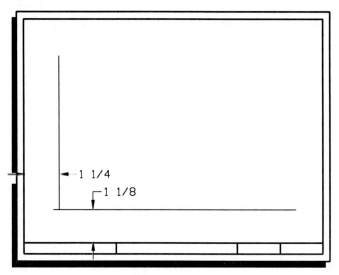

Fig. 3-36A Lay out the paper.

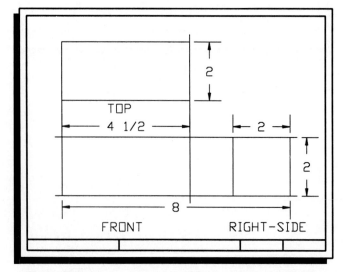

Fig. 3-36B Mark off horizontal and vertical dimensions.

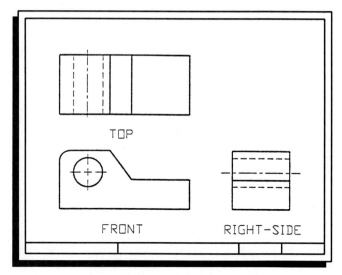

Fig. 3-36C Block in each view.

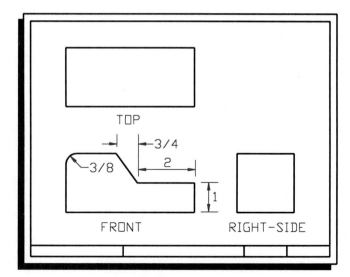

Fig. 3-36D Add features.

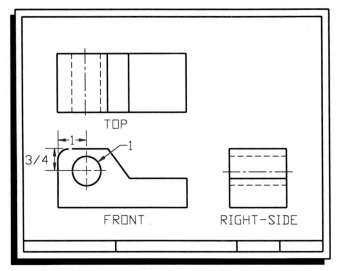

Fig. 3-36E Darken and finish.

Dimensioning

An object cannot be constructed from a picture alone. The views of an object show its shape, but dimensions and other shop notes are needed to supply additional information for construction.

Chapter 10, "Dimensioning," presents a detailed discussion of dimensioning conventions, rules, and practices that apply to computer-aided drafting. With few exceptions, they are the same for mechanical drafting. The major dif-ferences lie in the procedures. The mechanical drafter must pay strict attention to details such as arrowhead construction, lettering size and consistency, and dimension placement. Although these are also concerns of the CAD drafter, the computer takes care of consistency in all these areas. Once the drafter sets a procedure or style, the computer repeats the process with exacting efficiency.

ACTIVITY

To familiarize yourself with dimensioning styles used by mechanical drafters, look through drafting books or visit drafters in your area to look through drawings that have been dimensioned by hand.

1. Make a list of differences you see among the dimensioning styles.
2. Describe your favorite style and explain why it is your favorite.

Chapter 3 Review

1. Which traditional instruments does the drafting machine replace? What are the advantages and disadvantages of using a drafting machine?

2. Briefly explain the procedure for bisecting a line.

3. What is the difference between an inscribed hexagon and a circumscribed hexagon?

4. Which lettering style is most commonly used in engineering drawings?

5. What is orthographic projection? Why is it sometimes necessary?

Chapter 3 Problems

Using the principles you have learned in this chapter, recreate each of the following drawings, including the dimensions. Use the mechanical drafting instruments and techniques you have studied in this chapter.

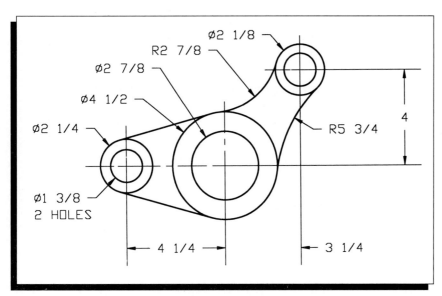

Problem 3-1

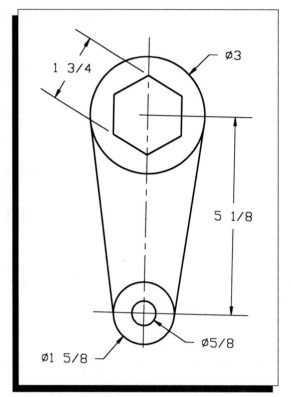

Problem 3-2

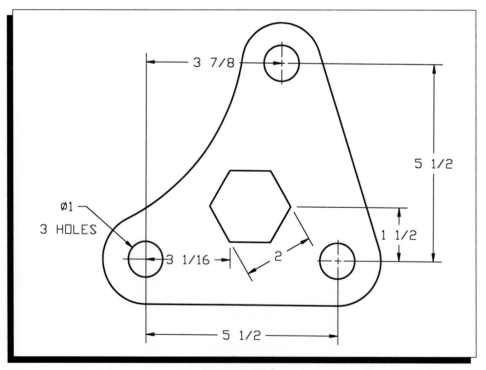

Problem 3-3

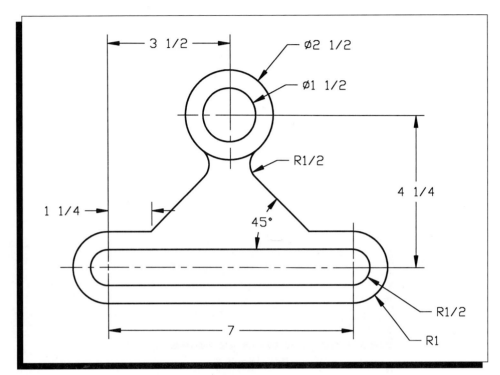

Problem 3-4

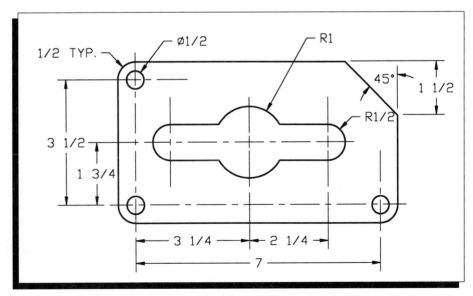

Problem 3-5

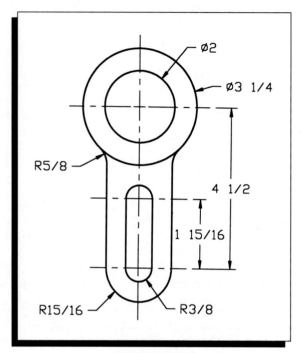

Problem 3-6

Using CAD

U.S. Navy Prints on Demand

Sailors on board ships in the United States Navy must read many drawings in order to repair and maintain the equipment. For example, one service manual takes 15 volumes for all the drawings and instructions for an engine. In the past, the Defense Printing Service used board-drafted images in their service manuals. When engineers made changes in designs or repair procedures, the plans had to be carefully redrawn; then all the volumes had to be reprinted. This required a great deal of time.

Ed Sargent works at the Bangor Submarine Base in Silverdale, Washington, where the Navy prints its manuals. Sargent, a "document conversion specialist," is busily converting drawings from the manuals into AutoCAD files so that even small changes can be made quickly on the computer without having to throw away out-of-date copies. Sargent can move lines and shapes, change scales and sizes, enter new dimensions, and add new drawings or delete old ones.

Converting the drawings to AutoCAD files is not so simple as it might sound. One method used, especially in the early days of CAD, is to trace the old drawings directly into AutoCAD. However, this process is very time consuming.

Ed uses another approach that is faster, but not so direct. He first electronically scans the drawings into a program called GTXRaster CAD. This program is essential; because, even though a scanner can import an image to the computer, it will not be in a form that can be used by AutoCAD. GTXRaster CAD does the job of converting the drawings into vectors—a form that matches AutoCAD.

Ed then works with AutoCAD to adjust or change the drawings as necessary and uses GTXRaster CAD to change the drawing back to the original format. The entire manual is then saved to another computer called a file server.

Now, when a submarine crew needs a new manual, they simply call the Defense Printing Service to order one overnight. The drawings go directly from the file server to a high speed printer that prints 8000 pages per hour. This electronic process is called Print on Demand (POD).

Ed Sargent is excited about his work and the fact that so much time is saved. "That's the beauty of it," he said. "We are saving so much money for the government!"

POD eliminates the need for storing warehouses of manuals. "We couldn't do it without the use of the scanner, GTX RasterCAD, the computer file server, and the high speed printer," Sargent says. "They are all tools that work together. Here, everything is electronic storage and electronic media. There is no need for paper anymore, except for the final output. And at 8000 pages per hour we can print entire service manuals on demand!"

Chapter 4

▶ Key Terms

graphics screen

cursor

Cartesian coordinate system

origin

World Coordinate System
 (WCS)

coordinate system icon

command line

screen menu

status line

pull-down menus

dialogue boxes

prototype drawing

alphabet of lines

layers

▶▶ Commands & Variables

GRID

LAYER

LIMITS

LINETYPE

LTSCALE

NEW

OPEN

ORTHO

SAVE

SNAP

UNITS

DDUNITS

Starting Drawings with AutoCAD

Objectives

When you have completed this chapter, you will be able to:

- begin an AutoCAD drawing.
- describe the standard screen layout, enter commands, and select options.
- create a custom prototype drawing.
- set drawing units, limits, Grid, Snap, and Ortho using the appropriate commands.
- create layers for drawing hidden lines and center lines and load their respective linetypes.
- save a drawing.
- start a new drawing using a customized prototype drawing.

Drawing is a method of visual communication. For drafters, drawing takes three forms: sketching, board drafting, and computer-aided drafting (CAD).

CAD can stand for either "computer-aided drafting" or "computer-aided design," depending on its primary use within an organization. When a company does both drafting and design work, its drafters and designers may use the acronym *CADD* for "computer-aided design and drafting."

Sketching is freehand drawing used to approximate ideas, while board drafting and CAD are accurate, precise drawing methods used to create detailed designs. Each method has its own tools and its own techniques.

When sketching and board drafting, drafters use hand tools and instruments to produce drawings on paper. CAD drawings, on the other hand, are created with a computer; then a printed, or hard, copy is made with a plotter or printer.

This chapter will introduce the operation of the AutoCAD computer-aided drafting program as operated on MS-DOS compatible computers. (AutoCAD is made by Autodesk, Inc., Sausalito, California.) It will also describe and illustrate the processes of setting up a prototype drawing and sheet layout.

Starting AutoCAD

AutoCAD may be run in a stand-alone or network environment. Stand-alone computers operate independently of each other, and each machine has its own software copies. Networked computers, however, are connected and share software. When you start the program, follow the instructions for the environment you are using.

The Graphics Screen

At start-up, AutoCAD proceeds directly to the **graphics screen,** which is the screen where drawing and editing occur.

> **Note** If you are using Release 10 or 11, the program begins with a Main Menu instead. Refer to Appendix B or C. See also the Introduction to this text.

The graphics screen contains five parts: the graphics area, the command line, the screen menu, the status line, and, behind the status line, the pull-down menus *(Figs. 4-1 and 4-2).* The **cursor,** which is controlled by the pointing device, allows you to locate and select the appropriate drawing commands and options.

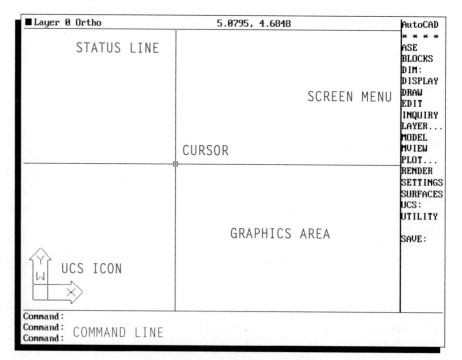

Fig. 4-1 The graphics screen.

The most common pointing devices are the mouse, digitizer, and trackball *(Fig. 4-3)*.

Note

The cursor might appear as crosshairs, a pick box, or another form depending on the command you select or the activity you are doing.

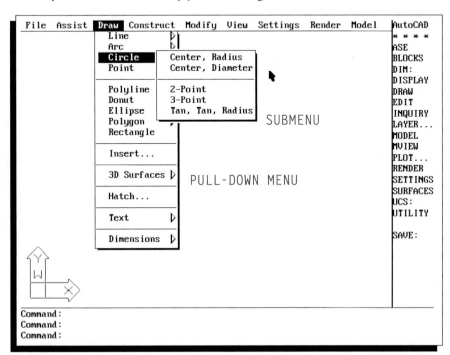

Fig. 4-2 Pull-down menu and submenu.

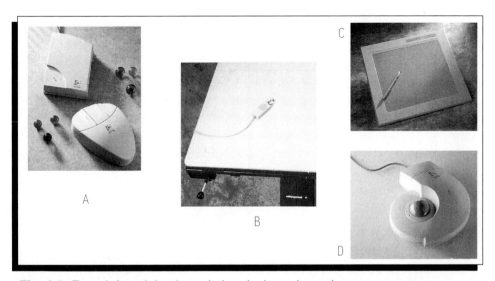

Fig. 4-3 From left to right, the pointing devices shown here are a mouse, a digitizer puck, a light pen, and a trackball. These and other pointing devices can be used to move the cursor and enter information in AutoCAD.

▶ The Graphics Area

The **graphics area** occupies most of the screen and is where drawing actually takes place. To understand the layout of this area, you should understand how AutoCAD drawing space is defined. AutoCAD is based on the **Cartesian coordinate system,** a point location system that uses a horizontal (X) axis and a vertical (Y) axis to assign values to specific points *(Fig. 4-4)*. If you have ever taken a geometry course, you will be familiar with this system.

The point at which the X and Y axes cross is (0,0). This point is called the **origin.** In AutoCAD's default display, the origin is located at the lower left corner of the graphics area. In other words, only quadrant I of the Cartesian coordinate system appears on the screen. AutoCAD refers to this default display area as the **World Coordinate System (WCS).**

The **coordinate system icon** appears in the lower left of the graphics area *(Fig. 4-2)*. This icon represents the orientation of the X and Y axes of the drawing. The arrow pointing to the right indicates that the X axis is horizontal, and the arrow pointing up indicates that the Y axis is vertical. The W appears because the drawing is in the World Coordinate System, which is AutoCAD's default coordinate system. Note that the icon does not plot or print on your drawing—it's there only for reference.

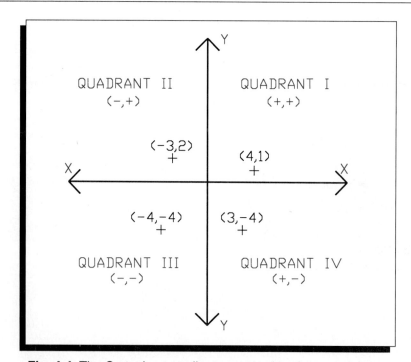

Fig. 4-4 The Cartesian coordinate system has four quadrants. Quadrant I is the only one in which both X and Y have positive values. In quadrant II, X is negative but Y is positive. In quadrant III, both X and Y are negative. In quadrant IV, X is positive but Y is negative. Quadrant I is AutoCAD's default display, referred to as the World Coordinate System.

▶ The Command Line

The **command line** is the main form of communication between AutoCAD and you. It actually takes up three lines at the bottom of the screen. The *Command:* prompt appears here, and this is where commands from the menus and keyboard are echoed, or repeated. AutoCAD "talks" to you and records your answers on the command line. Keep your eye on these lines for important information, prompts, and options.

▶ The Screen Menu

The **screen menu,** located at the right side of the screen, contains commands and menu options. The opening menu, which appears when you start AutoCAD, is called the "root menu." You can select commands and menu options by moving the cursor into the menu area with your pointing device and picking the command or option you want. Picking the word *AutoCAD* at the top of each menu cancels the command or option and returns you to the root menu. The CTRL-C key combination (hold the CTRL key down while you press the C key) also cancels commands and options and returns you to the root menu.

▶ The Status Line

The **status line** displays a number of items. The small square on the left shows the color of the active layer, followed by the name of the active layer. The blank space to the right displays various toggles such as Ortho, Snap, and Grid (discussed later in this chapter). The two numbers further to the right are the coordinate locations of the pointing device within the drawing area.

▶ Pull-Down Menus

The **pull-down menus** appear only when you move the graphics cursor off the drawing area into the status line *(Fig. 4-2).* The pull-down menus offer an optional method for entering commands or selecting options.

Command Entry

AutoCAD offers three methods for entering commands and selecting options: you can use the the screen menu, the pull-down menus, or the keyboard. Both the screen and pull-down menus require a pointing device. The pointing device is used to highlight and select a command or option.

Although numerous pointing devices are available, the 3-button mouse is a popular and efficient pointing device for AutoCAD. On a standard right-handed mouse, the left button is referred to as the pick button because it is used to select, or pick, an item on the screen. The center button is the object snap or OSNAP button, which you will learn to use later. The right button functions as an ENTER (↵) or RETURN key.

From this point, the pointing device will be referred to as a mouse for the sake of simplicity. If you are using a digitizer puck, trackball, or other pointing device, substitute the name of your device wherever the word "mouse" occurs.

▶ Screen and Pull-Down Menus

To activate the screen menu or pull-down menu, select the option or command with the pointing device. Then follow the operational sequence that appears in the command line at the bottom of the screen. Menu options that have ellipses (...) after them activate dialogue boxes when you select them. **Dialogue boxes** are pop-up boxes that let you perform operations by checking boxes or entering text *(Fig. 4-5)*. Pull-down menu options that have a small arrow after them open submenus when you select them.

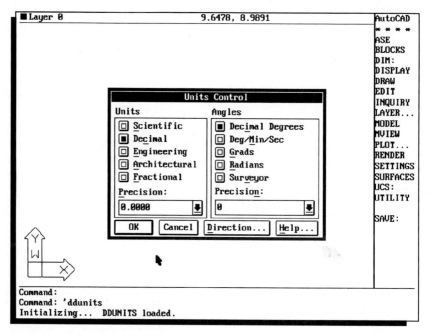

Fig. 4-5 Pop-up dialogue boxes allow you to set or change drawing specifications quickly and easily.

▶ The Keyboard

You may also enter commands and options on the command line directly from the keyboard. When you use the keyboard, you must use correct spelling and AutoCAD syntax. A *syntax* is a set of rules that applies to a given situation.

The AutoCAD syntax is the structure and the rules deduced therefrom that apply to the commands and options.

Even though pull-down and screen menus may seem easier to use at first, many people prefer to use the keyboard because it is a faster method of entry. Also, as a beginning AutoCAD student, you may learn more by using the keyboard. It may make the command sequences easier to follow and understand. The steps and sequences in this textbook will use whichever method is easiest for each individual case.

At the beginning of each chapter from this point, you will see a section called "Commands and Variables." This section lists keyboard commands, as well as the *variables,* or internal AutoCAD settings, that are covered in the chapter.

The function keys, located on the keyboard either above the keys or to their left, provide a shortcut for toggling some commands on and off. Also, the F1 key toggles between the text and graphics screens.

ACTIVITY

Follow the steps below to become familiar with the graphics screen.

1. Start AutoCAD. (The start-up procedure you follow depends on the type of computer system you have.)
2. Move the pointing device and note the movement of the cursor on the screen. What type of cursor appears?
3. Find the following components of the graphics screen: the graphics area, the command line, the screen menu, and the status line.
4. Move the cursor off the drawing area into the status line. Note the appearance of the pull-down menus.

Prototype Drawings

In a sporting event, before the game can begin, the rules must be agreed upon, the playing field set up, and the teams organized. The same holds true for CAD—certain parameters must be set up before you can begin drawing.

When you start a drawing in AutoCAD, the drawing screen is set up according to certain conditions. For example, the drawing might be set up to fit on a certain size paper when printed. It might be set for a certain

color when you draw, and so on. These default values are stored in a **prototype drawing** and become common to all drawings you create using that prototype drawing.

When you first start AutoCAD, the screen you see is set up from a prototype drawing named ACAD.DWG that comes with AutoCAD. You can use this default prototype drawing, change it so that it meets your needs, or create your own prototype drawing.

Custom Prototype Drawings

For experienced drafters, the ACAD.DWG prototype may be too general; if so, they may decide to create their own prototypes. They might prepare a simple prototype that establishes the units and sets values for limits, Grid, and Snap. You will learn more about these parameters later in this chapter. More advanced prototypes may add borders, title blocks, text styles, layers, and other variables. Drafters often have a separate prototype for each drawing size.

Planning the Drawing

Before you set up a prototype drawing, determine your needs and goals. Planning plays a crucial role in all CAD operations, and it should start at the beginning: with the prototype drawings. Will you be making mechanical drawings, electrical drawings, or architectural drawings? What plotter and paper sizes are available to you? How complex will your drawings be? Time spent in planning before drawing will pay off in less drafting time and lower frustration levels later.

Sheet Sizes and Layout

AutoCAD provides complete freedom from drawing size constraints. Drawings of any size can be made by adjusting the screen limits, as described later in the section "The LIMITS Command." Hard copies of the drawings, however, must conform to standard paper and equipment (plotter or printer) sizes, and it is often necessary to scale (reduce or enlarge) drawings to fit the available paper. A well-planned drawing will have limits set which accurately represent the finished drawing size. Fig. 4-6 lists the ANSI sheet letters with their respective sizes in inches, along with the International Standards Organization (ISO) sheet designations for metric sizes.

Some drafters also add a border and title strip to the prototype *(Fig. 4-7)*. Borders provide visual limits for a drawing and also provide a place to put essential information such as the name of the company, the date, and the drafter's name. Border sizes differ depending on the paper size used.

Creating a Prototype Drawing

In the activity on page 113, you will set up a simple prototype drawing with the UNITS, LIMITS, GRID, SNAP, ORTHO, LINETYPE, and LAYER commands. The commands you need to type are capitalized, and the information to be entered is in bold type. The prompts are in italics. The ⏎ symbol indicates a new line (press the ENTER or RETURN key).

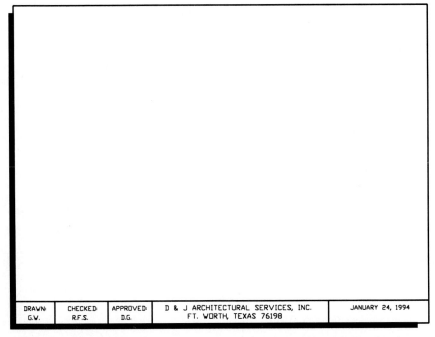

U.S. CUSTOMARY SERIES			ISO STANDARD	
Size	First Series	Second Series	Size	Third Series
A	8 ½ x 11 in.	9 x 12 in.	A0	841 x 1189 mm
B	11 x 17 in.	12 x 18 in.	A1	594 x 841 mm
C	17 x 22 in.	18 x 24 in.	A2	420 x 594 mm
D	22 x 34 in.	24 x 36 in.	A3	297 x 420 mm
E	34 x 44 in.	36 x 48 in.	A2	210 x 297 mm

Fig. 4-6 Standard ANSI and ISO drawing sheet sizes.

DRAWN: G.W.	CHECKED: R.F.S.	APPROVED: D.G.	D & J ARCHITECTURAL SERVICES, INC. FT. WORTH, TEXAS 76198	JANUARY 24, 1994

Fig. 4-7 A border such as this one provides visual boundaries for the drawing and provides a place for essential drawing information.

▶ The UNITS and DDUNITS Commands

The UNITS and DDUNITS commands establish the appropriate units of measure for a drawing. For example, do you want to measure in common fractions or decimal fractions? Will you measure angles in degrees or radians? The difference between the UNITS and the DDUNITS commands is that the DDUNITS command presents the Unit Control dialogue box, from which you can choose the units and precision of the units. (The precision determines the number of places to which AutoCAD calculates the math.) Refer back to Fig. 4-5 to see the Unit Control dialogue box. Fig. 4-8 shows the command sequence that results when you enter the UNITS command.

```
UNITS
Report formats:              (Examples)

   1.  Scientific            1.55E+01
   2.  Decimal               15.50
   3.  Engineering           1'-3.50"
   4.  Architectural         1'-3 1/2"
   5.  Fractional            15 1/2

With the exception of Engineering and Architectural formats,
these formats can be used with any basic unit of measurement.
For example, Decimal mode is perfect for metric units as well
as decimal English units.

Enter choice, 1 to 5 <2>: 2
Number of digits to right of decimal point (0 to 8) <3>: 3

Systems of angle measure:         (Examples)
   1.  Decimal degrees            45.0000
   2.  Degrees/minutes/seconds    45d0'0"
   3.  Grads                      50.0000g
   4.  Radians                    0.7854r
   5.  Surveyor's units          N 45d0'0" E

Enter choice, 1 to 5 <1>: 1
Number of fractional places for display of angles (0 to 8)
<0>:0

Direction for angle 0:
   East    3 o'clock = 0
   North  12 o'clock = 90
   West    9 o'clock = 180
   South   6 o'clock = 270
Enter direction for angle 0 <0>:0

Do you want angles measured clockwise? <N> N

Command:
```

Fig. 4-8 The UNITS command sequence.

Some AutoCAD commands switch the user automatically to a non-graphic text screen. You can toggle between the graphics screen and the text screen by pressing the F1 key.

ACTIVITY

As you go through the following sequence, AutoCAD presents choices in three main categories:
- report formats
- systems of angle measure
- direction for angle

1. *Command:* **UNITS**

When you enter the UNITS command, AutoCAD lists five choices under the heading "Report formats" and asks you to select the format for your drawing.

2. *Enter choice, 1 to 5 <default>:* **2** ↵

By entering 2, you have chosen to use decimal units in this drawing. Note that in the prompt line described above, the word "default" in brackets is used to represent a number or value that AutoCAD provides in the prompt line. You should see a number in the brackets. This number is either the system default value or the value you entered the last time you used this command. You can enter this number automatically by pressing the ENTER key.

3. *Number of digits to right of decimal point (0 to 8) <default>:* **3** ↵

This prompt appears only if you enter 1, 2, or 3 at the prompt line shown in step 2. If you had selected choice 4 or 5, the prompt would have been: *Denominator of smallest fraction to display (1, 2, 4, 8, 16, 32, or 64) <default>:* and you would choose the appropriate common fraction. Your choice for either prompt determines the degree of precision of your units.

4. *Enter choice, 1 to 5 <default>:* **1** ↵

Five choices are given under the heading "Systems of angle measure," and you are asked to select the system for your drawing. By entering 1, you have chosen to measure angles using decimal degrees.

5. *Number of fractional places for display of angles (0 to 8) <default>:* **0** ↵

Your choice determines the degree of precision of your angle units. You then have to define the direction for angle 0.

6. *Enter direction for angle 0 <default>:* **0** ↵

This choice orients a circle in degrees, grads, radians, or direction relative to a compass (East, North, West, and South) and to a clock face (3 o'clock, 12 o'clock, 9 o'clock, and 6 o'clock).

7. *Do you want the angles measured clockwise? <default>* **N** ↵

8. *Command:* Press the **F1** key to return to the drawing screen.

▶ The LIMITS Command

The LIMITS command sets the coordinates of the graphics area. In other words, it establishes the drawing size by marking the lower left-hand corner and the upper right-hand corner of the graphics area. AutoCAD drawings generally are made full size. This is essential if the AutoCAD drawing will be used with a CNC system. For example, if you are working with a 12" × 9" sheet of paper, you will set the lower left corner at 0,0 and the upper right corner at 12,9. If necessary, the drawing can be reduced when it is printed or plotted.

Note CNC stands for "Computer Numerical Control." A CNC system is a manufacturing system that accepts specially coded instructions from the design computer. AutoCAD files can be converted to CNC-acceptable format and used directly to manufacture the object shown in the AutoCAD drawing.

Many drawings are planned using a certain scale. For example, a drawing of a house will be used in the field by builders. It is not practical to draw the house at its full size. Architectural drawings are typically drawn at ¼" = 1', which means that an inch on the drawing equals four feet. To figure the drawing limits, you multiply each length of the drawing sheet by four. For example, a C-size sheet is 24" × 18". The limits for an architectural drawing on a C-size sheet are 0,0 and 96,72.

Because drawing limits define the size of the drawing, they usually depend on the following factors:

- the actual size of the object to be drawn
- the space required for dimensions, notes, and text
- the free space around each view of the object
- the border and title block areas

It may be helpful to make a sketch of the drawing to calculate the space needed.

Your Turn

ACTIVITY

The following procedure sets the drawing limits for a standard 12" × 9" A-size sheet. When you enter the values, always insert the X, or horizontal, value first. Separate the values with a comma but no spaces. The ZOOM command at the end of this sequence resets the screen to the actual limits you specified.

1. *Command:* **LIMITS** ↵
2. *Reset Model space limits: ON/OFF/<Lower left corner> <current value>:* **0,0** ↵
3. *Upper right corner <current value>:* **12, 9** ↵
4. *Command:* **ZOOM** ↵
5. *All/Center/Dynamic/Extents/Left/Previous/Vmax/Window/<Scale(X/XP)>:* **A** ↵

Entering A selects the All option. Similarly, AutoCAD allows you to select any command or option that has capital letters by entering only the capital letters.

Notice that "Scale(X/XP)" appears in angle brackets (<>). In AutoCAD, the option that appears in angle brackets is the currently selected option or the default option. In most cases, if you want to accept this current or default option, you can simply press the ENTER key.

Note

Grid, Snap, and Ortho

Once you have established the limits, you can set up the drawing aids Grid, Snap, and Ortho. The GRID command places a dotted grid on the screen at intervals that you designate. The grid appears only on the screen; it will not plot or print.

While Grid is a visible aid, Snap is an invisible grid that permits you to move your cursor to exact points. With the SNAP command, you can deter-

mine how far the cursor will move at each point.

The ORTHO command allows you to toggle the Ortho mode on and off. When Ortho is toggled on, it allows lines to be drawn only horizontally and vertically. Under most circumstances, no matter how you move the cursor, you cannot draw a diagonal line while Ortho is on.

Your Turn

ACTIVITY

The following procedure sets up Grid, Snap, and Ortho for your drawing.

1. *Command:* **GRID** ⏎
2. *Grid spacing (X) or ON/OFF/Snap/Aspect <current value>:* **1** ⏎

Note the grid that appears on the screen. Since you entered 1, the distance between each dot represents one unit.

3. *Command:* **SNAP** ⏎
4. *Snap spacing or ON/OFF/Aspect/Rotate/Style <current value>:* **0.25** ⏎
5. Move the cursor. Note that it snaps to a quarter of the distance between each dot.
6. *Command:* **ORTHO** ⏎
7. *ON/OFF <current setting>:* **on** ⏎

You will be able to see how this works when you draw lines in the next section.

Once they have been set, Grid and Snap may be toggled off and on with the function keys F7 and F9, respectively. Since Ortho has only off and on settings, you can toggle it with the F8 key at any time.

▶ The LINETYPE Command

Drawings are made with various kinds of lines. Different lines have different meanings that drafters must understand to express their ideas effectively. To standardize drawing communication, ANSI developed symbols for an **alphabet of lines** *(Fig. 4-9)*. The document describing the symbols is ANSI Y14.2M-1979(R1987).

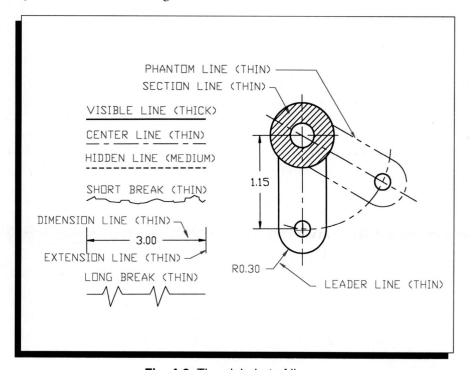

Fig. 4-9 The alphabet of lines.

The LINETYPE command activates line symbols that are stored in a library file. These lines include dashed, hidden, center, phantom, dot, dashdot, border, and divide *(Fig. 4-10)*. LINETYPE also contains provisions for creating lines that are not available in the library. Unless the drafter changes them using the LINETYPE command, all lines are drawn as continuous, or visible, lines.

AUTOCAD'S STANDARD LINETYPES

BORDER	DIVIDE
BORDER2	DIVIDE2
BORDERX2	DIVIDEX2
CENTER	DOT
CENTER2	DOT2
CENTERX2	DOTX2
DASHDOT	HIDDEN
DASHDOT2	HIDDEN2
DASHDOTX2	HIDDENX2
DASHED	PHANTOM
DASHED2	PHANTOM2
DASHEDX2	PHANTOMX2

Fig. 4-10 The standard library of linetypes supplied with AutoCAD.

As drawing sizes change, so do the relationships among the elements of the lines. Since Release 11, AutoCAD includes "2" and "X2" variations of all lines to halve and double the relationships. The LTSCALE (linetype scale) command adjusts the length of the elements and their spacing. Fig. 4-11 shows the effects of changing the scale factor on a center line.

ANSI line widths cannot be controlled within AutoCAD except as poly-lines. (Polylines are special lines, circles, and arcs which can be drawn with varying widths. You will learn more about polylines in Chapter 6.) However, you can change line widths when you create hard copy by changing the pens in your plotter. Plotters put lines on paper with felt tip, ballpoint, or wet ink pens that come in various line widths. Figs. 4-12 through 4-15 show examples of visible lines, hidden lines, center lines, and phantom lines.

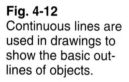

Fig. 4-11 The LTSCALE command alters the appearance of lines. As the LTSCALE setting becomes higher, the scale of the line elements becomes larger.

Fig. 4-12
Continuous lines are used in drawings to show the basic outlines of objects.

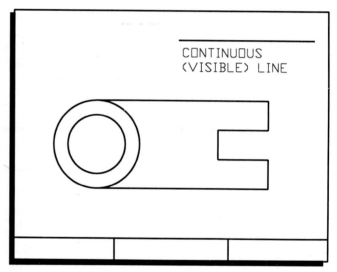

Fig. 4-13
Hidden lines are used to show lines that would be hidden from view if you were looking at the object.

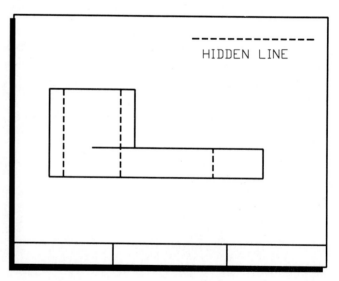

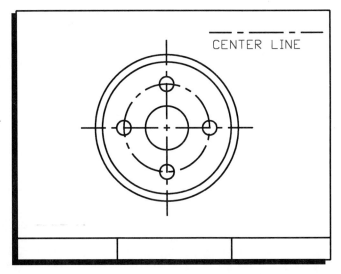

Fig. 4-14
Center lines are used to show imaginary lines through the center of an object or part of an object.

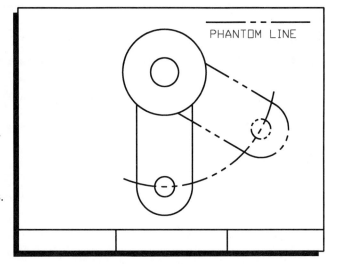

Fig. 4-15
Phantom lines are used to show lines that are not present in the real object. In this example, the phantom lines show a possible location of the object after it has been rotated around a central axis.

ACTIVITY

The following procedure will load the hidden and center lines.

1. *Command:* **LINETYPE** ↵
2. *?/Create/Load/Set:* **L** ↵
3. *Linetype(s) to load:* **hidden,center** ↵

Do not enter a space between the comma and the word "center." The Select Linetype File dialogue box appears. "ACAD" should appear after the word "File" at the bottom of the dialogue box *(Fig. 4-16)*. This is the default library file named ACAD (ACAD.LIN) which is where the line symbols are stored.

4. Pick **OK.**
5. *?/Create/Load/Set:* ↵

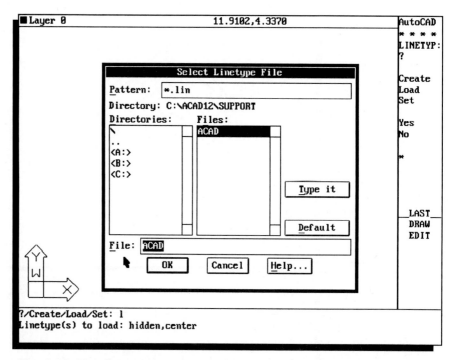

Fig. 4-16 The Select Linetype File dialogue box allows you to select a file that contains linetype specifications. AutoCAD's default linetype file is ACAD.LIN.

▶ The LAYER Command

Complex designs must frequently be broken down into their component parts. This can be done by using transparent overlays, which AutoCAD calls **layers** *(Fig. 4-17)*. Each layer contains specific information which can be viewed and/or reproduced on its own or in combination with other layers. All the layers together reflect the entire design.

Drafters generally create separate layers for views, sections, hidden features, dimensions, and notes. Each layer can have its own name, color, and linetype. You can make any layer current

for drawing, turn it off or on for visual impact, or freeze it so AutoCAD will ignore it. You can also plot layers separately so that you can use different pen widths or colors.

You can access the LAYER command either by entering the LAYER command or by using the pull-down menu with its associated Layer Control dialogue box. The dialogue box lets you perform operations by checking boxes or entering text. It is much easier to use than the LAYER command, which requires you to know the name of the layer you want to manipulate.

You can also access the Layer Control dialogue box directly by entering "DDLMODES" at the keyboard. Although Releases 10 and 11 use dialogue boxes for layer control, the boxes are different from the box in Release 12. Refer to Appendix B or C for details about Releases 10 or 11.

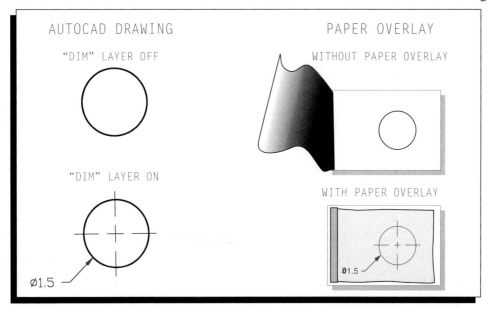

AUTOCAD DRAWING

"DIM" LAYER OFF

"DIM" LAYER ON

Ø1.5

PAPER OVERLAY

WITHOUT PAPER OVERLAY

WITH PAPER OVERLAY

Ø1.5

Fig. 4-17 AutoCAD's layers are similar to overlays on a drawing. You can display or hide each layer as you need it. In this example, the dimensions of an object are contained on a layer called DIM.

ACTIVITY

1. Activate the pull-down menus and pick **Settings, Layer Control.**

2. When the Layer Control dialogue box appears, the cursor is flashing in the blank bar under the word "New," as shown in Fig. 4-18. Type the layer name **HIDDEN,** and pick **New.** Note that HIDDEN is now listed as a layer name. It is currently set to "on," the color is white, and the linetype is "CONTINUOUS."

3. Create another layer by typing **CENTER** and picking **New.**

4. To change linetypes, highlight the **HIDDEN** layer by picking it. Then pick **"Set Ltype . . .".** The Select Linetype dialogue box appears.

5. Pick the hidden line (not the word HIDDEN). The linetype changes to HIDDEN.

6. Pick **OK.**

7. When the Layer Control dialogue box appears, click on the **HIDDEN** layer so that it will be cleared, or no longer selected.

8. Repeat steps 4 and 5 to change the CENTER layer linetype to CENTER. Refer to Fig. 4-19 for the completed dialogue box.

9. Pick **OK** to return to the command prompt.

Your Turn

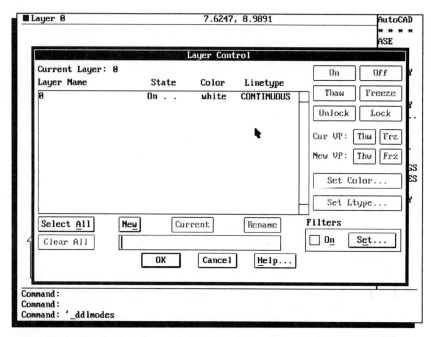

Fig. 4-18 The Layer Control dialogue box allows you to turn layers on or off, thaw or freeze them, set their characteristics, and create new layers. It also allows you to make a different layer "current."

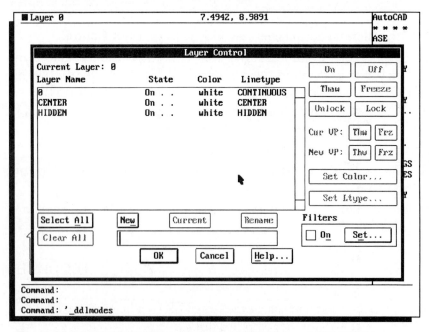

Fig. 4-19 In this illustration, the layers HIDDEN and CENTER have been created and their respective linetypes have been loaded.

▶ Saving the Prototype Drawing

You have now completed your prototype drawing. The final step is to name the drawing and save it. You should record the drawing name and the settings you have established for future reference.

Releases 10 and 11 use a different procedure for saving drawings. Refer to Appendix B or C.

There are several ways to save a drawing in AutoCAD. The SAVE command saves the drawing with the name you specified when you began the new drawing. If you did not specify a name at that time, you can specify one now.

The SAVEAS command allows you to save a copy of the current drawing using a new name that you specify. The END command saves the drawing with its existing name and ends the AutoCAD session.

ACTIVITY

1. *Command:* **SAVE** ↵

2. The Save Drawing As dialogue box appears with the flashing cursor in the "File:" box *(Fig. 4-20).* Type the path (drive and directory) where you want to save the drawing, followed by the name PROTO. For example, A:PROTO tells AutoCAD to save the drawing to drive A: with the name PROTO.

3. Pick **OK.** The drawing is saved with the name PROTO, and you are returned to the command prompt.

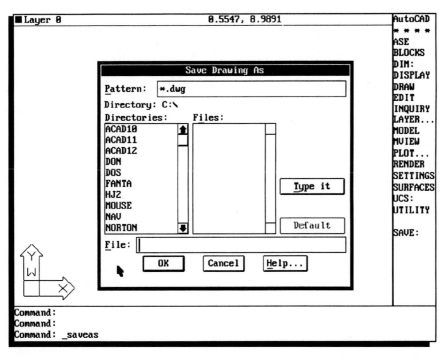

Fig. 4-20 The Save Drawing As dialogue box saves the drawing with the name you specify.

Drawing in AutoCAD

When you start a drawing session, you can begin a new drawing using AutoCAD's prototype or your own prototype, or you can open and continue with an existing drawing.

Note For Releases 10 and 11, Appendixes B and C describe the Main Menu and its options for beginning a drawing session.

Beginning a New Drawing

As mentioned earlier, when you start AutoCAD, the program goes directly to the graphics screen with a default prototype of ACAD.DWG. To accept AutoCAD's prototype for your new drawing, start drawing. However, if you want to use your custom prototype to create a drawing, you must load your prototype before you begin to draw. To load the prototype, you must begin a new drawing using the NEW command.

ACTIVITY

The following procedure replaces the ACAD.DWG prototype drawing with your custom prototype drawing.

1. *Command:* **NEW** ⏎

The Create New Drawing dialogue box appears *(Fig. 4-21)*.

2. Pick the box containing "acad" to the right of the "Prototype..." box and erase "acad" by using the DELETE or BACKSPACE key.

3. Type **PROTO** for the name of your prototype drawing. Be sure to include the path (disk drive and directory) if necessary.

Note You do not have to name a new drawing when you open it. To open a new drawing without naming it, press RETURN when the Create New Drawing dialogue box appears.

4. Pick the box to the right of the "New Drawing Name..." box and enter the name TEMP for your new drawing. Also, be sure to include the path (disk drive and directory) if necessary.

5. Pick **OK.**

6. Use the **SAVE** command to save the drawing.

You have now created a drawing based on the PROTO prototype. In the next section, you will reopen the drawing.

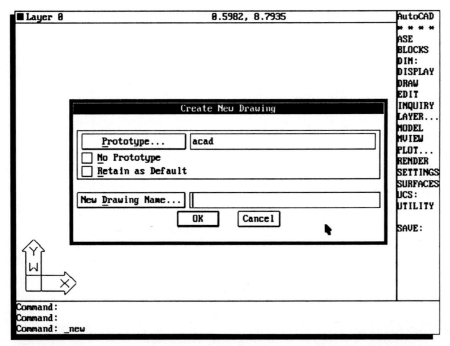

Fig. 4-21 The Create New Drawing dialogue box allows you to create a new drawing using the prototype of your choice. If you do not specify a prototype, AutoCAD uses its default prototype, ACAD.

Opening an Existing Drawing

After you have stored a drawing, you may want to work on it again in another drawing session. The following procedure shows you how to open an existing drawing using the OPEN command.

The procedure for opening an existing drawing is different in Releases 10 and 11 of AutoCAD. (Refer to Appendix B or C for the correct procedure to use in Releases 10 and 11.)

Note

Your Turn

ACTIVITY

1. *Command:* **OPEN** ↵

The Open Drawing dialogue box appears *(Fig. 4-22)*.

2. If the TEMP drawing appears in the "Files:" box, pick it. It will now show in the "File:" box.

3. If TEMP does not appear, you may need to change the directory or drive. If so, go to the "Directories:" box and click the mouse pick button on the directory or drive you specified when you saved the TEMP drawing. Then pick the OK button. A listing of the files in the selected directory or drive will appear in the "Files:" box. Pick **TEMP.**

4. When the drawing you wish to open appears in the "File:" box, pick **OK.**

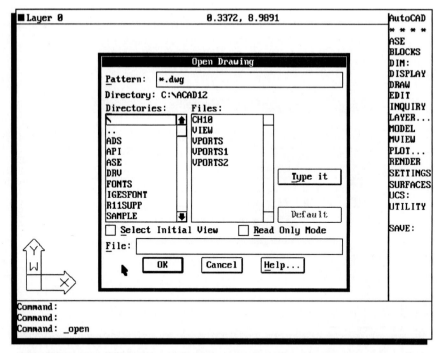

Fig. 4-22 The Open Drawing dialogue box allows you to open existing drawings within AutoCAD.

■ Chapter 4 Review

1. What is a dialogue box?
2. What is the purpose of a prototype drawing?
3. Describe three methods of entering a command in AutoCAD.
4. Why is it necessary to plan before you begin a drawing?
5. What does the UNITS command do?
6. Explain what a linetype is and how linetypes are used in AutoCAD.
7. How are hard copies of a CAD drawing produced? *(Fig. 4-23 A and B)*
8. What are layers in a CAD drawing? What is their purpose?
9. How do you save a drawing in AutoCAD?
10. How do you open an existing drawing?

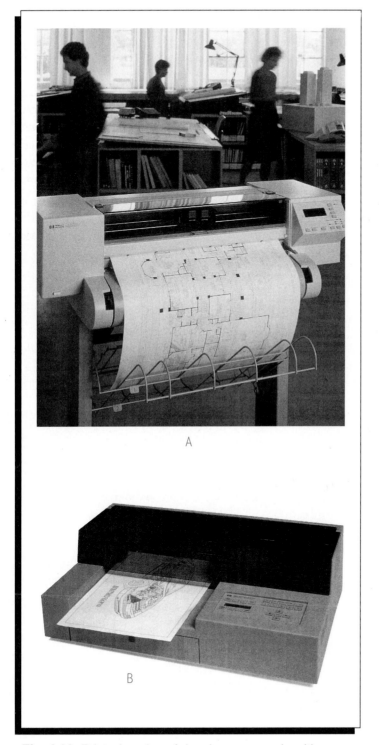

A

B

Fig. 4-23 Printed copies of drawings are made with
A) a plotter or B) a printer.

■ Chapter 4 Problems

Do the following problems to try the skills you learned in this chapter.

1. Obtain a copy of the *AutoCAD Reference Manual* from your teacher. This book is supplied with the AutoCAD software. Find the commands you learned about in this chapter and try other ways the commands can be used. You may want to do this as you go through the text; you may discover that you prefer to use the commands in ways other than those covered here. Read about the DDUNITS command and use it to set the units in the next problem.

2. Create a prototype drawing for use in architectural drafting. For architectural drafting, you should specify units of feet and inches. Set up the drawing as a C-size sheet at a scale of ¼" = 1'. Name your prototype and save it to a diskette for inspection by your teacher.

Using CAD

AutoCAD Used for Business and Recreation

Melanie Yamamoto, a self-taught AutoCAD user, could not imagine life without AutoCAD. She uses the program for business and recreation.

A graduate of the University of Hawaii with a degree in electrical engineering, Yamamoto first began using AutoCAD when she worked for the phone company as a transmission engineer. She used AutoCAD to set up a management system for installing, moving, and removing telephone equipment in the Central Office. The floor plans had specific layers to show the installers the location of existing, future, and removed equipment.

Yamamoto later worked as a CAD operator for an electrical engineering firm. She often worked with drawings of twenty different layers or more detailing lighting, electrical and communication outlets, appliances, existing and new wiring, as well as other engineering and architectural information.

Today, a self-employed Yamamoto uses AutoCAD for civil engineering and surveying applications. Since 1990, the Land Utilization Department for the city and county of Honolulu has required that all submitted maps of

land developments and subdivisions be in DXF format for their Geographical Information System (GIS). DXF is an AutoCAD file format that allows files to be shared with other CAD systems.

Because engineers and surveyors find that doing the annotations on their maps is tedious, time consuming, and not cost efficient, they send the drawings on a diskette to Yamamoto. Although she must be very accurate and diligent when checking the information in her client's drawings, Yamamoto finds the work to be very relaxing.

To fulfill the need for creativity, Yamamoto uses AutoCAD to design T-shirts, logos, brochures, and displays for presentations. "Since no drawing package can quite match the preciseness of realizing a design," Yamamoto usually imports the drawing into her desktop publisher. She depends on AutoCAD for creating highly geometric designs.

Yamamoto is impressed with the flexibility and creativity provided by AutoCAD. As the program's capabilities increase, so do the ways Melanie Yamamoto can utilize AutoCAD.

Chapter 5

► Key Terms

entities	editing
points	selection set
absolute coordinates	window
relative coordinates	crossing window
polar coordinates	concentric
blips	construction lines

►► Commands & Variables

LINE	ZOOM
CIRCLE	ERASE
ARC	TRIM
REDRAW	EXTEND
OFFSET	PLOT

Basic Drawing, Editing, and Plotting

Objectives

When you have completed this chapter, you will be able to:

- select points using absolute coordinates, relative coordinates, and polar coordinates.
- draw lines, circles, and arcs using AutoCAD commands.
- create parallel lines, circles, and arcs using the OFFSET command.
- create selection sets by various methods.
- clean up the screen using the REDRAW command.
- edit objects using the TRIM, ERASE, ZOOM, and EXTEND commands.

The saying "Someday we've got to get organized" is often followed by "Now we're organized. What are we going to do next?" These thoughts apply directly to computer-aided drafting. AutoCAD is running, the drawing is set up, and the prototype has been saved; you're organized. Now what are you going to do? The obvious answer is, "Now you draw." Chapter 5 introduces the basic concepts of drawing with the LINE, CIRCLE, and ARC commands, and editing with the jOFFSET, ZOOM, ERASE, TRIM, and EXTEND commands. The chapter also illustrates drawing cleanup with the REDRAW command. Finally, this chapter illustrates the procedures used to plot a drawing.

Drawing Entities

All AutoCAD drawings are made up of entities. **Entities** are predefined drawing elements you can place at specific coordinate locations by using an AutoCAD command. Some examples of entities you will use are points, lines, arcs, circles, text, dimensions, and polylines *(Fig. 5-1)*.

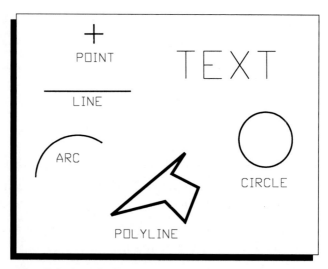

Fig. 5-1 AutoCAD supplies several pre-defined entities.

Using Coordinates

The fundamental unit of all CAD drawing is the point. **Points** describe screen locations and are used to place all other entities on the drawing screen. AutoCAD allows you to define points using three different types of coordinates: absolute, relative, and polar. In the exercises in this section, you will practice using all three. Refer to Fig. 5-2 as you read the following paragraphs about the different kinds of coordinates.

Points expressed using **absolute coordinates** have (X,Y) values that relate directly to the origin. The X value specifies how far the point is from the origin along the X axis. The Y value specifies how far the point is from the origin along the Y axis. For example, if you started a line at the origin (0,0) and entered the point (3,2), a line would extend from (0,0) to point (3,2) on the Cartesian coordinate system.

Relative coordinates relate the current point to the previous point. For example, the relative coordinate (4,2) identifies a point 4 units to the right on the X-axis and 2 units up on the Y-axis from the previous point. In AutoCAD, relative coordinate points are preceded by the @ symbol. In the preceding example, the point would be written *@4,2.*

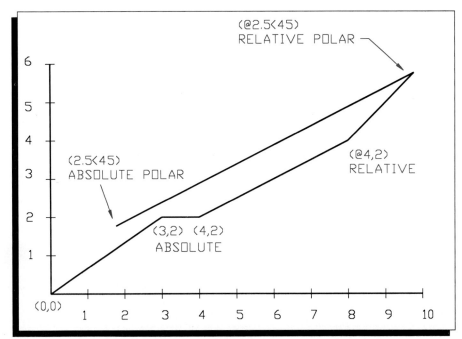

Fig. 5-2 Methods of locating points in AutoCAD.

Polar coordinates describe points at a distance and an angle from either the origin (absolute) or a previous entity (relative). They are usually expressed relatively, so the distance is preceded by the @ symbol. The angle is preceded by an angle symbol (<). For example, the polar coordinate *@2.5<45* places a point 2.5 units away from the previous entity at an angle of 45 degrees. The polar coordinate *3<45* places a point three units away from *absolute zero* at an angle of 45 degrees.

Using the Basic Drawing Commands

Since almost all objects can be defined with lines, circles, and arcs, the LINE, CIRCLE, and ARC commands form the basis of most AutoCAD drawings.

As you may remember from Chapter 4, AutoCAD provides three ways to issue commands. First, you can enter the command at the command line using the keyboard. The examples in this book are written for this method unless the text states otherwise. You can also use the mouse (or other pointing device) to select items from the screen menu or from the pull-down menus at the top of the screen. The command sequences you enter sometimes vary according to the method you use to enter them.

When you use the keyboard for entry, AutoCAD provides shorthand commands that allow you to enter a command by entering only the first few letters of the command name. For clarity, the examples in this book are written using the full command names. When you become more familiar with AutoCAD, however, you may wish to use the shorthand commands.

▶ The LINE Command

The LINE command constructs a line from one point to a second point. When accuracy is not required, you may choose the points using the mouse to place points on the screen "by eye." This is the fastest way of creating points, but it is the least accurate.

Make sure Snap is on when you use this method to avoid gaps and overlaps in your lines.

When accuracy is required, use the keyboard to enter absolute, relative, or polar coordinates for the points.

ACTIVITY

All of the exercises in this chapter will be performed on a single drawing. When you complete the exercises, you will have a drawing of a simple gasket.

The exercises in this chapter are designed to give you practical experience using the commands presented in this chapter. As you learn more about AutoCAD, you may find better, easier, or more practical methods of drawing similar objects.

To begin, start AutoCAD. Then follow the steps below to set up the drawing and to practice using the LINE command. As you perform steps 23 through 27, identify to yourself the type of coordinates used to specify the points. The result of this exercise is shown in Fig. 5-3.

1. *Command:* **NEW**

The Create New Drawing dialogue box appears.

2. Enter the name **GASKET** for the name of the new drawing. Be sure to enter the correct path name also.

3. Press **RETURN.**

AutoCAD sets the current drawing name to GASKET.

4. *Command:* **LINETYPE** ⏎

5. *?/Create/Load/Set:* **L** ⏎

6. *Linetypes to load:* **CENTER** ⏎

The Select Linetype File dialogue box appears.

7. Pick **OK.**

8. *Linetype CENTER loaded*
   ?/Create/Load/Set: ⏎

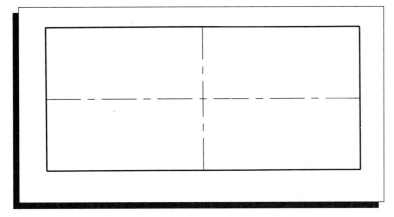

Fig. 5-3

9. Use the mouse to select the **Settings** pull-down menu and pick **Layer Control.**

10. Type **OBJECT** and use the mouse to pick the **New** button.

OBJECT appears in the Layer Name column at the top left of the dialogue box. Notice that all the layers are turned on, the color of each layer is white, and the linetype for each layer is continuous.

11. Using the mouse, pick the OBJECT layer name that just appeared.

The entire line becomes highlighted.

12. Pick the **Set Color...** button on the right side of the dialogue box.

13. When the Select Color subdialogue box appears, select **red** (you can type **red** or pick the red color at the top of the box). Then pick **OK.**

14. Use the mouse to pick the box below the New button. Then press the **BACKSPACE** key until OBJECT disappears from the box, and type **CENTER.**

15. Pick the **New** button to create the CENTER layer.

16. Change the color of layer CENTER to **yellow.**

Be sure layer OBJECT is not still highlighted, or its color will change to yellow also.

---

17. Pick the **Set Ltype...** button.

18. *Doubleclick* (press the mouse pick button twice) on the CENTER linetype. Be sure to pick the line, not the word CENTER.

Doubleclicking has the same effect as picking with the mouse and then picking the OK button. You may use this method in many cases where the OK button is displayed.

---

19. Highlight the OBJECT layer and pick the **Current** button. OBJECT should appear as the current layer at the top left of the dialogue box.

20. Check the chart on the Layer Control dialogue box. It should look like the one in Fig. 5-4. If it does, pick the **OK** button.

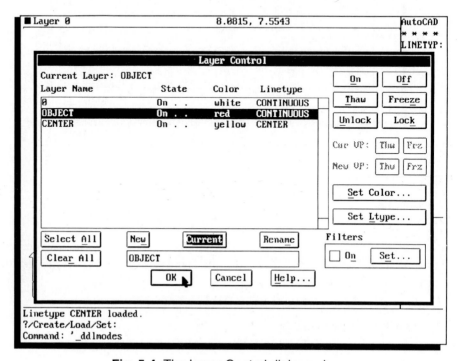

Fig. 5-4 The Layer Control dialogue box.

Now you can set up the drawing aids for your drawing.

21. Using the mouse, activate the pull-down menus at the top of the screen. Pick **Settings** and **Drawing Aids.**

The Drawing Aids dialogue box appears *(Fig. 5-5).* This is another way to control the Grid, Snap, and Ortho drawing aids.

22. Set up the following parameters:

   Ortho: **OFF** (not checked)

   Snap: **ON**

   X Spacing set to **0.5000**

   Y Spacing set to **0.5000**

   Grid: **ON**

   X Spacing set to **1.0000**

   Y Spacing set to **1.0000**

When you have set up these parameters, pick the **OK** button.

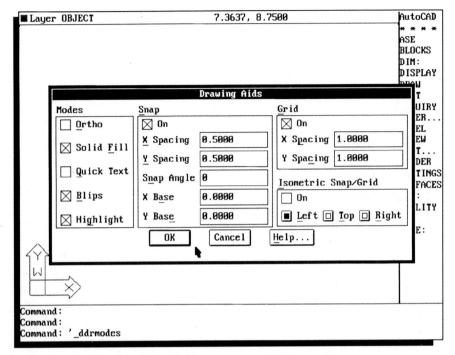

Fig. 5-5 The Drawing Aids dialogue box.

23. *Command:* **LINE** ↵
24. *From point:* **1.5,2** ↵
25. *To point:* **@0,4** ↵
26. *To point:* **@9,0** ↵
27. *To point:* **@4<270** ↵
28. *To point:* **C** ↵

A red rectangle 4 units high and 9 units long should now appear on your screen. If it does not, return to Step 23 and try again. What types of coordinates did you use? Step 24 specifies absolute coordinates. Steps 25 and 26 specify relative coordinates, and Step 27 uses relative polar coordinates. The **C** in Step 28 is short for Close. By pressing C, you tell AutoCAD to draw a line from the end of the last line segment you drew to the beginning of the first line segment you drew. This is an excellent method of creating a closed shape.

29. *Command:* **LAYER** ↵
30. *?/Make/Set/New/ON/OFF/Color/Ltype/Freeze/Thaw/LOck/Unlock:* **S** ↵
31. *New current layer <OBJECT>:* **CENTER** ↵
32. *?/Make/Set/New/ON/OFF/Color/Ltype/Freeze/Thaw/LOck/Unlock:* ↵
33. *Command:* **LINE** ↵

34. *From point:* With the mouse, create a vertical line that runs from the middle of the top of the rectangle to the middle of the bottom of it. To end the LINE command, press RETURN. The line should go through the exact center of the rectangle.

35. Create a horizontal line from the left side of the rectangle to the right side. It, too, should run through the exact middle of the rectangle.

Your drawing should now look like the one in Fig. 5-3. If you want to end your AutoCAD session at this point, be sure to save your work. You will need it for the next "Your Turn" exercises.

It is a good idea to save your work often, even though your work may not be finished. A momentary power outage could cause you to lose all the work you have done since the last time you saved it.

▶ The CIRCLE Command

When you enter the CIRCLE command, the following choices appear: *3P/2P/TTR/<Center point>:*. In the default method, you simply pick a center point for the circle and specify a radius. This is the method used in the exercises below.

The 3P (3 Points) option allows you to create a circle by specifying three points on the circle. After you specify the second point, an elastic circle appears on the cursor. The third point you pick gives the circle its size and final location.

The 2P (2 Points) option allows you create a circle by picking two points on the circle. When you pick the first point, an elastic circle appears on the cursor. The second point you choose sets the diameter and the location of the circle.

TTR stands for "tangent, tangent, radius." To create a circle using this method, you must specify two entities to which the circle is tangent. You then specify the radius to complete the circle.

ACTIVITY

The following exercises will give you a chance to try the CIRCLE command. Open the GASKET drawing you worked on in the previous "Your Turn" exercise. Then follow these step-by-step instructions. The result of this exercise is shown in Fig. 5-6.

1. Make the OBJECT layer current.

2. *Command:* **CIRCLE** ⏎

3. *3P/2P/TTR/<Center point>:* Using the mouse, pick the point at the intersection of the horizontal and vertical lines you drew through the middle of the rectangle.

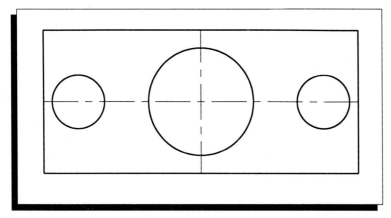

Fig. 5-6

4. *3P/2P/TTR/<Center point>: Diameter/<Radius>:* **1.5** ↵

A circle should appear in the rectangle as shown in Fig. 5-7.

5. *Command:* ↵

The CIRCLE command is re-entered.

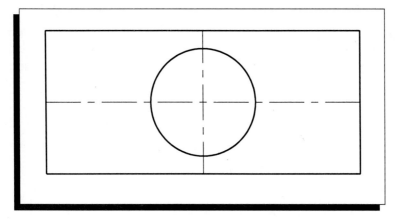

Fig. 5-7

As you have just seen, pressing RETURN at the *Command:* prompt re-enters the previous command. For example, when you need to draw several circles, you only need to enter the CIRCLE command once. For the remaining circles you can press the RETURN key. Note that this works only if you have not entered another command since you last entered the CIRCLE command.

Another helpful function of the RETURN key is to cancel operations already in progress. For example, suppose you begin a circle and then realize that you put the center point in the wrong location. To cancel the CIRCLE command and begin again, press RETURN twice. The first RETURN cancels the current circle operation and the second RETURN re-enters the CIRCLE command so that you can try again.

6. *3P/2P/TTR/<Center point>:* **2.5,4** ↵

7. *Diameter/<Radius>:* **0.75** ↵

8. *Command:* ↵

9. *3P/2P/TTR/<Center point>:* **9.5,4** ↵

10. *Diameter/<Radius>:* **0.75** ↵

Your drawing should now look like the one in Fig. 5-6. If you want to end your AutoCAD session at this point, be sure to save your work. You will need it for the next "Your Turn" exercises.

▶ The ARC Command

AutoCAD's default method of creating an arc requires three points: the starting point, a second point, and an end point. The second point may lie anywhere on the arc and helps define the shape of the arc.

The other methods of drawing an arc in AutoCAD are a little more complicated, but they allow the drafter greater control over the arc. These methods include various combinations of the arc's included angle, center, starting direction, end point, length of chord, and radius *(Fig. 5-8)*. (Remember to turn to the Pictorial Glossary at the back of this text for further information about any geometric terms you do not understand.) In addition to the above methods, an arc can be drawn as a continuation of a previous arc.

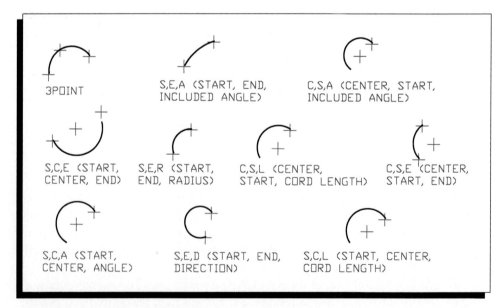

Fig. 5-8 AutoCAD provides several methods of creating arcs. Use whichever method is most appropriate for your current drawing.

ACTIVITY

The exercises below will give you a chance to try the ARC command. Open the GASKET drawing you worked on in the previous "Your Turn" exercise. Make sure that OBJECT is the current layer. Then follow these step-by-step instructions. The result of this exercise is shown in Fig. 5-9.

1. *Command:* **ARC** ⏎
2. *Center/<Start point>:* Pick the top left corner of the rectangle.
3. *Center/End/<Second point>:* **6,7** ⏎

Notice the elastic arc that appears at the cursor.

4. *End point:* Pick the top right corner of the rectangle.
5. *Command:* ⏎
6. *Center/<Start point>:* **10.5,6** ⏎
7. *Center/End/<Second point>:* **11.5,4** ⏎
8. *End point:* **10.5,2** ⏎

As you can see, you can use any combination of mouse picks and coordinate values to create an arc.

9. Use what you've learned about arcs to create the other two arcs shown in the drawing in Fig. 5-9.

The second point of the bottom arc is at absolute coordinates 6,1; the second point of the left arc is at absolute coordinates 0.5,4.

10. *Command:* **R** ⏎

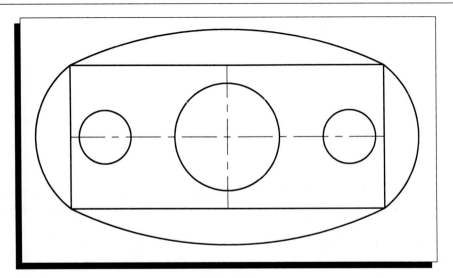

Fig. 5-9

R is the shorthand command for the REDRAW command. This is a general maintenance command that cleans up the screen by removing blips. (**Blips** are the small pick points left on the screen when you use the mouse pick button.) You may use the REDRAW command at any time during the step-by-step sequences in this text, even if it is not written into the sequence.

Your drawing should now look like the one in Fig. 5-9. If you want to end your AutoCAD session at this point, be sure to save your work. You will need it for the next "Your Turn" exercises.

Editing a Drawing

Editing is the process of making additions and/or changes to existing drawings. Basic editing commands include OFFSET, EXTEND, and TRIM. These commands can also be used in the original construction of a drawing.

In addition to these commands, editing frequently requires the use of general commands that allow you to erase objects and take a closer look at parts of the drawing. These commands will be introduced in this section.

Using Selection Sets

A **selection set** is a group of entities you pick for editing. When AutoCAD needs a selection set, the *Select objects:* prompt appears and the crosshairs change to a small selection box. You may respond to the prompt in a number of ways. First, you may select the objects directly by picking them with the mouse. In addition, you may select the objects by creating a rectangular area called a **window** around the objects. To define a window, you simply pick two opposite corners.

▶ Using Windows to Select Entities

AutoCAD recognizes two types of windows: those created from left to right and those created from right to left. When you create the window from left to right, all the objects that lie *entirely within* the window are selected *(Fig. 5-10A)*. When you create the window from right to left, the window appears with dotted lines. Any object it touches becomes part of the selection set, whether or not the objects lie entirely within the window. This type of window is called a **crossing window** *(Fig. 5-10B)*.

Objects that have been selected appear as dotted lines. When you pick entities one at a time using the mouse, the dotted lines help you keep track of what you have already selected.

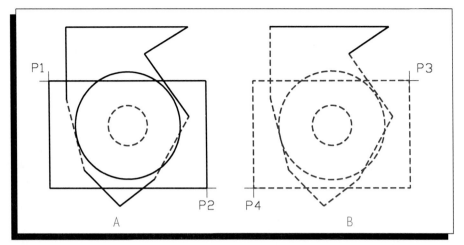

Fig. 5-10 A) When you create a window from left to right (P1 to P2) all entities that are entirely within the window are selected. B) When you create a window from right to left (P3 to P4), all entities even partially within the window are selected. This type of window is called a crossing window.

▶ Deselecting Entities

Occasionally, you may select an entity by mistake. If this happens while you are picking individual entities using the mouse, you can press **U** for "Undo." This command undoes the last selection you made without affecting previous selections.

If you are using a window to select entities and accidentally include an entity you don't want, the procedure is different. If you press **U**, the entire window disappears and none of the entities are selected. To avoid that, use the Remove feature to remove only the entities you don't want to include. Press **R** for "Remove." The *Select entities:* prompt changes to *Remove entities:* and any entity you select is removed from the selection set. When you are finished, press RETURN.

Using the Basic Editing Commands

After you have created a basic drawing such as the one you've worked on in this chapter, you may need to change it slightly or add details. Although the LINE, CIRCLE, and ARC commands are often useful for this, AutoCAD's basic editing commands can save you time. The basic editing commands are OFFSET, TRIM, and EXTEND. Two other commands you will frequently use—ZOOM and ERASE—are also included in this section.

▶ The OFFSET Command

The OFFSET command allows you to copy a line, circle, or arc exactly, but at a distance from the original. Offset lines are parallel to the original lines, and offset circles and arcs are concentric with the originals *(Fig. 5-11)*.

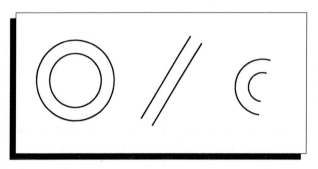

Fig. 5-11 The OFFSET command creates concentric circles and arcs as well as parallel lines.

When two circles or arcs are **concentric**, they share the same center point.

A practical example of the use of OFFSET is to create the interior and exterior walls of a house on a floor plan. Since the interior and exterior lines are parallel to each other, you can create the exterior lines first. Then offset the lines to create the interior walls.

ACTIVITY

The exercises below will give you a chance to try the OFFSET command. Open the AutoCAD drawing you worked on in the previous "Your Turn" exercise. Then follow these step-by-step instructions. The result of this exercise is shown in Fig. 5-12.

1. *Command:* **OFFSET** ↵
2. *Offset distance or Through <Through>:* **0.5** ↵

In this case, you entered an offset distance directly. If you had selected the "Through" option, AutoCAD would have asked you for a point through which the offset entity should pass.

3. *Select object to offset:* Pick the arc at the top of the drawing.
4. *Side to offset?* Pick the space below the arc to show that you want the arc offset to the inside of the object.
5. *Select object to offset:* Pick the arc on the right side of the drawing.
6. *Side to offset?* Pick the space on the inside of the object.
7. Continue following the prompts to offset the arcs on the bottom and left side of the drawing. After you complete the offset of the left arc, proceed to step 8.

8. *Select object to offset:* ⏎

The RETURN cancels the OFFSET command and returns you to the *Command:* prompt. Your drawing should now look like the one in Fig. 5-12. If you want to end your AutoCAD session at this point, be sure to save your work. You will need it for the next "Your Turn" exercises.

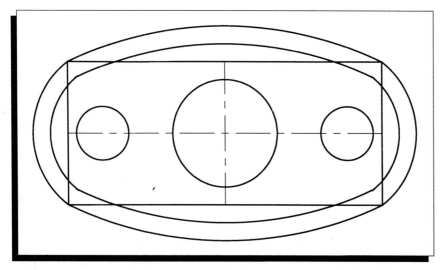

Fig. 5-12

▶ The ZOOM Command

As you work on a drawing, you may often need to "zoom in" on a particular part of the drawing to edit or to add details. AutoCAD's ZOOM command performs this function. Note that although the entities on the screen appear to move when you use ZOOM, they are not really moving. The ZOOM command simply changes the window through which you see the drawing on the screen.

The ZOOM command has several options. You can use a window to zoom in on a portion of the drawing. You can also change the size of the viewing window (the part of the drawing that appears on the screen). You can even move the viewing window around to see different parts of the drawing. The best way to discover the effect of the ZOOM options on a drawing is to use them. You will use some of the ZOOM options in the next "Your Turn."

▶ The ERASE Command

In many cases, CAD drafters create construction lines when they begin a new drawing. In CAD, **construction lines** are lines that are used as a reference or basis for other entities. The lines in the rectangle you created at the beginning of this chapter are good examples of construction lines. After the other entities are drawn, the construction lines can usually be removed. To do this, use the ERASE command. The ERASE command allows you to select one or more entities to erase from the screen.

ERASE is also useful when you make a mistake. If you create an entity that later you decide you don't want, you can remove it using the ERASE command. You will use the ERASE command in the next "Your Turn."

▶ The TRIM and EXTEND Commands

The TRIM and EXTEND commands are two of the most useful editing commands. TRIM allows you to trim lines, circles, and arcs that already exist on the screen. With the help of TRIM, you can incorporate parts of construction lines into your drawing by trimming away the unneeded parts. This reduces the time you spend setting up and creating the drawing.

When you use the TRIM command, AutoCAD first asks you to define the cutting edge—the edge to which

you want to trim the entity. Then you can pick any number of entities to trim to that edge. The only requirement is that the cutting edge must intersect the entity to be trimmed *(Fig. 5-13)*.

The EXTEND command does the opposite of the TRIM command: it extends existing entities to an existing boundary edge *(Fig. 5-14)*. (A boundary edge is a line, circle, or arc to which an entity will be extended.) You will use both of these commands in the following exercises.

Fig. 5-13 Follow this procedure to trim a line to another line: Make sure the lines cross. Enter the TRIM command and select the line to use as a cutting edge. Press RETURN. Select the line to be trimmed. Be sure to select the part of the line that you want to trim, rather than the part you want to keep.

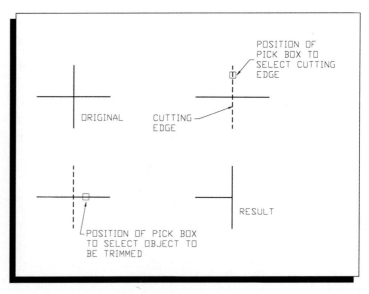

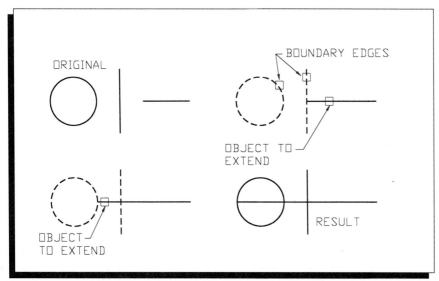

Fig. 5-14 Follow these procedures to extend a line to another line and to a circle.

ACTIVITY

The exercises below will give you a chance to try the TRIM and EXTEND commands. You will also use the ERASE and ZOOM commands in this sequence. Open the AutoCAD drawing you worked on in the previous "Your Turn" exercise. Then follow these step-by-step instructions. The result of this exercise is shown in Fig. 5-15.

1. *Command:* **ZOOM** ⏎

2. *All/Center/Dynamic/Extents/Left/Previous/VMax/Window/<Scale (X/XP)>:* **W** ⏎

Picking **W** selects the Window option of the zoom command. This allows you to use the mouse to specify the portion of the drawing you want to see more closely.

3. *First corner:* Pick near P1 as shown on Fig. 5-16.

4. *Other corner:* Pick near P2 as shown on Fig. 5-16.

AutoCAD zooms in on the portion of the drawing you specified *(Fig. 5-17)*. Notice that the ends of the inside arcs overlap slightly. You can use the TRIM command to clean up the intersections between the arcs.

5. *Command:* **TRIM** ⏎

If Ortho is still on, turn it off before performing the following steps.

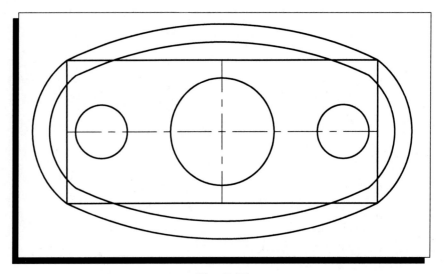

Fig. 5-15

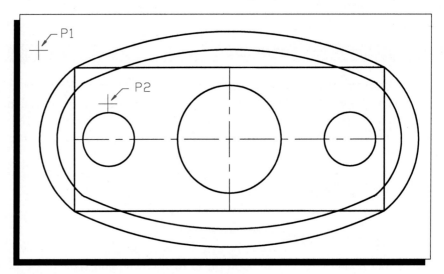

Fig. 5-16

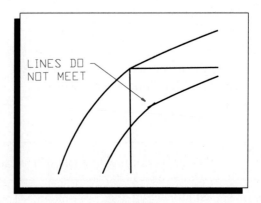

Fig. 5-17

6. *Select cutting edge(s)...*
   Select objects: Pick a point near the end of the left arc as shown in Fig. 5-18A.

7. *Select objects:* ⏎

8. *<Select object to trim>/Undo:* Pick the portion of the top arc to the left of the intersection of the two arcs, as shown in Fig. 5-18B.

9. *<Select object to trim>/Undo:* ⏎

10. *Command:* ⏎

11. *TRIM*

    Select cutting edge(s)...

    Select objects: Pick the top arc as shown in Fig. 5-18C.

12. *Select objects:* ⏎

13. *<Select object to trim>/Undo:* Pick the part of the left arc that extends above the top arc, as shown in Fig. 5-18D.

14. *<Select object to trim>/Undo:* ⏎

15. *Command:* **R** ⏎

The ends of the arcs now meet, creating a neater appearance.

16. *Command:* **ZOOM** ⏎

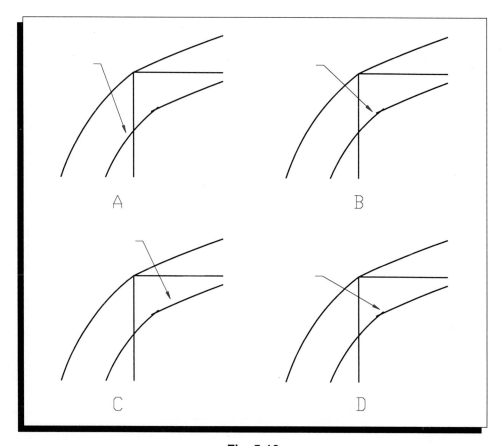

Fig. 5-18

17. *All/Center/Dynamic/Extents/Left/Previous/Vmax/Window/<Scale(X/XP)>:* **P** ⏎

By entering **P**, you selected the Previous option. AutoCAD zooms back to the previous view.

18. Repeat the above procedure for the other three arc intersections.

When you have finished, your drawing should look like the one in Fig. 5-15.

The rectangle you drew at the beginning of this chapter has now served its purpose. It was a construction prop that allowed you to create the arcs that make up the gasket. Now you can delete the rectangle from the drawing. Follow these steps. Fig. 5-19 shows the result of this exercise.

19. *Command:* **ERASE** ⏎

20. *Select objects:* Use the mouse to pick each of the four lines that make up the rectangle.

21. *Select objects:* When you have selected all four lines, press **RETURN.**

Now use the EXTEND command to extend the vertical and horizontal center lines to the edges of the object.

22. *Command:* **EXTEND** ⏎

23. *Select boundary edge(s)...*

    Select objects: Pick all four inside arcs.

When you need to extend (or trim) several entities, you can save time by selecting more than one boundary (or cutting) edge at a time. Then, at the next prompt, do all your trimming or extending in one pass.

24. *Select objects:* ⏎

25. *<Select objects to extend>/Undo:* Select each end of the horizontal and vertical lines, as shown in Fig. 5-20. When you have finished, press RETURN to cancel the command.

26. Save your work.

This completes the gasket drawing for this chapter. Your finished drawing should look like the one in Fig. 5-19. Since the center lines are drawn on their own layer, you can view or plot the drawing with or without the center lines. In the next section, you will learn how to plot your drawing to create a paper copy (also called a "hard copy").

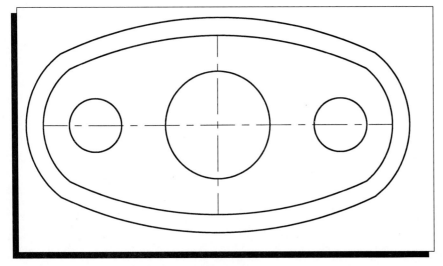

Fig. 5-19

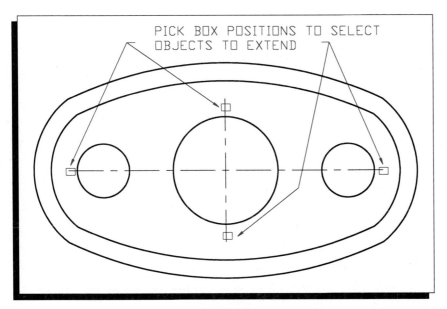

Fig. 5-20

Plotting a Drawing

The PLOT command accomplishes the final step: making a hard copy, or bringing the image on the computer screen to paper. Release 12 of AutoCAD allows you to use either a plotter or a printer, but it does not discriminate between the two. The PLOT command is used for both printing and plotting. This is not true of Releases 10 and 11. Refer to Appendixes B and C for more information about printing and plotting using these releases.

When AutoCAD is originally installed on a computer, the installer can configure the program to work with several plotting and printing devices. One of these devices becomes the default. However, the PLOT command allows you to change the default before you plot or print a drawing.

From this point, the text will refer to both printing and plotting as *plotting*, since the procedure is the same for both processes in Release 12.

When you enter the PLOT command, the Plot Configuration dialogue box appears *(Fig. 5-21)*. The dialogue box contains six sections. In the top left corner, the Device and Default Information section shows the name of the printer or plotter that is currently selected. The Device and Default Selection... button in this area allows you to select a different device if necessary.

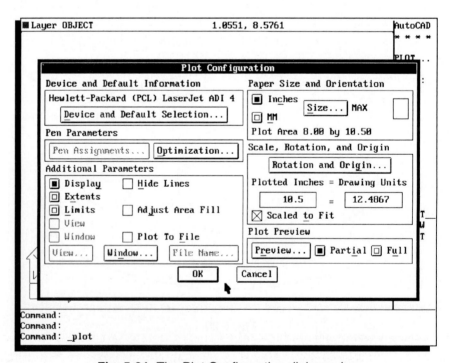

Fig. 5-21 The Plot Configuration dialogue box.

The Pen Parameters section controls the pens on multiple-pen plotters. The Pen Assignments... button allows you to associate specific colors with various line widths on plotters and many laser printers. The Optimization... button changes the way AutoCAD plots. (Refer to the *AutoCAD Reference Manual* supplied with the AutoCAD software for more information about optimization.)

The Additional Parameters section allows you to choose exactly what to plot. You can choose to plot:

- only the entities (or parts of entities) displayed on the screen (Display).
- all of the entities in the drawing, but none of the surrounding space that does not contain entities (Extents).
- the entire drawing area, even parts that do not contain entities (Limits).
- a named view of the drawing (View).
- a portion of the drawing defined by a window you create (Window).

You will learn more about views in Chapters 11 and 12.

For three-dimensional drawings, you can hide the hidden lines (lines that you would not ordinarily see in a real three-dimensional object) for the plot *(Fig. 5-22)*. You can also choose to plot to a file instead of a plotting or printing device.

The Paper Size and Orientation section allows you to choose the units in which the drawing will plot, as well as the size of the paper in the plotter. A line at the bottom of this area shows the active plotting area according to the options you have chosen.

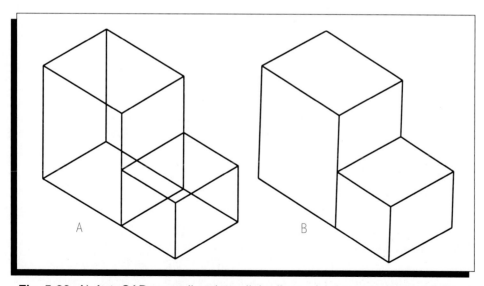

Fig. 5-22 A) AutoCAD normally prints all the lines of a three-dimensional object. However, you may not be able to see those lines if you are looking at the actual object. B) Choosing the Hide Lines option on the Plot Configuration dialogue box hides the lines that you would ordinarily not see.

The Scale, Rotation, and Origin section allows you to change the orientation of the drawing on the printed page. You can also specify how many AutoCAD units should be plotted to each inch of paper. If accurate scales are not necessary, you can pick the box next to "Scaled to Fit." When this box is checked, AutoCAD determines the size of the drawing automatically to fit it onto a single page.

The last section is Plot Preview. You can preview the plot in two ways. A partial preview shows you only how much space the current plot will take up on the page and where it will be positioned. The full preview shows you exactly how your plot will look on paper.

ACTIVITY

The exercises below will give you a chance to try the PLOT command. Open the AutoCAD drawing you worked on in the previous "Your Turn" exercise. Then follow these step-by-step instructions to plot your drawing.

1. *Command:* **PLOT** ⏎

The Plot Configuration dialogue box appears.

2. Look at the current plotter selection in the top left corner of the dialogue box. If the plotter you will use does not appear there, pick the Device and Default Selection button and choose your plotter from the list there. If your plotter does not appear on the list, consult your instructor. AutoCAD may need to be reconfigured for your computer.

3. In the Additional Parameters section, select Display.

If the box next to Display is highlighted, it is selected. If the box is not highlighted, pick it with the mouse to select it.

4. In the Paper Size and Orientation box, make sure the Inches box is checked.

5. In the Scale, Rotation, and Origin section, make sure the Scaled to Fit box is checked. The boxes below Plotted Inches = Drawing Units will show the final relationship of inches to drawing units. When you check this box, AutoCAD scales the drawing to fit the paper size you're using.

6. In the Plot Preview section, pick the box next to Full and then pick the Preview button to see what your plotted drawing will look like. If the entire drawing appears on the screen and you see no other problems, pick the OK button.

7. *Effective plotting area: 10.50 wide by 7.57 high*
Position paper in plotter.
Press RETURN to continue or S for hardware setup ⏎

Your plot should be finished within a few minutes, depending on your processor and plotter speeds.

As you go through this text, you may want to plot or print all of your drawings to keep in a portfolio.

■ Chapter 5 Review

1. Name the three types of coordinates you can use in AutoCAD to define points.
2. What are the three most basic drawing commands used in AutoCAD?
3. Suppose you have just completed a LINE command and AutoCAD is showing the *Command:* prompt. What is the fastest way to initiate another LINE command?
4. When you are using a window to select entities to edit, what is the difference between creating the window from left to right and creating it from right to left?
5. What is the purpose of the ZOOM command?
6. When you plot a drawing, what is the difference between plotting the *extents* of the drawing and plotting the *limits* of the drawing?

■ Chapter 5 Problems

Do the following problems to try the skills you have learned in this chapter. For each problem, recreate the drawing shown in this text as accurately as possible. Use all of the commands presented in this chapter at least once. You do not need to insert the text on your drawings. Save each problem to a diskette as a separate file. Name the problems PRB5-1, PRB5-2, PRB5-3, and so on.

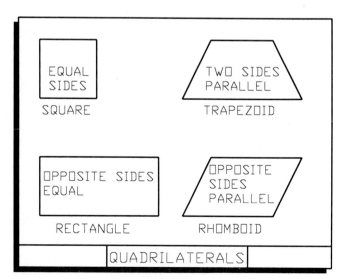

Problem 5-1

Problem 5-2

Problem 5-3

Problem 5-4

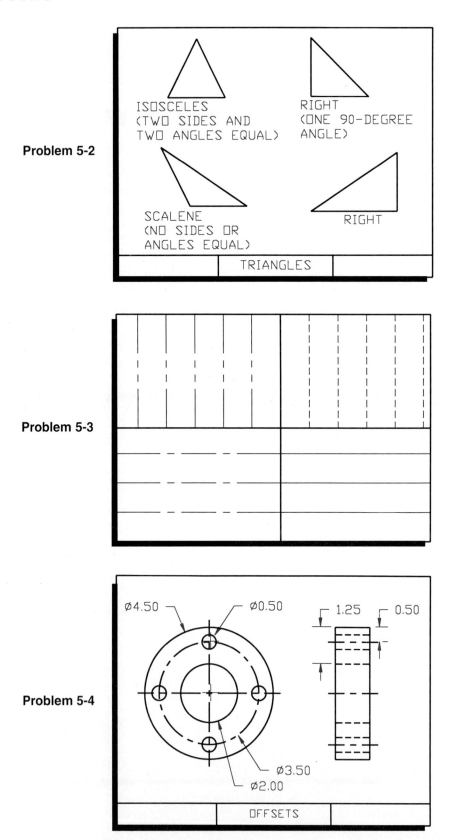

Problem 5-5

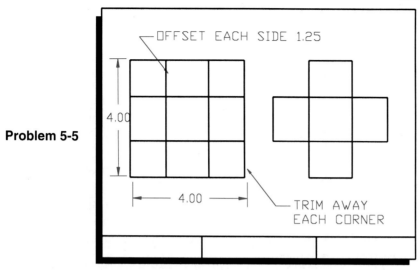

OFFSET EACH SIDE 1.25

4.00

4.00

TRIM AWAY
EACH CORNER

Problem 5-6

TRIM AND EXTEND
THE OFFSET LINES

4.00

@4<120

4.00

0.23 — OFFSET

157

Using CAD

AutoCAD Makes the World a Better Place

Larry Murphy is one of many civil engineers who uses AutoCAD, along with other application products, to make the world a better place in which to live. Murphy is a civil engineer for the firm of Anchor Engineering Services, East Hartford, Connecticut, whose primary focuses are environmental engineering and solid waste management.

As of October 1993, federal regulations for new landfills mandate that leachate (water that filters through a landfill and becomes polluted) must be collected and managed. Murphy's firm used AutoCAD to develop landfills that exceed the federal regulations. Their landfills collect the leachate in a double layer of synthetic liners. The leachate is then treated to remove metals and salts before it is sent to the municipal water treatment plant where it is treated again.

According to Murphy, Anchor Engineering also uses AutoCAD to develop plans to prevent pollution from stormwater run-off. Their goal is to minimize stormwater pollution at industrial recycling sites. Recyclers are required to contain the run-off from scrap metal and compost so that it won't pollute rivers and water supplies.

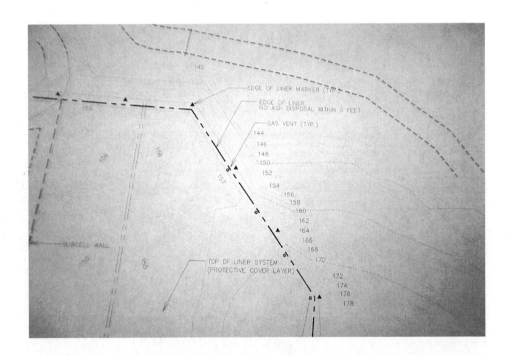

To develop a prevention plan, engineers use a map to create a site plan that shows run-off patterns and places where materials are stored. If no map of the site is available, they may use aerial photographs of the site taken by the Department of Environmental Protection or topographic quad maps from the USGS (United States Geological Survey). The photographs or quads are scanned and imported into AutoCAD.

Next, a map of the site is developed by using raster-to-vector conversion software such as CAD Overlay GS produced by Image Systems Technology, Inc. This product converts the scanned image (raster) to AutoCAD (vector) format. AutoCAD is then used to complete the map.

After the mapping is complete, the engineers use AutoCAD and third party applications such as COGO (Coordinate Geometry) and DTM (Digital Terrain Modeling) to create the site plan. Several companies make COGO and DTM products—Anchor Engineering uses the packages produced by Softdesk, Inc. The design created includes a system for cleaning the water of the unwanted pollutants. The system might include oil and water separators, a sediment pond, or grass lined swale.

According to Murphy, third party applications such as COGO, DTM, and CAD Overlay GS allow engineers to utilize the full power of AutoCAD. Used in conjunction with recycling and environmental management, these tools help engineers work to protect the future of our world.

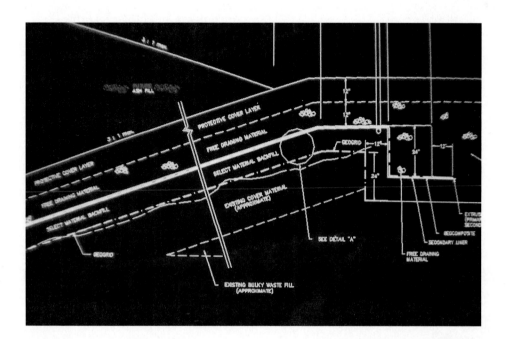

Chapter 6

▶

Key Terms

geometry
parallel lines
perpendicular lines
tangent
angle
vertex
circle
concentric circles
center line
regular polygons

triangles
quadrilaterals
object snap
aperture
pixels
running object snap
polygon
ellipse
polyline

▶ ▶

Commands & Variables

OSNAP
DDOSNAP
POLYGON

ELLIPSE
PLINE
PEDIT

Geometric Drawing

Objectives

When you have completed this chapter, you will be able to:

- understand the use of geometry in drafting.
- use the various object snap modes to snap to a specific place or point on an object.
- set and cancel a running object snap.
- use AutoCAD commands to construct geometric shapes.

Drafting is a form of visual communication. Although drawings are only representations of real objects, they should represent the real objects accurately. Drawings must be geometrically correct. Strictly defined, **geometry** is the study of the relationships of points, lines, angles, and figures in space. Drafters use these relationships to solve problems as they create drawings.

In addition to accuracy, speed is critical to successful drafting. For drafters, time is money. Drawings must be made quickly to meet deadlines and to satisfy customer requirements. But speed without accuracy is money wasted. Small errors and inaccuracies may appear minor on the surface, but incorrectly placed lines, distorted polygons, and misspelled words may lead to contract cancellations.

When you draw an object, you must decide the fastest and most accurate way to construct the object's geometry. Look at the wheel cover in Fig. 6-1. How many different shapes do you see? How can you quickly create these shapes to describe the wheel cover accurately? How would you begin to create the drawing?

With CAD, there are usually no unique solutions. The one you choose may depend upon the purpose of the drawing. For example, an assembly drawing shows how the different parts of an object fit together. A second drawing might be a technical illustration that shows what a finished object will look like *(Fig. 6-2)*. Also, as you gain experience and learn more commands, you will learn better ways to construct geometric shapes.

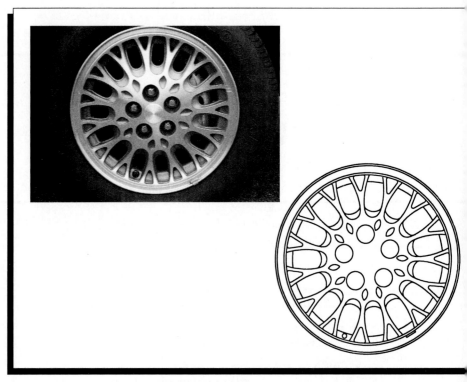

Fig. 6-1 How many different shapes can you see in this wheel cover?

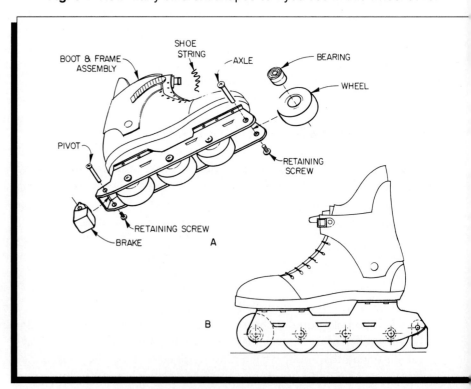

Fig. 6-2 A) This assembly drawing shows how the parts of an in-line skate fit toget
B) The technical illustration shows how the finished in-line skate should look.

There are two ways to draw with AutoCAD. One is to plod along with the basic commands you learned in Chapter 5. You will get the job done; but don't expect to hear from your customer again. The other is to use the power of AutoCAD to supplement and extend the basic commands. The commands and options presented in this chapter will enable you to create geometric shapes and relationships quickly and accurately.

Terminology

Although drafting is a visual medium, part of your credibility as a drafter depends on your knowledge of the language of drafting. The geometric terms which follow are part of that language. If you do not completely understand any of the terms, refer to the Picture Glossary at the back of this text.

- **Parallel lines [symbol ||]** never meet because they are an equal distance apart at all points.
- **Perpendicular lines [symbol ⊥]** meet in such a way that they form four 90-degree angles.
- A line, circle, or arc is **tangent** to another circle or arc if they touch at one point only. (Two lines cannot be tangent.)

- An **angle [symbol <]** is the circular measure of two lines that have one point in common, called a **vertex.**
- A **circle** contains 360 degrees measured in a counterclockwise (CCW) direction from the 3 o'clock or east position.
- **Concentric circles** have the same center point.
- A **center line** defines the center of an object or feature. Its symbol is a C with an L through it [℄].
- **Regular polygons** have three or more equal sides.
- Special polygons include **triangles,** which have three sides, and **quadrilaterals,** which have four sides.

ACTIVITY

1. Study the shapes and terms in Fig. 6-3. If there are any you are not familiar with, look them up in the Picture Glossary at the back of this text.

2. Write down the names of the parts of a circle. Draw a circle and indicate each of the parts in your drawing.

3. Draw obtuse, complementary, and supplementary angles. How do they differ?

4. Study the different types of polygons. Then have someone quiz you on them.

Your Turn

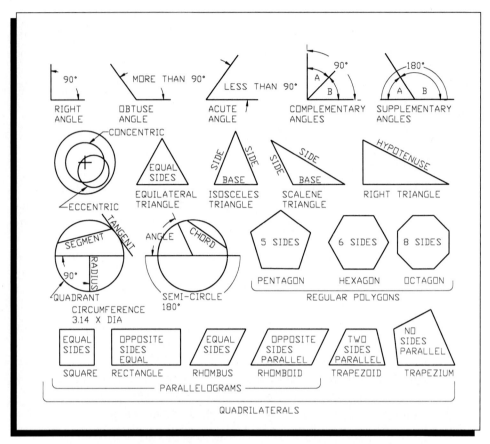

Fig. 6-3 Geometric shapes often needed in drafting.

Object Snap Modes

Object snap refers to the cursor's ability to snap to a specific place or point on an object. In AutoCAD, object snap is controlled by the OSNAP command. AutoCAD provides several object snap modes for use in a wide variety of situations. For example, you can bisect a line easily by snapping to the midpoint of the line.

The object snap modes are not commands in themselves. Rather, they augment commands such as LINE, CIRCLE, and ARC. Modes available within the OSNAP command include CENter, ENDpoint, INSert, INTersection, MIDpoint, NEArest, NODe, PERpendicular, QUAdrant, and TAN-

gent. Fig. 6-4 describes the use of each object snap mode.

The pull-down and screen menus contain the OSNAP modes. (They may also be accessed by pressing the middle button of a three-button mouse.) In many cases, however, you will find it more convenient to enter specific mode commands from the keyboard. You can enter object snap modes during construction by typing the first three letters of the mode in answer to a prompt such as *Line from point:* or *To point:*. For example, to draw a line that begins at the endpoint of an existing line, type END at the *Line from point:* prompt. This activates the ENDpoint object snap mode.

When a mode is selected, a pick box or **aperture** appears at the center of the crosshairs *(Fig. 6-5)*. To complete the object snap, place the point within the aperture and pick it. The crosshairs need not be directly on the point to be picked, but the point must be within the aperture's range.

When you are working in a crowded area of a complex drawing, you may need to make the aperture smaller to avoid selecting unwanted entities. You can change the size of the aperture by changing the number of **pixels,** or picture elements, with the APERTURE command.

Hot Tip

Object Snap Modes	
CENter	Snaps to the center of a circle or arc
ENDpoint	Snaps to the end of a line or arc
MIDpoint	Snaps to the midpoint of a line or arc
PERpendicular	Snaps to a line or arc at a 90-degree angle
INTersection	Snaps to the intersection of two entities
TANgent	Snaps a line, circle, or arc to its tangent point on another circle or arc
NEArest	Snaps to the nearest point on the selected entity
INSert	Snaps to the insertion point of a text or block entity
NODe	Snaps to a point, or *node*, that you have already created on the screen
QUAdrant	Finds the nearest quadrant point on a circle or arc (Quadrant points are points at which the entity intersects the X and Y axes: 0, 90, 180, and 270 degrees.)

Fig. 6-4

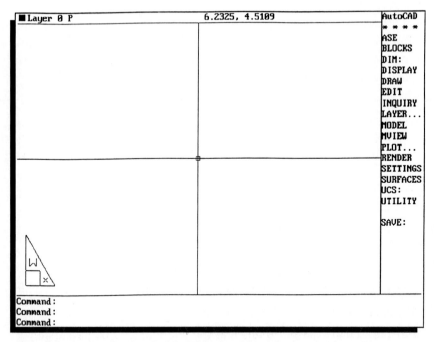

Fig. 6-5 Any entity that falls within the aperture, or pick box, is selected when you pick with the mouse.

ACTIVITY

Create the following geometric pattern using the object snap modes to supplement the commands you learned in Chapter 5. As you work, consider how difficult it would be if the object snap modes were not available. Think about each of the modes and how you could apply them to drawing other objects. Plot the drawing when you are finished, and look for visual inaccuracies. Are the proportions visually correct? Does the pattern appear to be in balance?

Before you begin, read through each of the exercises in this chapter. Would it be easier if you turned on the Grid and Snap? If so, set them to the proper intervals to help you perform the exercises.

1. Start AutoCAD and begin a new drawing named OBJSNAP.

AutoCAD does not require you to name a drawing when you first create it. To begin a new drawing without naming it, press RETURN when the Create New Drawing dialogue box appears. Note, however, that it is good practice to name your drawing when you first open it.

2. *Command:* **CIRCLE** ⏎

3. Place the center of the circle at point **3,4** and specify a radius of **1.**

4. Press **RETURN** to reenter the CIRCLE command. Place the center of the second circle at **6,4** and specify a radius of **1.**

Now place a line from the midpoint of one circle to the midpoint of the other by following this sequence:

5. *Command:* **LINE** ⏎

Now enter the CENter object snap mode.

6. *From point:* **CEN** ⏎

Note the appearance of the aperture.

   of Pick the circle on the left.

Be sure to pick the circle itself, not a location near the center of the circle.

7. *To point:* **CEN** ⏎

   of Pick the circle on the right.

8. *To point:* ⏎

Your drawing should look like the one in Fig. 6-6. Now use the MIDpoint object snap to bisect the line:

9. *Command:* **LINE** ⏎

10. *From point:* **MID** ⏎

   of Pick the line.

11. *To point:* **@3<90** ⏎

As you may recall from Chapter 4, this polar coordinate specifies a line 3 units long at a direction of 90 degrees, creating a vertical line.

12. *To point:* ⏎

13. *Command:* **LINE** ⏎

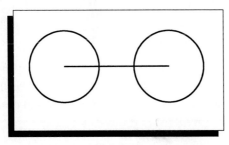

Fig. 6-6

Now try the INTersection object snap mode.

14. *From point:* **INT** ↵

    of Pick a point near the intersection of the horizontal and vertical lines.

15. *To point:* **@3<270** ↵

16. *To point:* ↵

Now use the ENDpoint and TANgent object snap modes to create lines connecting the top and bottom of the vertical line to the two circles. The TANgent mode places a line, circle, or arc tangent to another line, circle, or arc. Because a tangent requires at least one curved surface, two lines cannot be tangent. The tangential point will be the point closest to the aperture's pick point. Follow these steps:

17. *Command:* **LINE** ↵

18. *From point:* **TAN**

    of Pick a point near P1 on the upper left part of the left circle as shown in Fig. 6-7.

19. *To point:* **END** ↵

    of Pick the top of the vertical line.

20. *To point:* **TAN** ↵

    of Pick P2 near the upper right part of the right circle as shown in Fig. 6-7.

21. *To point:* ↵

22. Now repeat steps 17 through 21 to create similar lines connecting the two circles to the bottom end of the vertical line. When you finish, your drawing should look like the one in Fig. 6-8.

23. *Command:* **LINE** ↵

24. *From point:* **MID** ↵

    of Pick the first line at P1 as shown in Fig. 6-9.

25. *To point:* **PER** ↵

    to Pick the vertical line.

26. *To point:* **MID** ↵

    of Pick the second line near P2 as shown in Fig. 6-9.

27. Repeat this procedure to draw a line connecting the midpoints of the two slanted lines at the bottom of the object.

Hot Tip

You don't have to snap to the vertical line using the PERpendicular mode. You can snap directly from the midpoint of one line to the midpoint of the other.

28. On your own, create lines connecting the intersections of the lines you just created and the slanted lines with the centers of the circles.

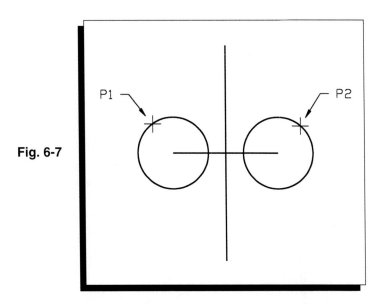

Fig. 6-7

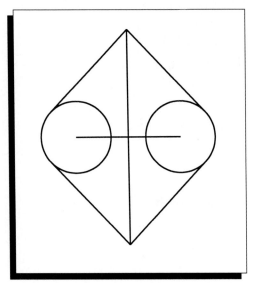

Fig. 6-8

Fig. 6-9

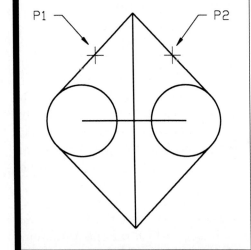

Your drawing should now look like the one in Fig. 6-10.

29. *Command:* **CIRCLE** ↵

30. *3P/2P/TTR/<Center point>:* **2P** ↵

31. *First point on diameter:* **END** ↵

    of Pick the top point of the vertical line.

32. *Second point on diameter:* **END** ↵

    of Pick the bottom point of the vertical line.

You will complete this exercise in the next "Your Turn" section when you practice the running object snap described below.

Fig. 6-10

Running Object Snap

Sometimes you need to repeat a specific object snap mode or modes frequently. To save time, you can set a continuous or **running object snap** to the modes you need. Set the modes by entering the OSNAP command, followed by the first three letters of the modes you want to set. Separate multiple modes with commas, but do not enter a space after the commas. If you want to set several modes, an even faster method is to enter the DDOSNAP command, which presents a Running

Object Snap dialogue box *(Fig. 6-11)*. Pick to select object snap modes you want to select. Then pick the OK button. You may also change the size of the aperture with the Running Object Snap dialogue box. When you set a running object snap by either method, the modes remain active until you turn them off. To cancel a running object snap, "unpick" the modes in the dialogue box or type OSNAP followed by NONE or OFF, or press the RETURN key (pick nothing).

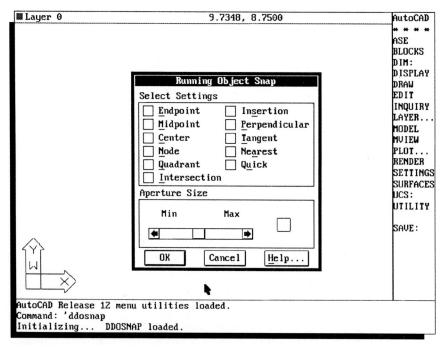

Fig. 6-11 The Running Object Snap dialogue box allows you to select or deselect several object snap modes at one time.

ACTIVITY

Your Turn

Return to the drawing you began in the previous "Your Turn" section. To finish the drawing, use the QUAdrant mode as a running object snap to create a square inside the large circle.

1. *Command:* **DDOSNAP** ⏎

The Running Object Snap dialogue box appears.

2. Pick the check box next to **Quadrant** and pick the **OK** button.

The QUAdrant object snap is now in the continuous, or running, mode.

3. *Command:* **LINE** ⏎

4. *From point:* Pick the large circle near the right side to select the quadrant point at 0 degrees.

5. *To point:* Continue counterclockwise around the circle, selecting near the 90, 180, and 270 degree points.

6. *To point:* **C** ⏎

Your drawing should look like the one in Fig. 6-12. When you placed the square within the large circle, you did not enter QUA each time to snap to the quadrant points. The QUAdrant running object snap mode let you snap to the quadrant points automatically.

The object snap modes can be used in many ways to help you create drawings accurately and rapidly. Use them routinely, even when they're not specified in the step-by-step procedures. You will improve your technique and drawing quality and increase your drawing speed.

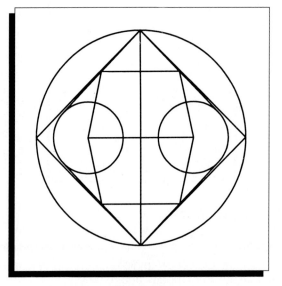

Fig. 6-12

Advanced Entity Construction

AutoCAD has several commands that let you create geometric shapes directly, without relying on the LINE, CIRCLE, and ARC commands. With these commands, you can make regular polygons, ellipses, and polylines.

The POLYGON Command

A **polygon** is a two-dimensional closed figure with three or more sides. A three-sided polygon is a triangle.

You can use the POLYGON command to create a regular polygon in two ways. With the default method, you are asked to pick the center point for the polygon. You may then inscribe the polygon within a circle or circumscribe it around a circle. When you specify the number of sides and the radius of the circle, AutoCAD has all the information needed to create the polygon. When you use this method, you never actually see a circle on the screen. AutoCAD does the calculations and draws the polygon for you.

The other method is to define one edge of the polygon using the Edge option of the POLYGON command. You define the size and location of the polygon by specifying the beginning and ending points of one side. You must also specify the number of sides. This method is useful when you need a polygon with sides equal to the edge of another part or entity.

ACTIVITY

This exercise will give you practice using the POLYGON command. You will begin by creating a circle so that you can see the relationship of the circle AutoCAD uses to the inscribed and circumscribed polygons.

1. *Command:* **CIRCLE** ⏎

2. *3P/2P/TTR/<Center point>:* **6,3** ⏎

3. *Diameter/<Radius>:* **1.5** ⏎

Now inscribe a six-sided polygon in the circle you just created. Make the center and radius of the polygon the same as those for the circle. Follow these steps.

4. Create a layer called **POLYGON** and set its color to green. Make POLYGON the current layer.

5. *Command:* **POLYGON** ⏎

6. *Number of sides <4>:* **6** ⏎

7. *Edge/<Center of polygon>:* **6,3** ⏎

8. *Inscribed in circle/Circumscribed about circle (I/C) <I>:* **I** ⏎

Notice that a rubberband polygon appears on the cursor, with the cursor attached to one vertex of the polygon. If you wanted to, you could answer the *Radius of circle:* prompt by picking a point using the mouse (or other pointing device). However, since you want to inscribe the polygon accurately in an existing circle, you'll specify the radius at the keyboard.

9. *Radius of circle:* **1.5** ⏎

The polygon appears inscribed in the circle *(Fig. 6-13A).* Now follow these steps to create another polygon. This time, the polygon will be circumscribed around the circle instead of inscribed in it.

10. *Command:* ⏎

11. *POLYGON Number of sides <6>:* ⏎

12. *Edge/<Center of polygon>:* **6,3** ⏎

13. *Inscribed in circle/Circumscribed about circle (I/C) <I>:* **C** ⏎

Notice that the rubberband polygon on the cursor is attached at the midpoint of one of the edges of the polygon.

14. *Radius of circle:* **1.5** ⏎

Fig. 6-13B shows the expected result. Look closely at the two polygons on the screen. What is the difference between polygons inscribed in circles and those circumscribed about circles? In an inscribed polygon, the radius of the circle equals the distance from the center of the polygon to each vertex (point) of the polygon. In a circumscribed polygon, the radius of the circle equals the distance from the center of the polygon to the center of each line segment (edge) that makes up the polygon.

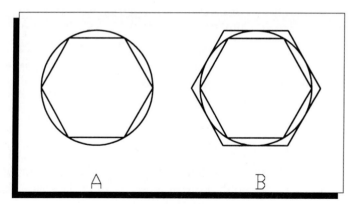

Fig. 6-13

Follow the steps below to try the Edge option of the POLYGON command. This option is useful when you need to match an edge of a polygon to an edge of an existing entity.

15. *Command:* **POLYGON** ⏎

16. *Number of sides <6>:* ⏎

17. *Edge/<Center of polygon>:* **E** ⏎

18. *First endpoint of edge:* **END** ⏎

   of Pick the upper left vertex of the circumscribed polygon (P1 in Fig. 6-14).

19. *Second endpoint of edge:* **END** ⏎

   of Pick the upper right vertex of the circumscribed polygon (P2 in Fig. 6-14).

Fig. 6-15A shows the expected result. Notice that the order in which you specify the endpoints is critical. AutoCAD builds polygons counterclockwise. If you had selected P2 first and P1 second, the resulting polygon would have been created in exactly the same place as (on top of) the circumscribed polygon on the screen *(Fig. 6-15B)*.

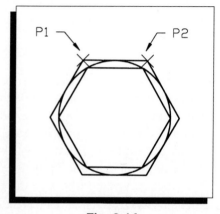

Fig. 6-14

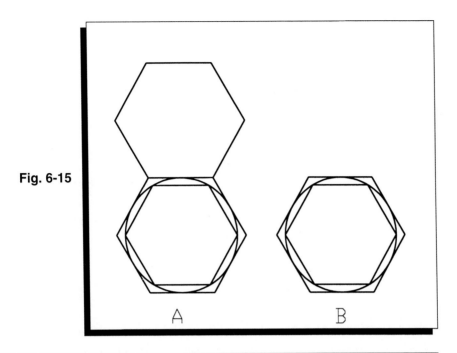

Fig. 6-15

A   B

The ELLIPSE Command

An **ellipse** is a circle viewed from an angle. It has two axes: the longer or major axis, and the shorter or minor axis *(Fig. 6-16)*. You can construct an ellipse by establishing both axes or by establishing the major axis and the angle of rotation around that axis *(Fig. 6-17)*. The default method is to specify the two axes. The second axis may be either the major or the minor axis. However, it always expands perpendicular to the first axis, regardless of where you move the crosshairs.

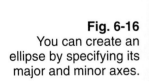

Fig. 6-16
You can create an ellipse by specifying its major and minor axes.

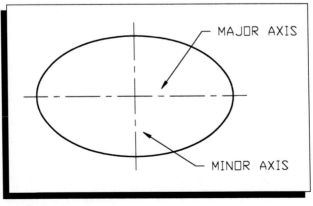

MAJOR AXIS

MINOR AXIS

In AutoCAD, ellipses are used mostly for isometric drawings and views. They are not usually used for two-dimensional auxiliary views.

Note

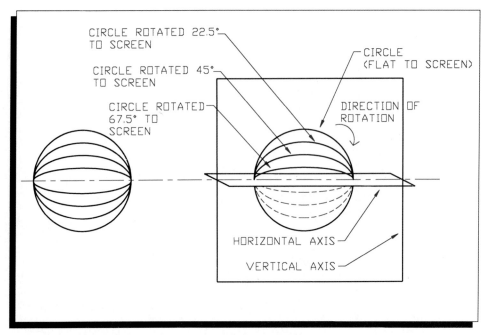

Fig. 6-17 Since AutoCAD views an ellipse as a circle that has been rotated into a plane that is not parallel to the computer screen, you can also specify an ellipse by identifying the major axis and the angle of rotation.

ACTIVITY

The purpose of this and the following "Your Turn" sections in this chapter is to allow you to gain experience using the many commands that are presented. You will not create an identifiable object as you did in Chapter 5. Rather, you will create a series of small drawings that consist mostly of isolated geometric shapes. It is important that you master the AutoCAD commands that allow you to create these shapes, since the geometric shapes are fundamental to drafting.

In this procedure, you will create several ellipses using the various options. As you create the ellipses, think about how you could use each method in an actual drawing. Then follow these steps. First is the axis endpoint option.

1. *Command:* **ELLIPSE** ⏎

2. *<Axis endpoint 1>/Center:* **3,5** ⏎

3. *Axis endpoint 2:* **8,5** ⏎

Move the cursor to see how the distance you choose for the other axis distance affects the shape of the ellipse. Notice that, no matter how you move the mouse, the second axis remains perpendicular to the first axis you defined.

4. *<Other axis distance>/Rotation:* Move the cursor to point **5.5,4** (as shown by the X,Y coordinate display at the top of the screen), and select that point using the mouse.

There will be times when you need to draw an ellipse so that its center is in a specific location. To do that, use the Center option of the ELLIPSE command, as shown in the following steps.

5. *Command:* **ELLIPSE** ⏎

6. *<Axis endpoint 1>/Center:* **C** ⏎

7. *Center of ellipse:* Pick a point near the center of the screen.

8. *Axis endpoint:* Move the cursor two units to the right and pick a point.

9. *<Other axis distance>/Rotation:* **1** ⏎

The Rotation option of the ELLIPSE command is a little more complicated. When you use this option, AutoCAD assumes that the first axis you specify is the major axis of the ellipse. Further, AutoCAD "sees" the ellipse as a circle that is turned at an angle to the drawing screen, so that it appears as an ellipse. When you specify an angle between 0 and 89.4 degrees, AutoCAD revolves the circle around the axis you specified, projects the circle onto the plane described by the X and Y axes, and draws the ellipse. Refer again to Fig. 6-17.

The ellipse does not exist as a circle in three-dimensional space. Instead, it is a two-dimensional object that is flat to the screen.

---

10. *Command:* **ELLIPSE** ⏎

11. *<Axis endpoint 1>/Center:* **2,4** ⏎

12. *Axis endpoint 2:* **2,5** ⏎

13. *<Other axis distance>/Rotation:* **R** ⏎

14. *Rotation around major axis:* **30** ⏎

This creates an ellipse that looks like a circle that is rotated 30 degrees from the screen around its major axis.

The PLINE and PEDIT Commands

A **polyline** is a connected series of line and arc segments that AutoCAD treats as a single entity. The PLINE command allows you to create polylines of specific widths and attributes. You can draw polylines in either two or three dimensions. They can be used as part of a geometric constructions or, because of their unique features, they can be constructions in themselves.

The procedure for editing polylines—changing their width, shape, and so on—is different from editing most entities. Polylines have their own editing command: PEDIT (Polyline EDIT).

The following procedure draws a polyline varying in width from 0 to 0.5 units, ending with an arc 0 units wide.

1. *Command:* **PLINE** ↵

2. *From point:* Pick a point at approximately P1 as shown in Fig. 6-18A.

At this point, you can set the beginning and ending widths of the first segment of the polyline.

3. *Current line-width is <current value>*

   Arc/Close/Halfwidth/Length/Undo/Width/ <Endpoint of line>: **W** ↵

4. *Starting width <current value>:* **0** ↵

5. *Ending width <0.0000>:* **0.5** ↵

6. *Arc/Close/Halfwidth/Length/Undo/Width/<Endpoint of line>:* Pick a point at approximately P2.

After each line or arc segment, you can change the width settings. To continue, follow these steps.

7. *Arc/Close/Halfwidth/Length/Undo/Width/<Endpoint of line>:* **A** ↵

8. *Angle/CEnter/CLose/Direction/Halfwidth/Line/Radius/Second pt/Undo/Width/ <Endpoint of arc>:* **W** ↵

9. *Starting width <0.5000>:* ↵

10. *Ending width <0.5000>:* **0** ↵

This will start the arc at the same width as the line and end it with a point. Since this is all part of the same polyline, the arc begins where the line stopped.

11. *Angle/CEnter/CLose/Direction/Halfwidth/Line/Radius/Second pt/Undo/Width/ <Endpoint of arc>:* Pick a point approximately P3.

12. *Angle/CEnter/CLose/Direction/Halfwidth/Line/Radius/Second pt/Undo/Width/ <Endpoint of arc>:* ↵

Fig. 6-18B shows the expected result (your drawing will not have the points marked).

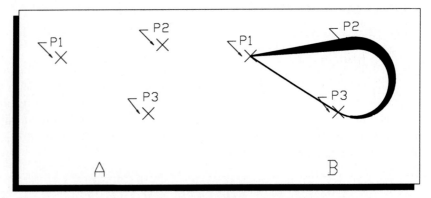

Fig. 6-18

The previous exercise was designed to give you practice in creating polylines, changing their widths, and so on. In drafting practice, an object that has polylines of this type would most likely be done in technical illustrations whose purpose is to show what the object looks like. Production drawings (used for manufacturing, assembly, or repair) usually contain polylines of a more consistent width or shape. The following exercise will give you a better idea of why polylines are often used instead of lines to create objects with complex shapes.

1. Find a picture of an object for which someone might need a dimensioned drawing. (Choose something fairly simple, but not too simple, so that you can grasp the concept without having to do hours of work.)

2. Enter the **POLYLINE** command and reproduce the outline of the object on the computer screen. Do not worry about setting the width or other characteristics of the polyline.

3. When you have completed the outline, press **RETURN** to end the POLYLINE command.

One of the biggest advantages of using a polyline instead of a line to create an object in AutoCAD is that the polyline is considered a single entity. If you need to move the entity, rotate it, or even delete it, you do not have to spend time selecting all of the individual line segments. By picking a point anywhere on the polyline, you can select the whole thing.

Before you continue with the steps below, be sure to save your drawing.

4. *Command:* **ERASE** ⏎
   Select objects: Pick a point anywhere on the polyline.

Notice that the entire polyline is selected.

5. *Select objects:* ⏎

The entire polyline is erased. (If you want to keep your drawing, use the UNDO command to undo the erase procedure.)

If you need to create a closed rectangular object to act as a single entity, you may wish to use the RECTANG command instead of the LINE or POLYLINE commands. The RECTANG command allows you to create rectangles quickly by specifying the location of the corners or the height and width of a rectangle. The resulting rectangle, like a polyline, is considered one entity by AutoCAD.

■ Chapter 6 Review

1. Why is geometry important to drafters?
2. Under what circumstances are two circles said to be concentric?
3. What is the purpose of the object snap modes?
4. Describe a method of creating a regular pentagon (five-sided figure).
5. What command would you enter to edit a polyline?

■ Chapter 6 Problems

To further your understanding of the material covered in this chapter, reproduce the following drawings using AutoCAD. Do not include the dimensions in your drawing.

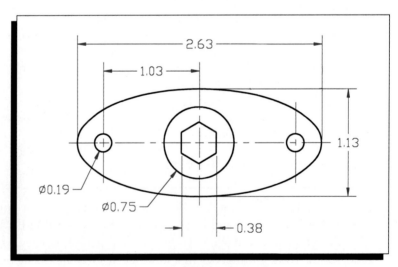

Problem 6-1

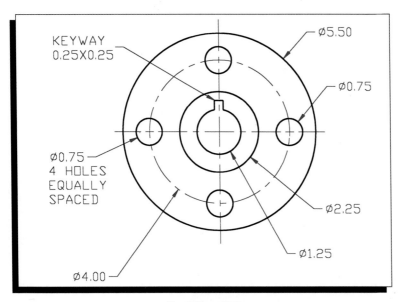

Problem 6-2

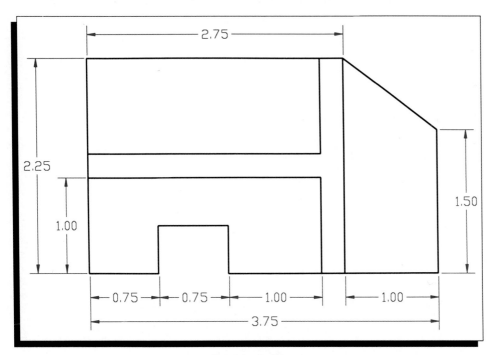

Problem 6-3

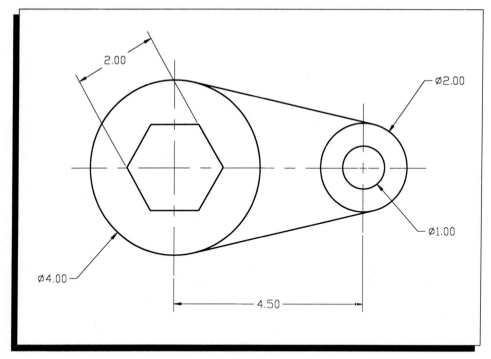

Problem 6-4

Using CAD

Jaws Knows Geometry

Richard Dreyfuss, who played in the box-office hit "Jaws," called the great white shark an "eating machine." Tourists who take the boat ride at the Jaws attraction at Universal Studios in Orlando, Florida, are given the thrill of their lives when they are "attacked" by a 32-foot long great white shark coming within three feet of their boat, fully convinced they will be the shark's next meal. Even though they know it is fantasy, they are terrified, and that's just what the designers wanted.

Mark Woodbury is one of the principal members of the Universal Studios Florida Jaws attraction design team. Woodbury said, "We tried to create a shark that was larger than any shark that had ever been caught—one which would scare the living daylights out of you." The design team created what Woodbury called "an envelope of experience" to explain what they wanted people to see and feel on the boat ride.

The designers put themselves in the place of a visitor to Universal Studios who would take part in a Jaws experience. They observed killer whales in a tank and saw how these large animals propelled themselves. They needed their shark to move and thrash about like a real shark. They wanted to create a terrifying ride.

After deciding on the "envelope of experience," the design concepts were broken down to their simplest components. This is where AutoCAD entered the picture.

Cliff Jennings, who works for Oceaneering Technologies, Inc. (Otech), was part of a team that worked six months to bring the designers' ideas into reality. They had to analyze the shark's motions, break them down into the simplest terms, and understand the geometry involved. "What you see in the ride is just the tip of the iceberg," said Jennings. After the engineers and designers made big decisions, Jennings and the Otech team "designed the machinery and drive systems" that moved the giant shark. "This is where AutoCAD helped us with the geometry," he said. "We would rotate an assembly through certain arcs to make sure there weren't any interferences and integrate it with where the boat was travelling and to make sure safety zones and other factors would be taken into account."

Jennings and the OTECH team had to maintain good communications among all the designers. "It was a challenge to break things down to discrete tasks and work together to draft something that had never been done before," Jennings said. "We had about 20 drafters, and everybody was working on a little piece of the shark at the same time." Everyone had to follow basic rules which allowed all the drawings to flow directly into a unified AutoCAD database.

Woodbury said, "First you imagine the terrifying experience and then you have to figure out how to build it, and anytime you try to build something big—a skyscraper or a shark—you have to break it down to its basic geometry."

The designers, engineers, and drafters relied on AutoCAD's geometric design capabilities for even the smallest parts to enable the 32-foot Jaws to terrify and thrill guests at Universal Studios every day of the year.

Photo Courtesy of Universal Studios

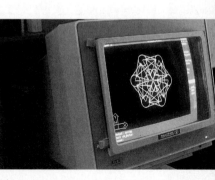

Chapter 7

▶

Key Terms

mirror line   round
fillet   grips

▶ ▶

Commands & Variables

DIVIDE	COPY
REGEN	MOVE
SETVAR	ARRAY
PDMODE	MIRROR
PDSIZE	FILLET
MEASURE	CHAMFER
DIST	SCALE
BREAK	STRETCH

Editing Geometric Shapes

▶ ▶ ▶

Objectives

When you have completed this chapter, you will be able to:
- use AutoCAD commands to edit and manipulate entities and objects on drawings.
- use AutoCAD's grips to edit the shape and size of basic entities.

E ven geometrically correct figures may need to be changed. Frequently they must be copied, moved, divided, scaled, or generally adjusted. Speed and accuracy again become important factors. What a waste of time to have to draw three copies of the same figure in different locations! The COPY command will duplicate the figure for you wherever and whenever you wish. Think of the chances for error if you were to draw a figure twelve times equally spaced radially around a common center. The ARRAY command will do it accurately with only a few keystrokes.

Basic Editing Commands

AutoCAD provides numerous commands to help you edit and manipulate your geometric shapes. The OFFSET command, which was introduced in Chapter 5, is one example. There are many more, however. In this chapter you will discover other commands, such as the COPY and ARRAY commands mentioned above, that can help you make the most of AutoCAD's power to construct drawings.

The DIVIDE Command

DIVIDE breaks an object into equal parts. You can use the DIVIDE command to inscribe a polygon within an existing circle. Divide the circle into a number of parts equal to the number of points on the polygon. Connect each divide point with a straight line.

When you use the DIVIDE command in its default mode, you cannot see the division points. To view the divisions, you can use the SETVAR command (which SETs the VARiables) and set the PDMODE (Point Display MODE) variable to display the points the way you want them to appear *(Fig. 7-1)*. In practice, the SETVAR command is often bypassed. You can type PDMODE at the command prompt and enter its new value.

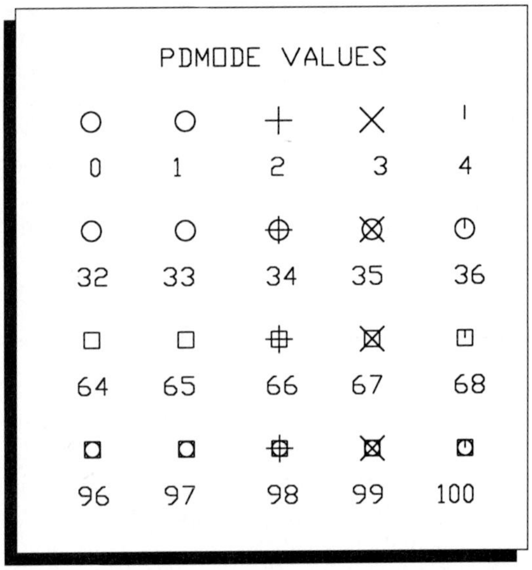

Fig. 7-1 The PDMODE variable allows you to display points in a variety of ways. The display method you choose depends mostly on the drawing.

PDMODE VALUES

○	○	+	×	│
0	1	2	3	4
○	○	⊕	⊗	⊙
32	33	34	35	36
□	□	⊞	⊠	⊡
64	65	66	67	68
▢	▢	⊞	⊠	▢
96	97	98	99	100

Note

Depending on the scale of your drawing, you may need to adjust the size of the points. You can make the points larger or smaller by setting the PDSIZE variable.

Your Turn

ACTIVITY

The following procedure inscribes a regular hexagon in a circle. You may want to compare this procedure with the procedure used in manual drafting.

1. *Command:* **CIRCLE** ↵
2. *3P/2P/TTR/<Center point>:* Pick a point near the center of your screen.
3. *Diameter/<Radius>:* **3** ↵

To inscribe the hexagon in the circle, you will need to know the endpoints of the line segments. A simple way to do this is to use the DIVIDE command to divide the circle into six equal parts. Follow these steps:

4. *Command:* **DIVIDE** ⏎

5. *Select object to divide:* Pick the circle.

6. *<Number of segments>/Block:* **6** ⏎

The circle is now divided into six equal parts. To see the division points on the circle, change the PDMODE.

7. *Command:* **PDMODE** ⏎

8. *New value for PDMODE <0>:* **3** ⏎

The value of PDMODE changed to display an X at point locations, although you still cannot see the change. To display a change in this mode setting, you will need to regenerate the drawing using the REGEN command. REGEN is different from the REDRAW command. REDRAW just refreshes the screen, but REGEN recalculates all of the internal vectors used by the AutoCAD program to display drawings. Therefore, REDRAW will not display the change in the PDMODE variable, but REGEN will.

9. *Command:* **REGEN** ⏎

You should now be able to see the points at the divisions on the circle *(Fig. 7-2)*. To complete the hexagon, follow these steps:

10. *Command:* **OSNAP** ⏎

Object snap modes: **NOD** ⏎

Using the NODe running object snap will help you place the endpoints of the line segments precisely at the nodes (division points) on the circle.

11. *Command:* **LINE** ⏎

12. *From point:* Pick near one of the points on the circle.

13. *To point:* Continue picking the points until you have completed the hexagon.

14. Press **RETURN** to end the LINE command.

The hexagon is now inscribed in the circle *(Fig. 7-3)*. If it is necessary to remove the points from sight, you can do so by changing PDMODE back to 0.

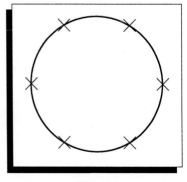

Fig. 7-2

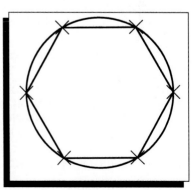

Fig. 7-3

Note

If the drawing is going to be used with Computer Numerical Control (CNC), the points must be erased entirely instead of merely hidden from view. Computer Numerical Control is a system in which the drawing is used directly to control the computer-operated machinery that manufactures a part.

To erase the points, *do not* change PDMODE back to zero. Instead, use the ERASE command and select the points individually. You may need to set PDSIZE to a larger value to be able to pick the points. An easier way is to use a window to select the whole drawing. Then use the Remove option of the ERASE command to remove the circle and polygon from the selection set.

The MEASURE and DIST Commands

The MEASURE command is similar to DIVIDE. It places marks at equal distances along an entity. However, the MEASURE command does not divide the entity into equal parts. Instead, you specify the length of each part, and MEASURE marks off parts of that length. Any extra space along the entity is left after the last mark. MEASURE is useful when you need to construct additional entities at specific intervals along an existing entity. Fig. 7-4 demonstrates the difference between the MEASURE command and the DIVIDE command. Don't forget to set the PDMODE to a visible value as you did with DIVIDE.

The DIST command tells you the distance from one point to another. It does not mark points at specific distances as the MEASURE and DIVIDE commands do. When you specify a starting point and a stopping point, the DIST command tells you the total distance between the two points.

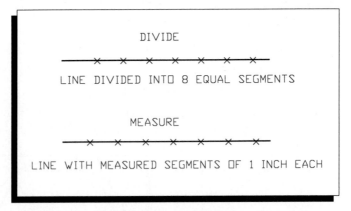

Fig. 7-4 The DIVIDE command divides a line, circle, or arc into a number of segments you choose. The MEASURE command measures segments of a *length* you choose along a line, circle, or arc. If the distance doesn't divide evenly into the length of the line or the circumference of the circle, the MEASURE command leaves a short segment at the end of the entity.

DIST also displays additional information that is most useful when you are working with three-dimensional drawings. The values shown are the angle in and from the XY plane and the delta (change) in X, Y, and Z from the first location to the second.

ACTIVITY

Be sure the PDMODE variable is still set at 3, as it was for the DIVIDE command. The following procedure marks a line every 1.25 units *(Fig. 7-5)*. Note the short space after the last mark.

1. Draw a line that extends most of the way across your screen.
2. *Command:* **MEASURE** ⏎
3. *Select object to measure:* Pick the line you drew in step 1.

Note that if you pick the line at the right of center, the measurement will start from the right. If you pick the line at the left of center, the measurement will start from the left.

4. *<Segment length>/Block:* **1.25** ⏎

You should see the points at 1.25-unit intervals. Look at the intervals on each end of the line. Does one seem shorter than the other? Use the DIST command to find out.

5. *Command:* **DIST** ⏎

Do not enter the entire word *DISTANCE*. If you do, AutoCAD assumes that you want to know the current value of the Distance variable. This variable is set when you use the DIST command, but you cannot change it directly. If you enter DISTANCE at the command prompt, the current setting of the Distance variable appears, followed by "(read only)."

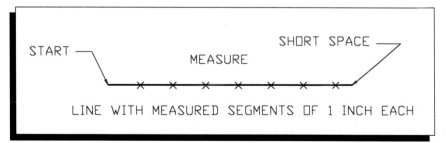

Fig. 7-5 The MEASURE command applied to a line. Notice the short space or segment to the right of the line. If you had told AutoCAD to begin measuring at the right side of the line, the short space would be on the left side instead.

6. *First point:* **END** ⏎

 of Pick one end of the line.

7. *Second point:* **NOD** ⏎

 of Pick the point closest to the end of the line you chose in step 6.

 The distance information should appear. Is the distance 1.25? If not, AutoCAD started measuring from the other end of the line. This is the distance remaining after the line was measured into as many 1.25-unit segments as possible.

8. Repeat steps 5 through 7 for the other end of the line.

The BREAK Command

The BREAK command divides a single entity into two separate entities at a point you specify. BREAK can also break a line, circle, or arc so that part of the entity is erased. When you choose two points on an entity, the part of the entity between the points disappears.

ACTIVITY

The following procedure will help you learn more about the BREAK command.

1. Draw two lines and a circle approximately as shown in Fig. 7-6. Be sure that the circle runs through both lines. Do not draw the arrows or text.

2. *Command:* **BREAK** ⏎

3. *Select object:* Pick the top line at P1 on Fig. 7-6.

AutoCAD assumes that your pick point in step 3 is the first point for determining the break.

4. *Enter second point (or F for first point):* **F** ⏎

If you had entered a second point here, AutoCAD would have removed the part of the line that lies between the two points. However, you can be more specific than that by choosing to enter the first point precisely.

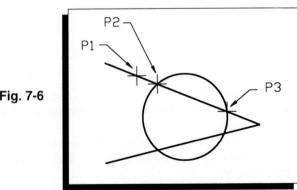

Fig. 7-6

5. *Enter first point:* **INT** ⏎

6. *of* Pick the intersection of the circle and the top line (P2 in Fig. 7-6).

7. *Enter second point:* **INT** ⏎

   of Pick the other intersection of the circle and top line (P3 in Fig. 7-6).

The part of the line that overlapped the circle should disappear *(Fig. 7-7)*.

To break a line at only one location (make it into two entities without making any of it disappear), you can use the @ symbol. When you enter this symbol at the *Enter second point:* prompt, it places the second break point at the same location as the first.

8. *Command:* **BREAK** ⏎

9. *Select object:* Pick the lower line.

10. *Enter second point (or F for first point):* **F** ⏎

11. *Enter first point:* **INT** ⏎

    of Pick the left intersection of the lower line and the circle.

12. *Enter second point:* **@**

Although you cannot immediately see any change, the line is now broken at that intersection. To see how this affects the line, erase the part of the line that extends to the left of the circle.

13. *Command:* **ERASE** ⏎

14. *Select objects:* Pick the lower line to the left of the circle.

Only the part of the line that extends to the left of the circle is highlighted, because it now exists as a separate entity from the rest of the line.

15. *Select objects:* ⏎

The highlighted part of the line disappears *(Fig. 7-8)*.

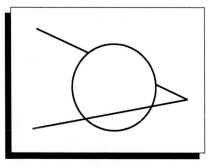

Fig. 7-7

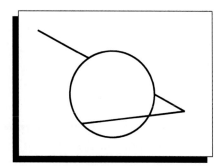

Fig. 7-8

The COPY and MOVE Commands

The COPY command lets you copy one or more entities to another location. When you enter the COPY command, you are asked to identify a base point on the original entity. The base point is used to locate a single copy accurately.

If you wish to make more than one copy and put the copies in different locations, select the Multiple option after you enter the COPY command *(Fig. 7-9).* You can draw an item, such as a fastener, once and then copy it to as many places as you wish.

The MOVE command is similar to the COPY command, except that the existing entity moves to a new location. It no longer exists at the original location.

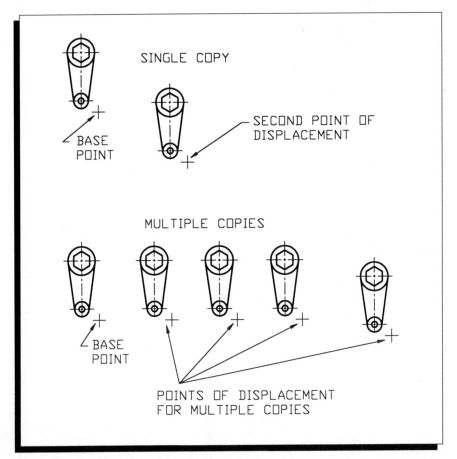

Fig. 7-9 AutoCAD defaults to a single copy of the objects you select. By selecting the Multiple option, however, you can make as many copies as you need without reentering the COPY command. Every time you pick a location, a new copy appears.

ACTIVITY

Follow these steps to practice the COPY and MOVE commands.

1. In the upper left corner of your screen, create a rectangle about 1 unit high and 1.5 units wide.

2. *Command:* **COPY** ⏎

3. *Select objects:* Use a crossing window to select all of the lines that make up the rectangle.

Recall from Chapter 5 that to create a crossing window, you pick opposite corners of the window from right to left.

4. *Select objects:* ⏎

5. *Base point or displacement>/Multiple:* **END** ⏎

   of Pick the lower left corner of the rectangle.

The entities you selected now appear at the cursor. You can place them by picking with the mouse or by using the keyboard to specify coordinates.

6. *Second point of displacement:* Pick a point anywhere on the screen.

A copy of the rectangle appears at that point. Now use the MOVE command to move the copy to a different location.

7. *Command:* **MOVE** ⏎

8. *Select objects:* Use a crossing window to select the lines that make up the copy of the rectangle.

9. *Select objects:* ⏎

10. *Base point of displacement:* Pick the lower left corner of the rectangle.

11. *Second point of displacement:* Pick a point at a different location on the screen.

The copy of the rectangle moves to the destination you specify. In the next steps, you will explore the COPY Multiple option.

12. Erase the copy of the rectangle from the screen, but leave the original rectangle.

13. *Command:* **COPY** ⏎

14. *Select objects:* Select the lines that make up the rectangle.

A fast way to select all of the objects on a drawing screen is to respond to the *Select objects:* prompt with the word **ALL**. All of the entities on the screen become highlighted.

15. *Select objects:* ⏎

16. *<Base point or displacement>/Multiple:* **M** ⏎

17. *Base point:* Pick the bottom left corner of the rectangle.

18. *Second point of displacement:* Pick several locations on the screen.

The rectangle appears at each point you pick. Notice that the *Second point of displacement:* prompt reappears after each copy.

19. *Second point of displacement:* ⏎

The ARRAY Command

The ARRAY command is similar to the Multiple option of the COPY command because it copies one or more entities several times. However, the ARRAY command repeats the copy a specified number of times in a regular pattern. There are two basic types of arrays: rectangular and polar *(Fig. 7-10)*. When you select a rectangular array, AutoCAD prompts you for the number of rows and columns in the array. When you select a polar array, the next prompt asks where the center point of the array will be. Then you can choose the number of items in the array and the angle to fill. A polar array can extend for less than 360 degrees *(Fig. 7-11)*. You can also rotate the objects as they are arrayed or maintain all of the copies in the original orientation *(Fig. 7-12)*.

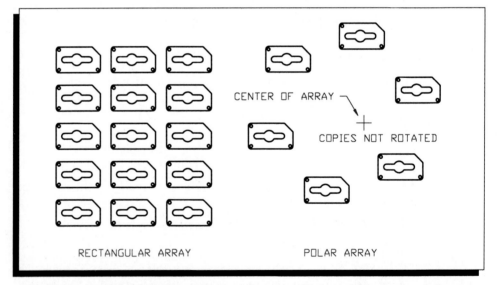

RECTANGULAR ARRAY

CENTER OF ARRAY

COPIES NOT ROTATED

POLAR ARRAY

Fig. 7-10 AutoCAD permits two types of arrays. A rectangular array copies an object or objects into rows and columns. A polar array copies them in a circular pattern. The type you choose depends on the individual drawing you are creating.

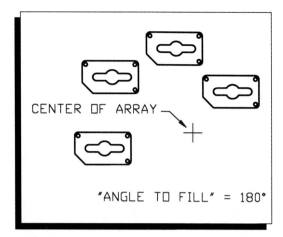

Fig. 7-11 A polar array does not necessarily have to extend 360 degrees (a full circle). This array extends for 180 degrees.

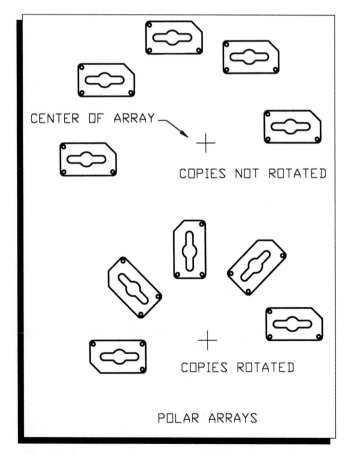

Fig. 7-12 When you choose a polar array, you must decide whether you want each copy of the object rotated.

The following exercise creates a polar array of 12 arrows around a circle starting at the 12 o'clock position.

1. Draw a circle 5 units in diameter.

2. Use the CENter and QUAdrant object snap modes to draw a line from the center of the circle to the upper or 90-degree quadrant.

3. Draw two lines from the intersection of the circle and the line to form an arrowhead *(Fig. 7-13A)*.

4. *Command:* **ARRAY** ↵

5. *Select objects:* Pick each of the three parts of the arrow.

6. *Select objects:* ↵

7. *Rectangular or Polar array (R/P) <R>:* **P** ↵

8. *Center point of array:* **CEN** ↵
   of Pick the circle.

9. *Number of items:* **12** ↵

The number you specify at the *Number of items:* prompt includes the original object. Entering 12 at this prompt actually makes 11 copies of the arrow.

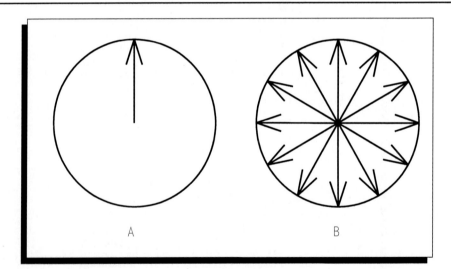

Fig. 7-13

10. *Angle to fill (+=ccw, -=cw) <360>:* ↵

11. *Rotate objects as they are copied? <Y>* ↵

The arrow is arrayed around the center point of the circle *(Fig. 7-13B)*.

12. Try this sequence again with a different circle, but at the *Angle to fill* prompt, enter 180 degrees. Notice that AutoCAD arrays all twelve arrows into only half of the circle.

The MIRROR Command

The MIRROR command copies an object as a reflection of the original. The mirror image can be at any angle to the original, depending on the mirror line you specify. The **mirror line** is the imaginary line or axis around which the original entities are rotated to create the mirror image *(Fig. 7-14)*.

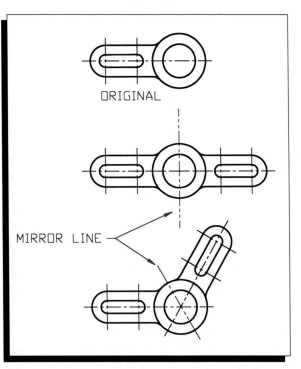

Fig. 7-14 When you use the MIRROR command, the placement of the mirror line is very important. The object is mirrored on the opposite side of the line.

ORIGINAL

MIRROR LINE

ACTIVITY

Follow these steps to mirror two objects.

Your Turn

1. Draw the arc with the line through it as shown in Fig. 7-15A.
2. *Command:* **MIRROR** ⏎
3. *Select objects:* Select the arc and line.
4. *Select objects:* ⏎
5. *First point of mirror line:* Pick a point near the top of the mirror line shown in Fig. 7-15B.

When you pick the first point of the mirror line, the mirror image of the object appears on the cursor and swings as you move the cursor before you pick the second point. This image shows how the final mirror will look at any given angle of the mirror line. Keep in mind that the line is an imaginary line; AutoCAD will not place it permanently on the screen. When you choose the second point, the mirror image appears according to your placement of the imaginary line. The imaginary line itself does not need to be very long, since it merely establishes the angle of the mirror image.

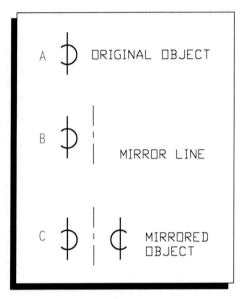

Fig. 7-15

 You may want to turn Ortho on before you pick the second point.

6. *Second point:* Pick a point near the other end of the mirror line shown in Fig. 7-15C.

7. *Delete old objects? <N>* ⏎

You would answer *Yes* to this question only if you were replacing the old object with the mirrored one.

The FILLET and CHAMFER Commands

Drafting terminology normally refers to a small radius of an inside corner as a **fillet** and a small radius of an outside corner as a **round.** For simplicity, AutoCAD combines both fillets and rounds with the command FILLET. FILLET uses an arc of a specified radius to connect two lines, circles, or arcs, or combinations of these.

The CHAMFER command is similar to the FILLET command. While a fillet rounds an outside corner, a chamfer cuts it sharply with a line segment *(Fig. 7-16).*

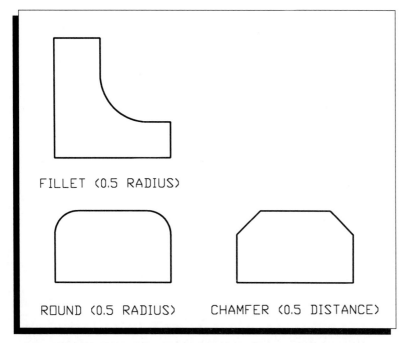

Fig. 7-16 The FILLET and CHAMFER commands are similar in that they both "cut" the corners of an object. However, their methods of doing so are different. The AutoCAD FILLET command executes both fillets and rounds. The CHAMFER command executes chamfers.

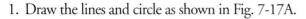

ACTIVITY

Follow the procedure below to set the radius to 0.25 and fillet the objects.

1. Draw the lines and circle as shown in Fig. 7-17A.

The space between the circle and vertical line must be within the 0.25 fillet radius, but the distance between the lines is not important. The FILLET command will extend or trim them to suit the required radius.

---

2. *Command:* **FILLET** ⏎
3. *Polyline/Radius/<Select first object>:* **R** ⏎
4. *Enter fillet radius <current value>:* **0.25** ⏎
5. *Command:* **FILLET** ⏎
6. *Polyline/Radius/<Select first object>:* Pick the horizontal line.
7. *Select second object:* Pick the vertical line.
8. *Command:* **FILLET** ⏎
9. *Polyline/Radius/<Select first object>:* Pick the vertical line.
10. *Select second object:* Pick the circle. Your results should look like the drawing in Fig. 7-17B.

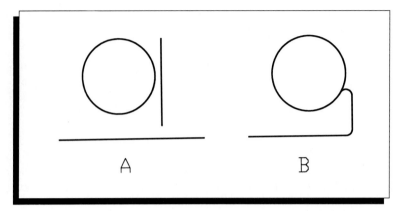

Fig. 7-17

Now begin with a clean drawing screen and follow these steps to set the chamfer distances and to chamfer two lines.

1. Draw two lines as shown in Fig. 7-18A.

Hot Tip Don't worry about the distance between the lines. The CHAMFER command will adjust it accordingly.

2. *Command:* **CHAMFER** ⏎

3. *Polyline/Distance/<Select first line>:* **D** ⏎

The Distance option allows you to select the distances before you create a chamfer.

4. *Enter first chamfer distance <current value>:* **0.5** ⏎

The chamfer distance is the distance the line will be drawn from each edge. Unlike the fillet radius, the chamfer distances can be set independently *(Fig. 7-18B).*

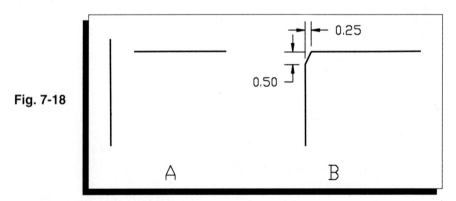

Fig. 7-18

5. *Enter second chamfer distance <current value>:* **0.25** ⏎

6. *Command:* **CHAMFER** ⏎

7. *Polyline/Distance/<Select first line>:* Pick the vertical line.

8. *Select second line:* Pick the horizontal line.

The chamfer should appear. Your drawing should look like the one in Fig. 7-18B (without the dimension marks).

The SCALE Command

You already know that if you want to make an object smaller on your screen, you can use the ZOOM command. However, the ZOOM command does not change the actual size of the object. It merely changes your view of the object so that the object appears smaller. To reduce or increase the actual size of the object, you can use the SCALE command. By default, you enter the percentage of the original to which you want to scale the object. For example, to reduce the size to a quarter of the original size, you enter 0.25 (for 25 percent). To increase the size to twice the original size, you enter 2 (for 200 percent).

ACTIVITY

Follow these steps to reduce the size of a triangle by 50 percent (½ or 0.5) of its original size.

1. Use the POLYGON command to draw a triangle approximately as shown in Fig. 7-19A.

2. *Command:* **SCALE** ⏎

3. *Select objects:* Pick the triangle.

4. *Select objects:* ⏎

5. *Base point:* Pick the lower left corner of the triangle.

6. *<Scale factor>/Reference:* **0.5** ⏎

The size of the triangle changes to half of its original size. To check this, you can repeat the procedure. This time, however, use the DIST command to measure the base of the triangle before and after the scaling operation. The second distance should be half the first *(Fig. 7-19B).*

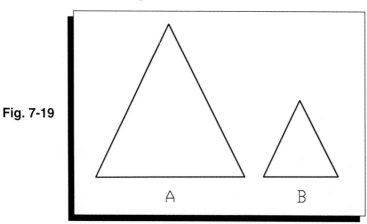

Fig. 7-19

A   B

Using Grips

Sometimes you may need to change the shape of an object on the screen. There are many complicated ways to do this, but AutoCAD supplies a very simple method called grips *(Fig. 7-20)*. **Grips** are small blue boxes that appear at intervals along an entity. You activate them whenever you select the entity *before* you enter a command. When you pick one of these boxes, it turns solid red. You can then move your mouse to a different location on the screen and pick again. The box you selected moves to the new location.

Grips also offers a quick way of copying entities on the screen. When you activate the grips and pick a grip point, the STRETCH command is entered automatically. The prompt changes to *<Stretch to point>/Base point/Copy/Undo/eXit:*. If you pick a new point, you choose the default, "Stretch to point," and the *Command:* prompt returns. However, if you choose the Copy option, you can make unlimited copies of the selected entity. This is similar to the COPY Multiple option, except that it is often faster.

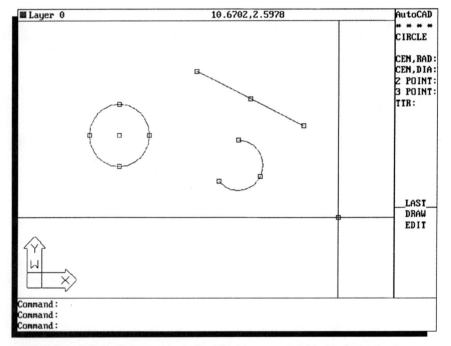

Fig. 7-20 Grips allow you to edit objects more quickly by "gripping" a point on an entity and moving it with the mouse to a new location.

ACTIVITY

The best way to understand grips is to use them. Follow these steps to change a line and a circle using grips.

1. Create a circle on the drawing screen. The size does not matter.

2. Without entering a command, use your mouse to pick the circle.

Notice the small blue boxes that appear at each of the quadrant points and in the center of the circle. These are grips.

3. Use the mouse to pick the grip in the center of the circle.

The grip box turns red.

4. Move the cursor around the screen to see the rubberband circle that is now attached to the cursor. Then pick a new location on the screen.

The entire circle moves to that location.

5. Now pick one of the grips along the edge of the circle so that it turns red.

6. Move the cursor to see its effect on the circle.

Notice that the center of the circle remains in the same place, but the diameter of the circle changes when you move the cursor.

7. Pick a new location for the grip to change the diameter of the circle.

8. Erase the circle from your screen and draw a short line.

9. Without entering a command, pick the line. The grips should appear.

10. Pick the grip at one end of the line so that the grip turns red.

11. Move the cursor to a different location and pick again to change the location of the endpoint.

Notice that the *other* endpoint of the line is unaffected by this procedure.

12. Now pick the grip in the middle of the line.

13. Move the cursor and notice that the entire line moves.

When you select the middle grip, the entire line is affected. Selecting either endpoint of the line affects only that endpoint.

14. ** *STRETCH* **

    <Stretch to point>/Base point/Copy/Undo/eXit: **C** ↵

This activates the Copy option.

15. Pick several different locations on the screen.

Notice that a copy of the original line appears at each location.

16. To end the sequence, press **X**.

Chapter 7 Review

1. What is the difference between the DIVIDE and MEASURE commands?
2. What is the purpose of the PDMODE variable?
3. When might the PDSIZE variable need to be changed?
4. Describe at least two ways to create multiple copies of an entity.
5. What is the difference between a fillet and a chamfer?
6. What are grips and how are they activated?

Chapter 7 Problems

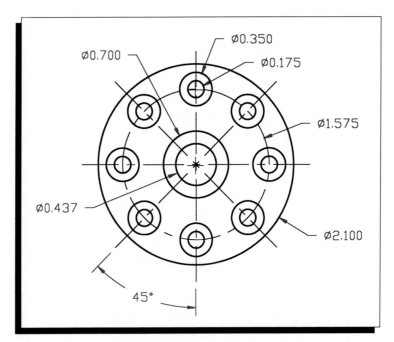

Problem 7-1

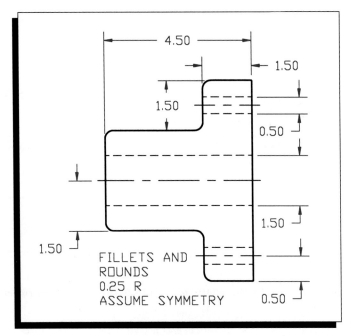

Problem 7-2

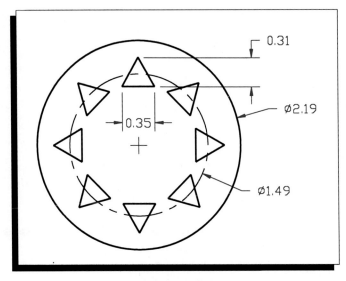

Problem 7-3

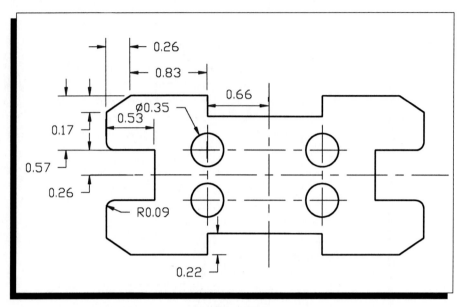

Problem 7-4

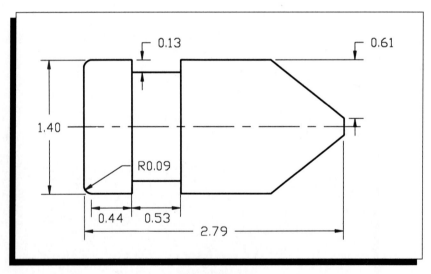

Problem 7-5

Using CAD

Off-Road Bicycles Designed with AutoCAD

"Bicycle design hasn't changed since the beginning of the century," says Bill Townsend, co-owner of Bushido Bicycles in California. However, Townsend has found ways to put his knowledge of race car design into his unique, custom-built, off-road bicycles. Bushido builds bicycles for top athletes and serious amateurs.

Even though Townsend says that most modern bicycle designs are traditional, his designs for bicycles are radical even by today's advanced concepts. The bicycles are all built with a full suspension system which Townsend explains in terms of "what goes where and why, what happens when this moves, and where do these other pieces fit." Townsend uses geometry to give Bushido bicycles their speed and excellent handling characteristics.

Each Bushido off-road bicycle is unique. When Townsend designs a bicycle for a customer, he must know the person's size, weight, and riding preferences to determine the steering and seat tube geometry. "It's like having a suit made instead of buying one off a rack," he said. The intended terrain the bicycle will travel and the type of handling the customer expects determine the bottom bracket height and wheel base. "Then all that remains is to connect center lines," said Townsend. "AutoCAD's geometry calculator is handy to determine lengths and angles. With this information, it is a simple matter to cut tubing to the proper length and angle."

Townsend explained that "everything we design is based on geometry. We can whip things out pretty quickly for an individual this way. And with AutoCAD we can give them a color rendering of what their bike will look like." The price of an off-road Bushido bicycle starts at $6,000. Only a handful are made each year.

Townsend also designs bicycle components for world-wide manufacturers. He produces detail drawings to give the manufacturers exact fits and tolerances. "AutoCAD speeds up the process for me," said Townsend. "I can do the design for a company in Switzerland, FAX it to them, and in two days I'll have their red-line copy back. We can answer their questions, make changes, and in four hours they have the new design." Townsend enthusiastically calls it "world-wide instantaneous results."

Using AutoCAD, said Townsend, "a designer has no drafting board, just infinite physical space which places no limit to your imagination—you can do anything here!"

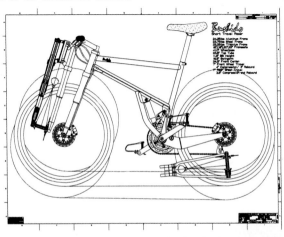

Chapter 8

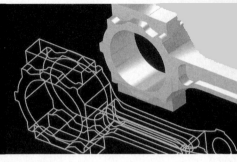

► Key Terms

multiview drawings
orthographic
  projections
coincide
foreshortened
visualization

model space
paper space
viewports
thickness
plan view

►► Commands & Variables

TILEMODE
VPORTS
MVIEW
PSPACE
MSPACE

VPLAYER
THICKNESS
ELEV
VPOINT

Views and Techniques of Drawing

Objectives

When you have completed this chapter, you will be able to:

- visualize objects as orthographic projections.
- lay out the six faces of an object in their proper relationships.
- select views for drawing.
- place and align views properly.
- create orthographic projections to include visible lines, hidden lines, center lines, circles, and arcs.
- use AutoCAD to create a multiview drawing.

Looking through two eyes, you see objects with binocular vision, which allows you to see in three dimensions. Even when you close one eye, your brain interprets real-life objects as three-dimensional (3D). When objects become pictures, however, they lose their depth. The objects may be photographed or drawn with depth, but depth is only an illusion. The surface upon which they are displayed is two-dimensional, and the object is flat. This chapter discusses **multiview drawings:** two-dimensional representations of three-dimensional objects.

Views

Over the years, many attempts have been made to view two-dimensional pictures in three dimensions. Fig. 8-1 shows a Victorian stereoscope which was the forerunner of the View-Master® 3D viewer in Fig. 8-2. Both viewers use the same principle of offsetting two similar images and viewing one image with each eye. The two images fool the brain into thinking the objects in the two-dimensional pictures are three-dimensional.

Fig. 8-1 This is an example of a Victorian stereoscope, circa 1901; it was one of the first devices that created a three-dimensional image from two-dimensional pictures.

Fig. 8-2 The three-dimensional viewer works much the same way as the Victorian stereoscope.

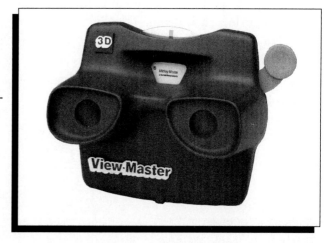

Two-Dimensional Objects

Although most of our world is three-dimensional, we do have some two-dimensional objects, or at least objects that we think of as two-dimensional. A gasket, a sheet of paper, wallpaper, and a rug are examples of items that can be described in terms of length and width only. Depth may be a consideration, but it can be specified and need not be drawn. Objects such as those in Fig. 8-3 can be fully described in only one view.

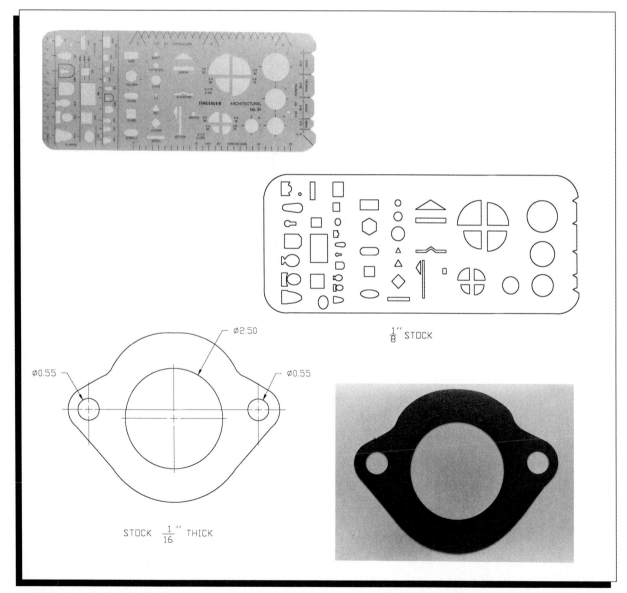

Fig. 8-3 A gasket, rug, shear blank, and stencil are all items that can be described in one view. (A shear blank is a blank sheet of metal that can be sheared (cut) and stamped to create parts.)

Orthographic Projections

When depth becomes a visual consideration, it takes more than one view to show an object completely. Multiview drawings, called **orthographic projections**, show three-dimensional objects by projecting two or more views of the object onto a flat surface. The flat surfaces are called planes of projection.

A column or cone can be described with two views: one to show its length and another to show its diameter *(Fig. 8-4)*. As objects grow in complexity, two views may not be enough to describe their features fully. In these cases, multiple views of the object are necessary. For example, think about the features of a computer monitor. You view the monitor from the front, but the controls may be on the left side or the right side. The bottom sits on a swivel base, and to plug it in, you need to see the rear. To describe a monitor fully with its various features, you need multiple views *(Fig. 8-5)*.

Fig. 8-4 Examples of the two views needed to describe cones and cylinders.

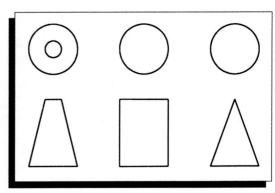

Fig. 8-5 A computer monitor and each of its faces in their proper position on the glass box.

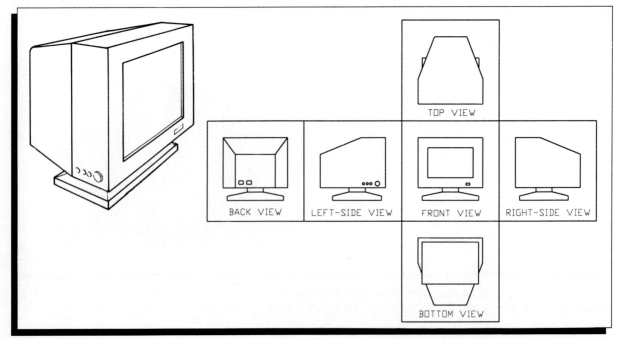

▶ Lines

When you create multiview drawings, you will need to consider various lines. Continuous (visible) lines, hidden lines, center lines, and coincidental lines are examples of the linetypes you will encounter.

Center lines indicate axes of symmetry, or imaginary lines that run through the middle of an object *(Fig. 8-6)*. Therefore, drawings are frequently created around them. Center lines also locate features such as holes, and they provide a base line for dimensioning the features.

Hidden lines represent surfaces that cannot be seen in a view because they are hidden by other features *(Fig. 8-7)*. Hidden lines are generally drawn in all views in which they appear; however, in some complex drawings, the experienced drafter may omit them.

When two or more features appear at the same location or point, they are said to **coincide** *(Fig. 8-8)*. If a visible line coincides with a hidden line or a center line, show the visible line. If a hidden line coincides with a center line, show the hidden line Two lines **coincide** when one line lies directly on top of the other so that only one line is visible.

AutoCAD maintains a library of lines that includes both the center line and the hidden line. Refer to Chapter 4, Fig. 4-6, for the ANSI alphabet of lines and Fig. 4-7 for the AutoCAD linetypes. Chapter 4 also introduced the LINETYPE command for loading lines, the LAYER command for creating layers to contain specialized lines, and the LTSCALE command for changing the relationships between the various line elements. Review that information now if you do not remember it, since this chapter will use those commands.

AutoCAD does not allow you to control the placement of the various line elements. It may look nice if two short center line dashes cross to form the center of a circle. However, in practice, this will probably not be the case. Hidden lines will begin and end at the locations you specify, but the ends of the lines may not look exactly as you might want them to. (Note, however, that when you dimension the circle, AutoCAD places a small cross in the center of the circle. The cross will be visible if the center lines are not present.)

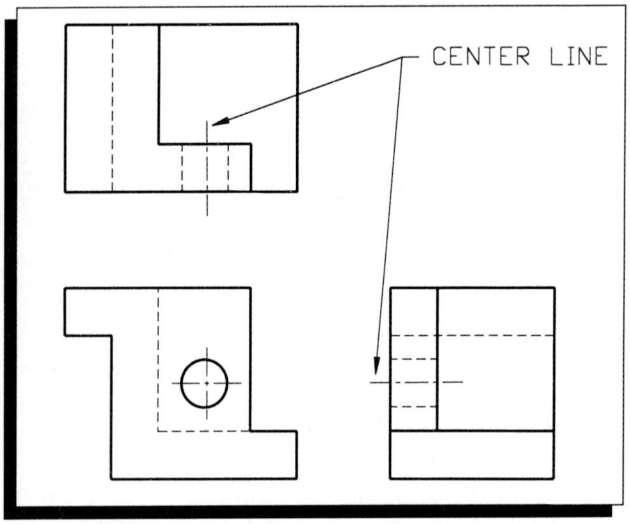

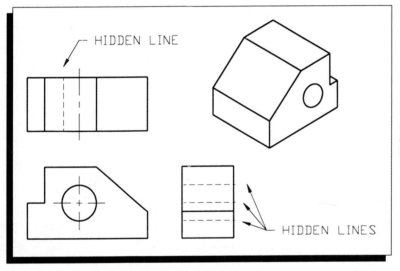

Fig. 8-6 Different lines are hidden in different views on a multiview drawing.

Fig. 8-7 Axes of symmetry are imaginary lines through an object that show that it is symmetrical.

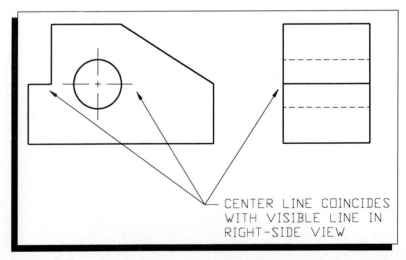

Fig. 8-8 When two lines coincide, they look as though only one line were present.

▶ Surfaces

The way a surface is drawn depends on its shape and its position with respect to the viewing plane (the side it is viewed from). Surfaces that are parallel to the viewing plane appear in their true size. Surfaces that are perpendicular (at 90 degrees) to the viewing plane appear as lines *(Fig. 8-9)*.

Surfaces that are at an angle other than 90 degrees to the viewing plane appear **foreshortened**; that is, they appear shorter than their true length. When surfaces are foreshortened, it is necessary to obtain their true size by referring to more than one view *(Fig. 8-10)*.

Curved surfaces do not show as curves in all views *(Fig. 8-11)*. Cylinders and cones appear as polygons when viewed parallel to their axes. They appear round only when you look at them in cross section. Spheres appear as circles in all views. Fig. 8-12 shows examples of curved surfaces in various views and in combination with other curved surfaces and with flat surfaces.

Notice that when you draw the top view of a curved object, you cannot show the curve at every point. Instead, drafters insert "lines of surface separation" to show when a sharp surface change occurs. No lines of surface separation appear with tangents and connected arcs. Refer again to Fig. 8-12.

Holes are reverse — or negative — cylinders, cones, and spheres. In other words, they are cylindrical, conical, or spherical shapes that are missing, or removed, from a solid object.

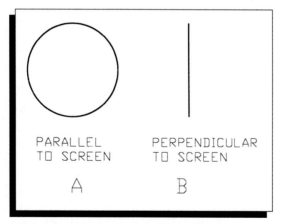

Fig. 8-9 A) When a surface is parallel to the viewing plane, it appears in its true size and shape. B) When a surface is perpendicular to the viewing plane, the surface appears as a line.

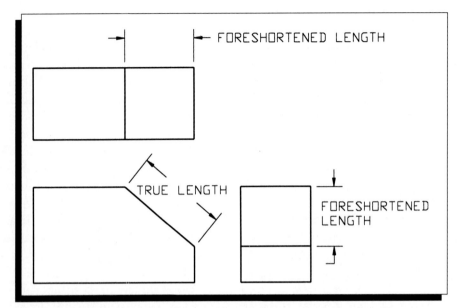

Fig. 8-10 More than one view may be necessary to determine the true size of a surface that appears foreshortened.

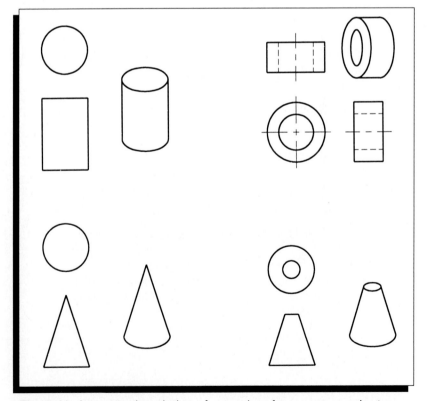

Fig. 8-11 Accurate description of curved surfaces may require two or three views.

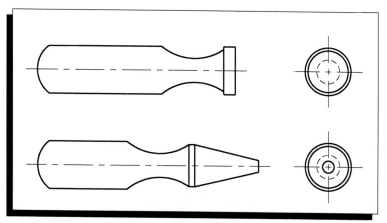

Fig. 8-12 Curved surfaces often occur in combination with other types of surfaces. Notice that lines of surface separation occur in the top view only when an abrupt or major change occurs in a curved surface.

▶ The Glass Box

Visualization, the ability to see in the mind's eye what an object looks like, is the key to drafting. Three-dimensional objects must be turned so that the drafter can fully describe the necessary features of the three dimensions: width/height, width/depth, and depth/height.

All objects can be reduced to a general rectangular form which has six sides: front, rear, top, bottom, left, and right. Any irregularities, that is, surfaces that are curved, offset, or at an angle, are projected to a uniform flat surface for viewing. The result is as if the object were placed inside a hinged glass box and all the surfaces were projected to the glass walls *(Fig. 8-13)*. The intersections of surfaces in different planes appear as lines. As the box opens, *(Fig. 8-14)*, each of the six sides revolves onto a flat surface *(Fig. 8-15)*. The process is similar to making the cereal box in Fig. 8-16, only the cereal box starts out as a flat pattern and is folded to make the box.

The glass box always opens the same way, and the views are always placed in the same relationship to each other. The front and back views show the width and height, the top and bottom views show the width and depth, and the left and right side views show the depth and height. Each view shows only two of the three dimensions. In lieu of a glass box, the object itself can be turned to reveal the various surfaces *(Fig. 8-17, p. 220)*.

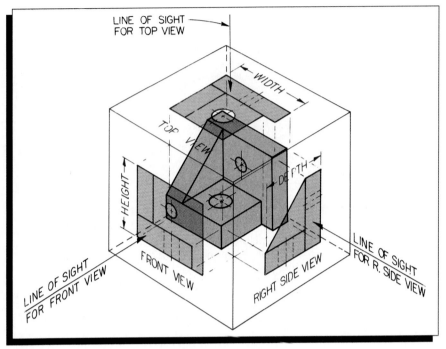

Fig. 8-13 Visualizing an object inside a glass box can help you determine its proper view.

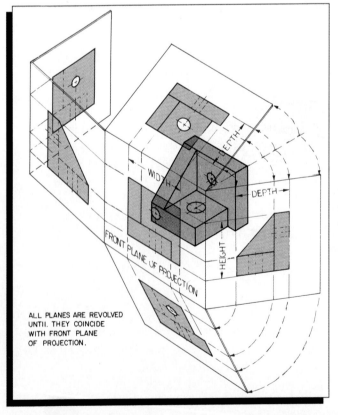

Fig. 8-14 As the glass box begins to open, you may start to see some similarity to an orthographic layout.

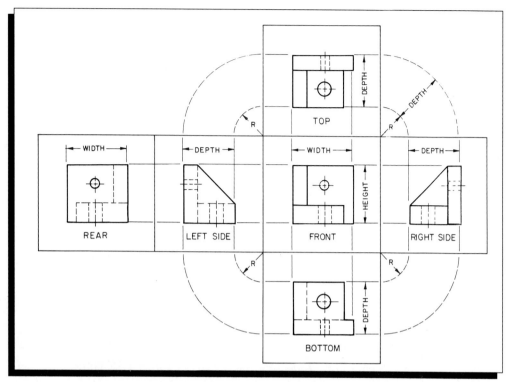

Fig. 8-15 The glass box opens to reveal the orthographic layout.

Fig. 8-16 A cereal box pattern is folded to create the finished box. This is just the opposite of the glass box model.

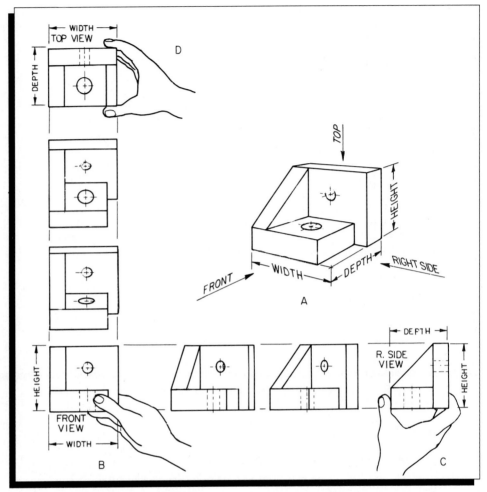

Fig. 8-17 As an alternative to using the glass box model, you can rotate the object manually, if it is not too large.

ACTIVITY

As you begin to practice visualizing the various views of objects for orthographic projection, it may be helpful to have a paper mockup of the glass box model. Follow these steps to create the one shown in Fig. 8-5.

1. Start AutoCAD and begin a new drawing named **GLASSBOX**. Set up the drawing as a 9" by 12" sheet, and set Snap to **0.125**.

Since you will be forming the box when you finish, it is important that all the sides be equal. To accomplish this, you can begin by dividing the AutoCAD screen into rectangles of equal size.

Be sure Snap and Ortho are on to make this job easier. **Hot Tip**

2. Using the **LINE** command, draw lines that begin and end at the points described below.

From	To
0,6	12,6
0,3	12,3
3,0	3,9
6,0	6,9
9,0	9,9

Your screen should now be divided as shown in Fig. 8-18.

3. Use the **TRIM** command to trim away the top and bottom thirds of the line that now extends from 3,0 to 3,9.

4. Use the **LINE** command and the **END**point object snap to close in the remaining squares.

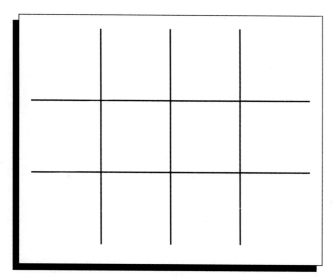

Fig. 8-18

After you complete this operation, your drawing should look like the one in Fig. 8-19 — it should look like the two-dimensional layout of the glass box.

5. Zoom in on the center box, which will contain the front view.

6. On your own, create the front view of the computer monitor as shown in Fig. 8-20. You may need to refer again to Fig. 8-5 to see the details of the monitor for this and the following steps. Again, Snap will help you with this job, but you will need to turn Ortho off to create the diagonal lines of the base.

7. Now zoom in on the space for the bottom view of the object.

8. Draw the bottom view, being careful that the edges of the bottom of the monitor align exactly with the edges shown in the front view. Be sure to include the rectangle that represents the base of the monitor.

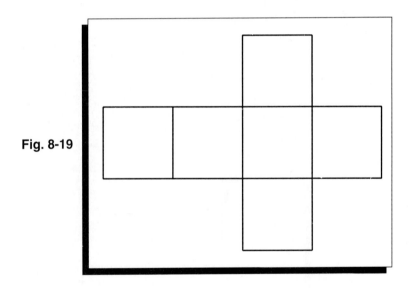

Fig. 8-19

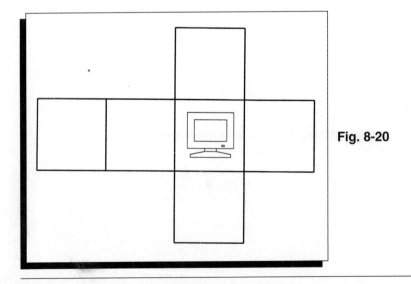

Fig. 8-20

Now you can use the MIRROR command that you learned in Chapter 7 to your advantage. Notice that by using the MIRROR command, you can create the basic lines in some of the views without having to draw them individually. This is one of the advantages of using CAD methods.

9. Use the **ZOOM** command to view the entire drawing.

10. *Command:* **MIRROR** ⏎

11. *Select objects:* Use a window to select all of the lines you created in the bottom view of the object (do not include the walls of the box, however).

12. *Select objects:* ⏎

13. *First point of mirror line:* Select a point in the center of the center box; for example, you might choose the point **6,4.5**.

14. *Second point:* Select another point that bisects the center box; for example, **9,4.5**.

15. *Delete old objects?* <N> ⏎

Notice that AutoCAD placed the mirrored copy perfectly in the square you will use for the top view. However, you still have a little work to do to complete the top view. If you think about it, you will realize that not all of the swivel base is visible from the top view. Refer again to the finished box in Fig. 8-5 if you need help understanding this concept. To show only the parts of the swivel base that you would ordinarily see from the top view, you need to trim away some of the lines in the swivel base.

16. Use the TRIM and BREAK commands to trim away the invisible, or hidden, lines of the swivel base.

If you used the RECTANG or POLYLINE command to create the swivel base, you may need to use the BREAK command before you can use the TRIM and BREAK commands effectively.

Note

Your finished top view should look like the one in Fig. 8-5.

17. Zoom in on an area of the drawing that includes the front view and the square that will hold the right-side view.

18. Using the **LINE** command, create the right-side view as shown in Fig. 8-5. As you create this view, be sure that the top, bottom, and swivel base of the monitor align with the same features in the front view.

19. Once again, use **ZOOM** to view the entire drawing.

20. Use the **MIRROR** command to mirror the right-side view into the box for the left-side view.

When AutoCAD prompts you for a mirror line, indicate an imaginary line that runs vertically through the exact middle of the front view.

21. To finish the left-side view, zoom in on the view and create the control knobs. (You may need to turn off Snap first.)

22. Finally, zoom in on the square that will hold the back view.

23. Create the back view, making sure that the lines for the front and back of the monitor align with the front and back in the other views.

24. Plot your glass box and cut it out.

Your glass box model may be more useful to you if the different views are labeled, as they are in Fig. 8-5. You may either write in the labels at this time or wait until after you have completed Chapter 9, "Text," and use AutoCAD to label the views.

25. Assemble the glass box model and secure it with tape.

Layout of Views

When planning the layout of a drawing, you should include only those views necessary to describe the object. Most objects can be fully described without using all six views. In fact, most objects can be described clearly with one, two, or three views. When drawing three views of the object, drafters usually select the front, top, and right-side views, by convention. They may use additional views or alternate views as required.

Primary View

The front view, the primary view of an object, is generally the most important view, and it is usually drawn in the object's most stable position. Common objects such as buildings, tables, or appliances should be drawn in their normal upright position — the way people are used to seeing them. Objects that have no particular orientation may be drawn in any convenient or logical position; however, long objects such as bolts, screws, pencils, and shafts should be drawn horizontally.

Other Views

Since the front view is the primary view, all the other views should be aligned with it. For example, center lines as well as object lines should match. The use of the glass box visualization technique can help you keep the views aligned and positioned properly.

Visualization

"I just can't see it!" is a common complaint among beginning drafters. All drafters need the ability to visualize spacial relationships, and this ability can be improved with practice. Fig. 8-21 shows a bracket, a projection of the bracket's surfaces, and the resulting three-view drawing.

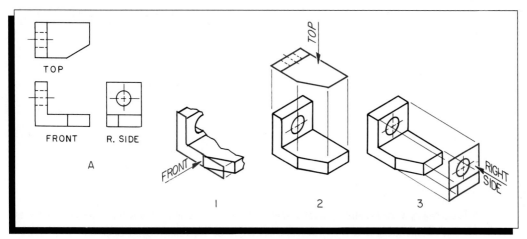

Fig. 8-21 This is the proper way to visualize a bracket.

ACTIVITY

Follow the steps below to practice visualizing objects from a drafter's point of view.

1. Practice visualizing the objects in Fig. 8-22. Sketch the front, top, and right-side views, and compare your results with Fig. 8-23 on the next page. Draw the views with AutoCAD.

2. Now reverse the process. From the orthographic drawings in Fig. 8-24 (page 228), sketch the pictorials, and compare your sketches with those in Fig. 8-25. Draw the views with AutoCAD. (Refer to Chapter 2 if you need a review of isometric sketches.)

Practice visualization constantly. As you look at objects in the three-dimensional world, visualize them in two dimensions. Sketch a mental picture; then move around to check your results.

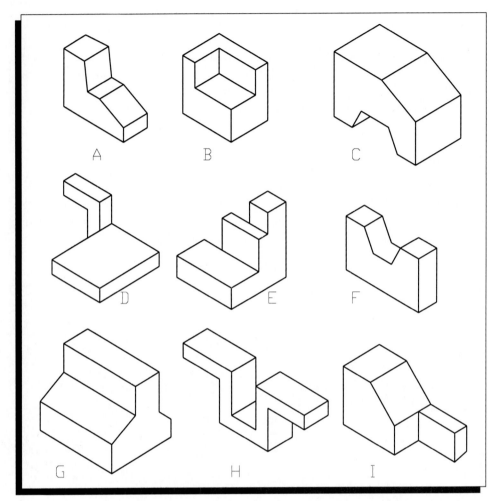

Fig. 8-22 Isometric drawings for visualization practice.

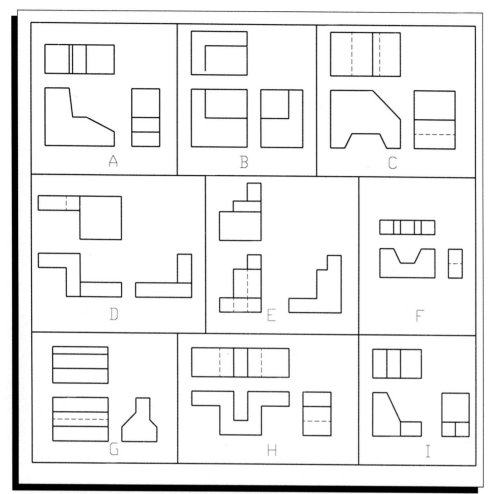

Fig. 8-23 Answers for visualization practice examples in Fig. 8-22.

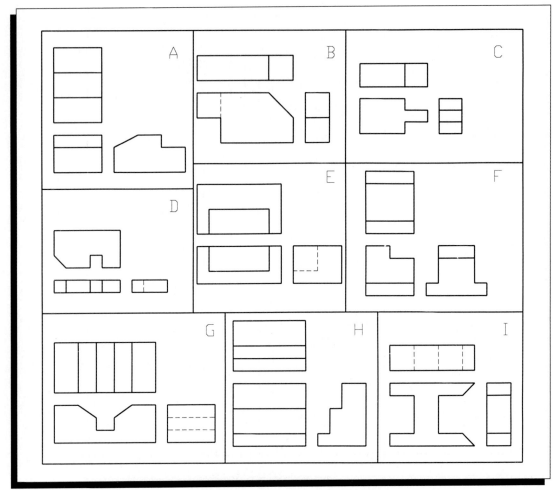

Fig. 8-24 Orthographic drawings for visualization practice.

Multiview Drawings with AutoCAD

As you have read, orthographic projections come from three-dimensional objects. Inside the glass box is a solid part. Each face of the part becomes two-dimensional when it is projected against a flat surface. In AutoCAD, you can create two-dimensional views by projecting three-dimensional drawings against your two-dimensional computer screen.

Imagine your computer monitor as a glass box. You are outside the box, and the object is inside your monitor. You can see all sides of the object *(Fig. 8-26A)*. Actually, your monitor is more like a box with one glass side. Only the side of the object that is parallel to your screen is visible through the glass. To view the various surfaces of the object inside your

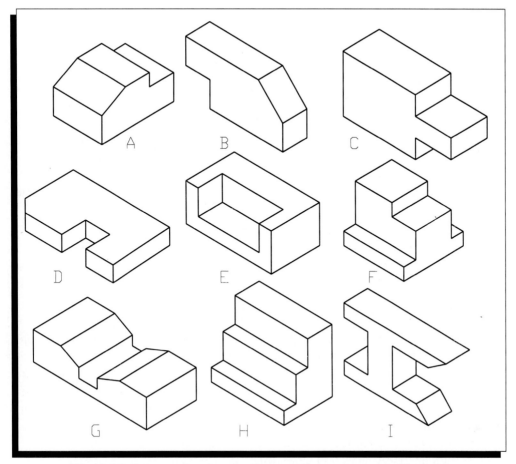

Fig. 8-25 Answers for visualization practice examples in Fig. 8-24.

monitor, you must turn the object so that the appropriate side is parallel to your screen *(Fig. 8-26)*.

Now imagine that your monitor screen is divided into four parts. Each part becomes a separate screen, and a different view can be shown in each screen. In the lower left quadrant you can have the front view; in the lower right quadrant, the right-side view; in the upper left quadrant, the top view; and in the upper right quadrant the pictorial view. Sounds like the standard layout for a three-view drawing, doesn't it?

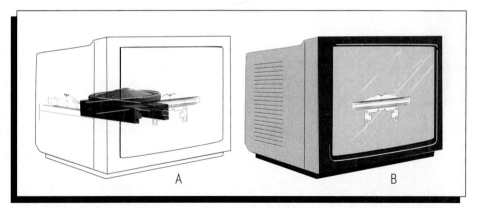

Fig. 8-26 A) If your monitor were a glass box, you could see all sides of the object. B) Your computer monitor can also be considered a box with one glass side.

Model Space and Paper Space

AutoCAD has two types of space: model space and paper space. **Model space** is the standard drawing space. When you start AutoCAD, you enter model space by default. Prior to Revision 11, model space was the only space available. **Paper space** lets you lay out and annotate (add notes and labels to) your drawing. It also allows you to plot two or more views of your drawing — something you cannot do in model space. In paper space, the icon in the lower left corner of your screen changes to a triangle, and the letter "P" appears in the status line *(Fig. 8-27)*.

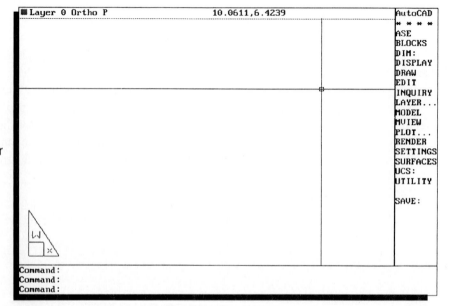

Fig. 8-27 The paper space icon appears when you enter AutoCAD's paper space.

Setting Up Paper Space

If you plan to create a multiview drawing, you will probably need to use paper space. Therefore, you must set up your drawing for paper space. To do so, you need to understand TILEMODE. TILEMODE is a system variable that, when set to 0, lets AutoCAD work in paper space.

By default, TILEMODE is set to 1, the setting required for model space. When TILEMODE is set to 1, it is turned on, and the VPORTS command can be used. In this mode, AutoCAD Revisions 12 and 11 are compatible with earlier versions for working with viewports.

To activate paper space, you must turn TILEMODE off by setting it to 0. The VPORTS command is disabled, and the MVIEW command takes its place. In addition, the PSPACE, MSPACE, and VPLAYER commands are activated. PSPACE switches you from model space to paper space, and MSPACE switches you from paper space to model space. VPLAYER is similar to the LAYER command, but it allows you to control the layers in each viewport individually.

You can draw objects in both model space and paper space. You can move between them without changing the objects. However, you must edit each object in the same type of space in which you created it. You may plot in model space or in paper space. If you plot in model space, you will plot only the active view. If you plot in paper space, you can plot all the views at once.

Viewports

Viewports divide your screen into multiple viewing areas *(Fig. 8-28)*. You can see one, two, three, or four parts of your drawing at one time. AutoCAD's default is one viewport. If you are working in model space (with TILEMODE set to 1), the VPORTS command controls the viewports. Whatever you do in one viewport will appear in all of the viewports — unless you freeze the current layer in the other viewports using the VPLAYER command. In paper space, the MVIEW command controls the viewports.

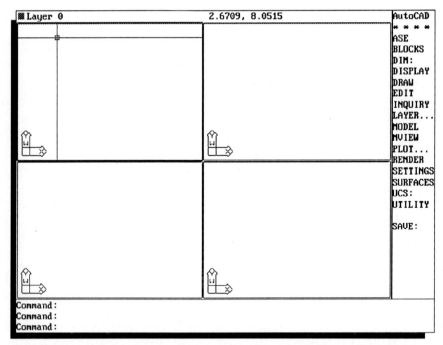

Fig. 8-28 Setting up several viewports on the AutoCAD screen allows you to view an object from several viewpoints at once. On this screen, four viewports are active, showing the three-dimensional object from various viewpoints.

ACTIVITY

The following procedure will set up four viewports in paper space. To help you understand the two types of space and their differences, this procedure also steps you through some simple creation. You will draw in both model space and paper space.

1. Start AutoCAD and begin a new drawing named VIEWS.
2. Create the following layers:
   OBJECT   color white
   VPORTS   color blue
3. Make **VPORTS** the current layer.
4. *Command:* **TILEMODE** ⏎
5. *New value for TILEMODE <1>:* **0** ⏎

The screen responds with "Entering Paper space. Use MVIEW to insert Model space viewports. Regenerating drawing." The paper space icon appears in the lower left corner of the screen. The letter "P" appears in the status line following the current layer name.

6. *Command:* **MVIEW** ⏎

7. *ON/OFF/Hideplot/Fit/2/3/4/Restore/<First Point>:* **4** ⏎

8. *Fit/<First Point>:* **F** ⏎

The screen responds with "Regenerating drawing." Two blue lines divide the screen into four viewports placed to fit by AutoCAD.

9. Make the **OBJECT** layer current.

10. *Command:* **MSPACE** ⏎

Throughout this exercise, you may use the shorthand commands MS and PS for MSPACE and PSPACE, respectively.

The viewports created in paper space continue into model space. You have four separate screens, each with its own UCS icon. The crosshairs appear in the active screen, which has a darker border than the other screens.

11. Change the active screen by moving your pointing device and picking one of the inactive screens.

12. Draw a rectangle in your active screen. It appears in all four screens.

13. Plot your rectangle.

14. *Command:* **PS** ⏎

The rectangles are in paper space.

15. Plot the paper space screen.

16. Compare your model space and paper space plots.

Notice that when you plot in model space, only the active viewport plots. When you plot in paper space, all of the viewports plot.

17. Draw a circle in one of the paper space viewports. Notice that it only appears where you have drawn it.

18. Change to model space. The circle still appears only where it was drawn.

19. Try to erase the circle.

You cannot do it. You drew the circle in paper space, not in model space, so you cannot edit it in model space.

20. Change to paper space and erase a rectangle.

That does not work either. The rectangles were not drawn in paper space.

21. Erase the circle.
22. Change to model space and erase a rectangle.
23. Make sure all four viewports are empty. Then save the drawing. You will need it later in this chapter.

Three-Dimensional Objects

Now that you are familiar with model space, paper space, and viewports, it is time to put a three-dimensional object on the screen. The simplest kind of three-dimensional (3D) object is a two-dimensional object with thickness. **Thickness** is the same as height *(Fig. 8-29)*.

A box is a simple 3D object. A box with a cylinder sitting on top of it is more complex *(Fig. 8-30)*. A box has thickness which starts at the bottom of the box and goes up. The cylinder also has a thickness which starts at the bottom of the cylinder and goes up. The problem is, the bottom of the cylinder is not at the same location as the bottom of the box. The bottom of the cylinder starts at the top of the box. The location of the bottom of a 3D object is its elevation. If you draw a box with a thickness of 2" and an elevation of 0" and you draw a cylinder with a thickness of 1"

and an elevation of 2", you will have a 1"-high cylinder sitting on a 2"-high box *(Fig. 8-31)*.

AutoCAD has a system variable for thickness. You can set this variable directly by entering THICKNESS at the command prompt, or you can enter it as part of the ELEV (Elevation) command. (An object's elevation is the level at which the bottom of the object begins.) In most cases, you will need to change both the thickness and the elevation, so it is easier to use the ELEV command to do both at once.

When you draw a 3D object using the ELEV command, you will be looking down on the object from above. AutoCAD calls this the **plan view**. You cannot see the thickness of the object until you change your position of viewing by using the VPOINT command.

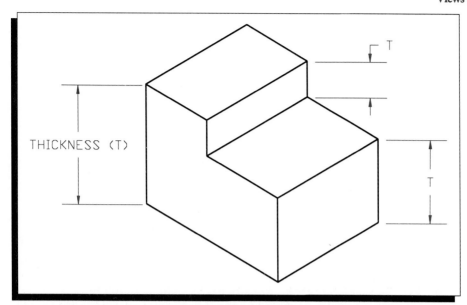

Fig. 8-29 The thickness of an object is the same as its height.

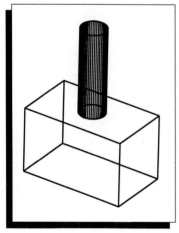

Fig. 8-30 To achieve a box with a cylinder sitting on top of it, you have to draw the box at one elevation and thickness setting, then change the settings and draw the cylinder. In AutoCAD, this is best done from the top or plan view.

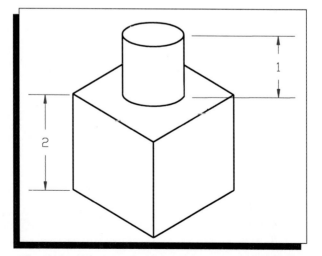

Fig. 8-31 By changing the elevation and thickness, you can create objects that appear in three-dimensional space.

ACTIVITY

The following procedure uses the viewports which you created in the preceding "Your Turn." You will draw a 3D object in model space and view it as a top view (plan view), front view, right-side view, and pictorial view *(Fig. 8-32)*.

1. Start AutoCAD, open your **VIEWS** drawing, and make the **OBJECT** layer current.
2. Enter model space if you are not already there, and make your upper left viewport active.
3. *Command:* **ELEV** ⏎
4. *New current elevation <Current>:* **0** ⏎
5. *New current thickness <Current>:* **3** ⏎
6. *Command:* Draw a rectangle approximately as shown.
7. Use the **POLYGON** command to place a hexagon within the rectangle approximately as shown, with the same elevation and thickness as the rectangle.
8. *Command:* **ELEV** ⏎
9. *New current elevation <0>:* **3** ⏎

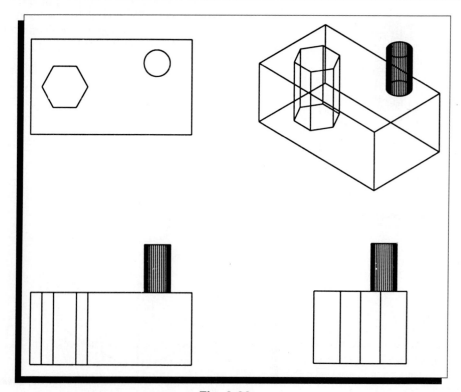

Fig. 8-32

10. *New current thickness <3>:* **2** ↵

11. Draw a circle approximately as shown.

12. Make the upper right view active.

13. *Command:* **VPOINT** ↵

14. *Rotate/<View point> <Current>:* **1,-1,1** ↵

The numbers you entered in step 14 refer to AutoCAD's internal compass and axes tripod *(Fig. 8-33)*. The compass consists of a center point that represents the north pole, an inside circle that represents the equator, and an outer circle that represents the south pole. The three numbers you entered refer to those three items, respectively. By entering various combinations of 1 and 0, you can view the object from all of the basic views:

    0,0,1 = plan (top) view
    0,0,-1 = bottom view
    1,0,0 = right-side view
    -1,0,0 = left-side view
    0,1,0 = back view
    0,-1,0 = front view

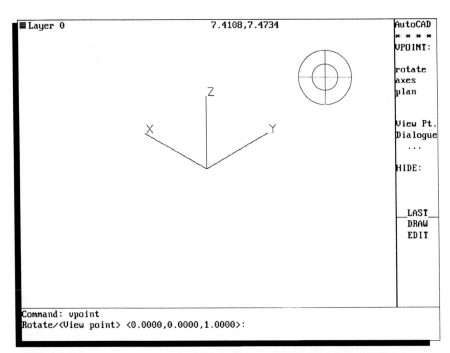

Fig. 8-33 When you enter the VPOINT command and press RETURN twice, AutoCAD displays this compass and axes tripod. By moving the mouse, you can position the cursor, which represents your point of view. When you pick a point, the drawing changes to show the point of view you selected.

The screen responds with "Regenerating drawing." Now you have a 3D view.

15. *Command:* **ZOOM** ↵

16. *All/Center/Dynamic/Extents/Left/Previous/Umax/Window/<Scale (X/XP)>:* **1** ↵

The ZOOM response of 1 sets the view to the full scale you defined when you set up the drawing limits.

Hot Tip If you want to remove the hidden lines, type HIDE at the *Command:* prompt. To replace the hidden lines, type REGEN (for REGENERATE).

---

17. Make the lower left viewport active.

18. *Command:* **VPOINT** ↵

19. *Rotate/<View point> <Current>:* **0,-1,0** ↵

Note If you had pressed RETURN in step 19 instead of entering the numbers, a graphic representation of AutoCAD's compass and tripod would have appeared on your screen, and you could have chosen your viewpoint by moving the mouse.

---

20. *Command:* **ZOOM** ↵

21. *All/Center/Dynamic/Extents/Left/Previous/Umax/Window/<Scale (X/XP)>:* **1** ↵

22. Make the lower right viewport active.

23. *Command:* **VPOINT** ↵

24. *Rotate/<View point> <Current>:* **1,0,0** ↵

25. *Command:* **ZOOM** ↵

26. *All/Center/Dynamic/Extents/Left/Previous/Umax/Window/<Scale (X/XP)>:* **1** ↵

Some of the rotated views may not fit entirely within their viewports, so they may appear to be cut off. To fix this problem, you can change the sizes of the various viewports. This can be done only in paper space. It is not uncommon for viewports to overlap one another in AutoCAD.

27. Switch to paper space.

28. Click on one of the viewport lines that appears to cut off a view of the object. The AutoCAD grips will appear.

29. Grab the grip marker with your mouse and move it so that the entire view of the object appears on the screen.

30. Repeat this process for any other viewports that cut off views of the object.

31. Freeze the VPORTS layer and plot your drawing.

Freezing the VPORTS layer made it disappear, which means AutoCAD simply ignores it. You could also make the VPORTS layer disappear by turning the layer off. Freezing is usually the preferred method, however, because AutoCAD will regenerate a drawing faster if layers are frozen instead of turned off.

Views generated with AutoCAD in the above manner do not necessarily follow the rules of multiview drawing. Hidden lines do not appear as hidden lines. View alignment may or may not be accurate. Center lines in one view may look strange in another. AutoCAD provides methods of correcting these problems, but they are beyond the scope of this textbook. For more information, refer to the *AutoCAD Reference Manual.*

■ Chapter 8 Review

1. What is a multiview drawing?
2. What are the six possible views of an object?
3. Which is the primary view of most objects?
4. What do two lines look like when they coincide?
5. What is a foreshortened surface?
6. When you are plotting a multiview drawing in AutoCAD, what is the practical difference between being in model space and being in paper space?
7. To what value should TILEMODE be set if you are going to work in paper space?
8. Explain the advantage of using AutoCAD to create a multiview drawing.

■ Chapter 8 Problems

To apply the drafting skills presented in this chapter, create the following drawings using AutoCAD. You may draw each view separately or create a 3D object and display the appropriate views. The dimensions are there for your reference only. You do not need to include them in your drawings.

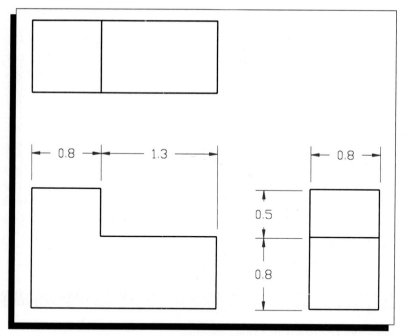

Problem 8-1

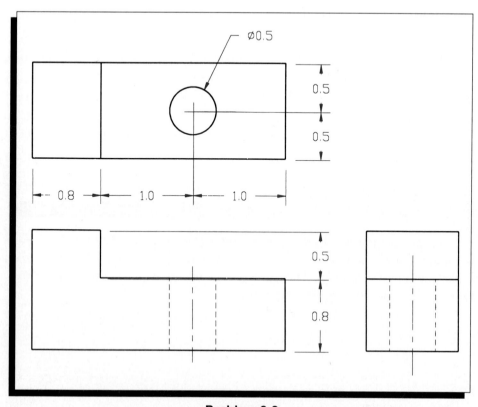

Problem 8-2

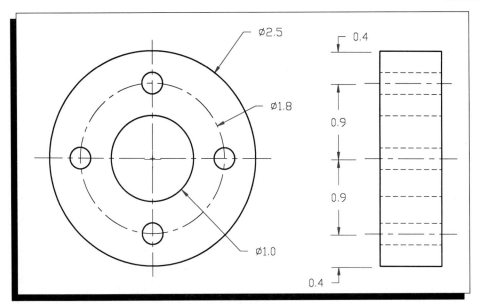

Problem 8-3

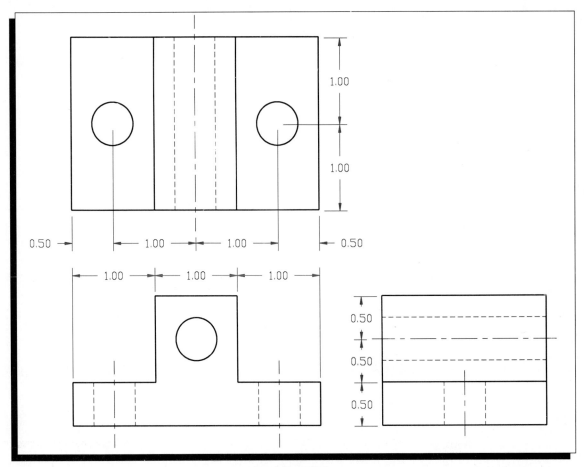

Problem 8-4

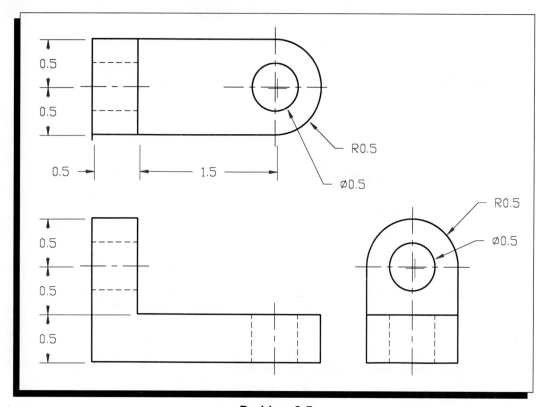

Problem 8-5

Using CAD
AutoCAD Provides Data for Public Health and Safety

The Northern California Power Agency (NCPA) owns and operates two geothermal power plants (geothermal is heat from the earth) in the geysers of northern California. In giant fractures deep within the earth, water boils into steam. NCPA operates 69 wells in Lake and Sonoma Counties that tap the underground steam energy to produce electricity.

After a decade of extensive use, however, steam production has decreased. To improve the productivity of the steam wells, NCPA injects treated waste water from nearby communities into the underground steam reservoirs. They use AutoCAD to help monitor the injection program.

NCPA also closely monitors the frequent seismic (earthquake) activity around the injection wells. These small earthquakes measure less than 3.0 on the Richter scale but can often be felt by area residents.

The entire steam field, including the 69 steam wells and seven injection wells, has been drawn into AutoCAD. Engineers and geologists have superimposed data about underground movement of the injected water and the resulting seismic activity onto the AutoCAD drawings of the steam field area. These maps allow the engineers to better understand the reservoir. AutoCAD plan and section views help them make decisions about how best to sustain the geothermal reservoir beneath the steam wells.

Lake County Energy and Resource Manager Mark Dellinger believes that the injected waste water is a benefit to the energy resources of northern California because it helps sustain geothermal production. In addition, it allows community development to occur because waste water is being disposed of in an environmentally superior manner. Dellinger said, "We need seismic monitoring of the area to provide data for public health and safety." AutoCAD can analyze and make this data visible, says Dellinger. "Up until now the public hasn't had the ability to see seismic data generated in the steam fields."

The AutoCAD drawings are done by NCPA's Bud Schroeder in paper space so that both the plan views and section views of the wells can be shown simultaneously. Schroeder produced the drawings that help NCPA explain the complex interactions that occur in the steam fields.

Because Schroeder used AutoCAD to clearly represent drawings and data at the same time, NCPA has been able to illustrate that the life and productivity of their steam fields in northern California can be increased with augmented water injection.

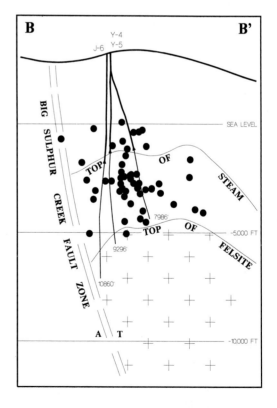

Drawing Courtesy of Bud Schroeder

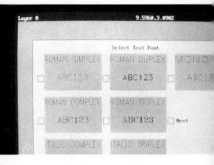

Chapter 9

▶

Key Terms

text   obliquing angle
font   justification
style   overscore
width factor   underscore

▶▶

Commands & Variables

STYLE
TEXT
DTEXT
DDEDIT
CHANGE

Text

▶ ▶ ▶

Objectives

When you have completed this chapter, you will be able to:

- enter text on a drawing using the TEXT and DTEXT commands.
- justify text using various justification options.
- set text fonts and styles.
- enter special symbols in text.
- modify text using the DDEDIT and CHANGE commands.

Pictures convey ideas and concepts. However, drafters must add letters, numbers, characters, and symbols to describe objects, give directions, explain ideas, and list dimensions. **Text** is a general term for the letters, numbers, characters, and symbols used for visual communication in CAD.

Board drafters letter their drawings by hand. This is slow, time-consuming work. CAD makes lettering easy to read, fast, and consistent in both size and style. CAD also makes it easier to modify text and change fonts and styles. Like board drafters, CAD users follow standard conventions and rules that govern text appearance and placement on drawings.

Fonts and Styles

Fonts and styles refer to the character, shape, size and other features that determine the visual impact of the text. A **font** is a complete set of characteristics of a certain typeface. It is a typeface design. Fig. 9-1 shows examples of several AutoCAD fonts. A text **style** is a variation of a font. In AutoCAD, a specific style retains the basic features of the font, but varies other features such as width, height, and slant *(Fig. 9-2)*.

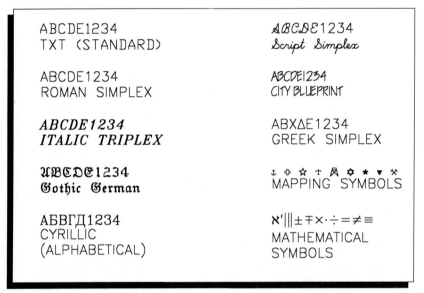

Fig. 9-1 AutoCAD supplies a variety of fonts and styles to meet the needs of most drafters and designers.

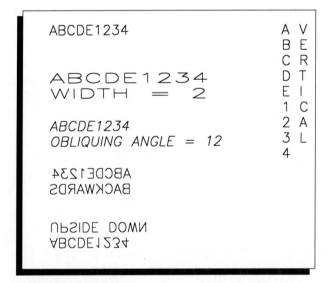

Fig. 9-2 AutoCAD provides several variations of each font. The variations you choose make up the style you use.

Text Standards

How do you know what fonts and styles to use on a given drawing? Individual company standards and ANSI standard Y14.2M-1979(R1987), "Line Conventions and Lettering," should control how text appears on your drawings. The following general rules apply to text placement.

1. All text on a given drawing should be of the same height, a minimum of 0.125" (⅛ inch), except that titles, captions, drawing revisions, drawing numbers, etc., may be in larger sizes, generally between 0.188" (³⁄₁₆ inch) and 0.25" (¼ inch).
2. You may use slanted letters, but never mix slanted letters with vertical letters. The recommended slant is 68 degrees.
3. Text may be upper case or mixed upper and lower case, following accepted rules of capitalization.
4. Use simple fonts and styles for clarity and ease of reading.
5. Do not mix fonts within a single drawing.
6. In addition to company policy and ANSI standards, follow your own common sense. You will occasionally find that you must break one or more of the rules above. Remember that none of these rules are absolute. You may break them sometimes, but only with valid reason.

AutoCAD Fonts and Styles

AutoCAD stores its library of fonts under the general heading Styles, even though the word "styles" actually refers to variations of fonts. This is because once you have selected a font in AutoCAD, you can vary its style to suit your specific drafting needs.

AutoCAD provides a variety of fonts, including Roman, Cyrillic, Greek, and Gothic, as well as symbols for mathematics, mapping, meteorology, astronomy, and music. AutoCAD defaults to Original TXT font and the STANDARD style.

There are various ways to set the current style in AutoCAD. First, you can enter the STYLE command. AutoCAD asks you for the new style name. The current style appears as the default. Unless you know the style name you need, however, this can be a difficult procedure. A much easier way to access the fonts and styles is to use the pull-down menu so that you can see your options. The following procedure changes the font and sets the style.

ACTIVITY

Begin a new drawing in AutoCAD and name it TEXT. Then follow these steps to change the current font and style in AutoCAD.

1. Activate the pull-down menus and pick **Draw, Text, Set style...**

The Select Text Font dialogue box appears *(Fig. 9-3)*. There are 38 fonts and symbols listed by name on the left and by alphanumeric design on the right.

Only the first 20 choices appear on the first "page" of this dialogue box. To see the remaining fonts, you can pick the Next button near the bottom of the dialogue box. Another page of fonts appears. To return to the original page, pick the Previous button.

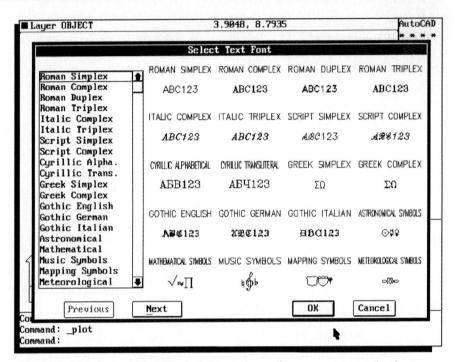

Fig. 9-3 The Select Text Font dialogue box allows you to see the fonts available before you choose one.

2. Select **Script Complex** by picking either the words on the left or the illustration on the right. (When you pick either one, they are both highlighted.)

3. Pick **OK.**

4. *Font file <txt>: scriptc Height <0.0000>:* ↵

This prompt confirms that you have selected "scriptc," or Script Complex, for your new font and asks for a height. The default height is 0. When you accept this default, AutoCAD prompts you for a height each time you insert text on the drawing. If you specify another height now, the height will be fixed, and you will not be able to enter a height at the time you insert the text on the drawing. In most cases, it is better to accept the default value of 0.

5. *Width factor <default>:* **1** ↵

The **width factor** is the expansion or compression of the characters. The default, or standard width, is 1. Values greater than 1 expand the characters, and values less than 1 compress the characters *(Fig. 9-4).*

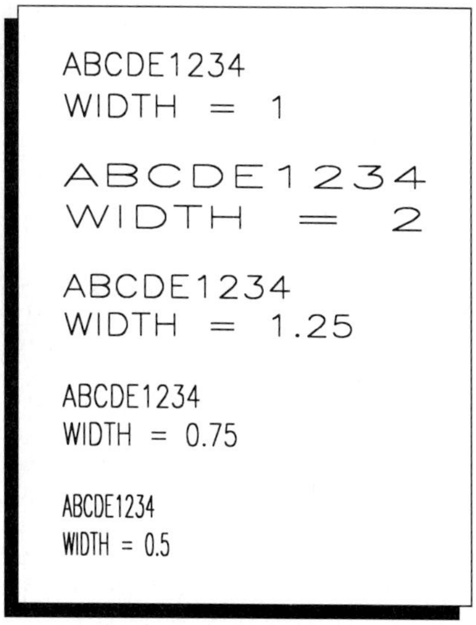

ABCDE1234
WIDTH = 1

ABCDE1234
WIDTH = 2

ABCDE1234
WIDTH = 1.25

ABCDE1234
WIDTH = 0.75

ABCDE1234
WIDTH = 0.5

Fig. 9-4 The width factor
has an impact on the
appearance of text
in a drawing.

6. *Obliquing angle <default>:* 0 ⏎

The **obliquing angle** is the slope of the characters. A slope of 0 creates vertical characters. A positive value creates a positive slope; that is, the characters slant to the right. A negative value sets a negative slope, and characters slant to the left *(Fig. 9-5)*.

Now that you have specified how the characters will appear, AutoCAD asks the following questions about each line of text. These prompts do not refer to the individual letters.

7. *Backwards? <N>* ⏎

8. *Upside-down? <N>* ⏎

9. *Vertical? <N>* ⏎

10. *SCRIPTC is now the current text style.*
    Command:

You have now set up a new style for use in your drawing. Save the drawing, even though you have not done any drawing or text work yet. By saving the drawing, you are saving the style you just created as the current style. You will use this drawing in the next "Your Turn."

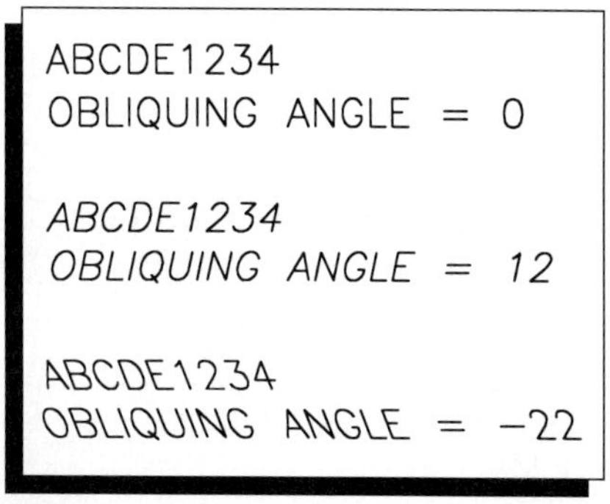

Fig. 9-5 A positive obliquing
angle makes text slant to
the right; a negative obliquing
angle makes it slant to the left.

ABCDE1234
OBLIQUING ANGLE = 0

ABCDE1234
OBLIQUING ANGLE = 12

ABCDE1234
OBLIQUING ANGLE = −22

Text Entry in AutoCAD

AutoCAD provides two commands for placing text on drawings: TEXT and DTEXT. The two commands are similar, except that DTEXT is easier to use in many cases. Using either command, you can enter most text directly.

Some symbols and style options, however, require special codes. This section will explain how to enter all types of text using TEXT and DTEXT.

The TEXT Command

The TEXT command is the most basic command for entering text. When you enter this command, the prompt changes to *Justify/Style/<Start point>:*. Notice that the default is *Start point*, which means you simply pick a point on the screen to begin the text. The other options allow you to change the justification and style before you begin entering text. (Justification is explained in the following section.)

If you choose the default and pick a starting point, AutoCAD prompts you for the height and rotation angle of the text. You can set these two characteristics independently each time you use the TEXT command. Then the prompt asks for the text. As you type the text, it appears at the command line. When you press RETURN, the text appears on the screen.

▶ The Justify Option

Justification refers to the alignment of the text on the screen. It also sets the location of the starting point in relation to the text. Therefore, the Justify option allows you to place text on the screen relative to specific points. The justification options give you a large degree of freedom to place text exactly where you want it on the screen, relative to entities that already appear on the drawing.

AutoCAD allows you to justify text in any of 14 different ways *(Fig. 9-6)*. Both the Align and Fit options insert the text between two points that you specify. However, Align maintains the text's horizontal and vertical proportions. Fit just stretches or shrinks the text without changing its height. The

Center and Right options justify the text horizontally from the center and from the right side respectively. Middle justifies the text both horizontally and vertically from the middle of the text string.

The other nine options are listed at the prompt by their initial letters. TL, TC, and TR refer to Top Left, Top Center, and Top Right, respectively. The point you pick to begin the text on the screen will be located at these positions in the line of text you enter. ML, MC, and MR refer to Middle Left, Middle Center, and Middle Right, and BL, BC, and BR refer to Bottom Left, Bottom Center, and Bottom Right, respectively.

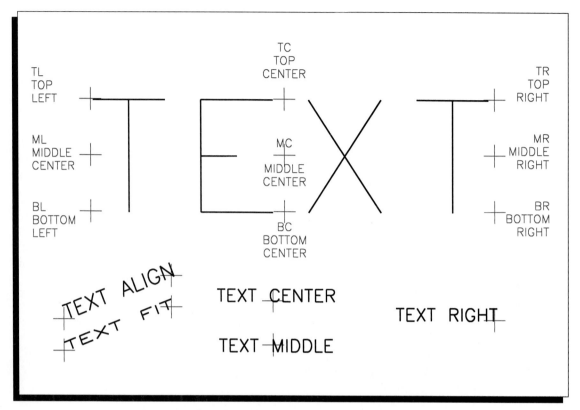

Fig. 9-6 AutoCAD's text alignment options.

▶ ## The Style Option

The Style option of the TEXT command works in exactly the same way as the STYLE command. If you choose the style option, AutoCAD asks for the style name you want. The current style name appears as the default.

The DTEXT Command

DTEXT stands for "Dynamic TEXT." All of the options of the TEXT command are also present in the DTEXT command; the two work in exactly the same way. However, the DTEXT command offers some important advantages over the TEXT command.

First, DTEXT allows you to see the text on the screen *as you are entering it.* This gives you the advantage of seeing the text in place while you can still go back and change it. This feature also makes it easier to know how much text will fit in a given space.

Next, you can enter multiple lines of text with one DTEXT command. The TEXT command allows you to enter only one line at a time—when you press RETURN, the command ends. In DTEXT, when you press RETURN, the cursor moves to the beginning of the next line automatically. To end the DTEXT command, press RETURN twice.

Special Symbols

The keyboard does not contain all of the characters you may want to use as text on your drawing. You can create special symbols by using a double percent sign (%%) followed by a character that represents the symbol you wish to use. If you wanted to, you could create the entire American Standard Code for International Interchange (ASCII) character set in this manner. However, since these are readily accessible on the keyboard, this is not usually necessary.

In addition to the ASCII character set, AutoCAD has its own special symbols that use the following control characters:

- %%C draws the diameter symbol (φ)
- %%D draws the degree symbol (°)

- %%P draws the plus/minus tolerance symbol (±)

For example, to create the text "25°," type the control sequence 25%%D.

There are two ways to create the percent symbol (%) in AutoCAD. In many cases, you can just enter it from the keyboard. However, when the percent symbol will be followed by a symbol that requires a control sequence, you must also precede the % with the control sequence %%. The following examples may help you understand this concept.

To enter this text:	Type this:
45%	45%
45% ±5	45%%%%P5

Note When you use control sequences with DTEXT, the control sequences appear in the text as you enter it. The special characters do not appear until you press RETURN.

You can also use control sequences to underscore and overscore text. An **underscore,** or underline, is a line that runs beneath the text. The control sequence for an underscore is %%U. An **overscore,** or overline, is a line that runs above the text. The control sequence for it is %%O.

These two control sequences are different from those used to create special characters; they act as toggles to turn the feature on and off. For example, the first time you enter %%U, the underscore feature is turned on. All text from that point will be underscored until you enter %%U again to turn the feature off *(Fig. 9-7)*. For example, if you entered "%%OThis text is overscored,%%O %%Uthis text is underscored, %%O and this text is both overscored and underscored.%%U%%O" at the *Text:* prompt, the result would look like the text in Fig. 9-7.

This text is overscored, this text is underscored,

and this text is both overscored and underscored.

Fig. 9-7 The effect of the underscore and overscore toggles in AutoCAD.

ACTIVITY

Open the TEXT drawing you created in the previous "Your Turn." Then follow these steps to compare the TEXT and DTEXT commands. First, use the TEXT command to enter your name and the name of your school.

1. *Command:* **TEXT** ⏎
2. *Justify/Style/<Start point>:* **J** ⏎
3. *Align/Fit/Center/Middle/Right/TL/TC/TR/ML/MC/MR/BL/BC/BR:* **C** ⏎
4. *Center point:* Pick the starting point for your text in the upper middle portion of the screen *(Fig. 9-8).*

The point you pick with the crosshairs will determine the center of the first line of text.

5. *Height <Default value>:* ⏎

Instead of accepting the default value, you could have selected a new value. You could also have selected a height visually by moving the crosshairs on the screen. The distance you moved the mouse (or other pointing device), regardless of direction, would become the new text height.

6. *Rotation angle <Default value>:* ⏎

The default value of 0 places your text horizontally on the drawing. When you enter a new value, the text rotates in a counterclockwise direction, pivoting on your starting point *(Fig. 9-9).*

7. *Text:* Use the %%U code to underscore your full name. Remember to use the %%U command both before and after your name to turn the underscore on and off. When you have completed this line of text, press **RETURN.**
8. *Command:* ⏎

As you may recall, pressing RETURN reenters the previous command.

9. *Justify/Style/<Start point>:* ⏎

By pressing RETURN at this prompt, you direct AutoCAD to align this text directly under the text you created using the previous TEXT command. The alignment also defaults to the alignment you chose at the previous text command, so the text you enter now will be centered directly beneath your name.

10. *Text:* Type the name of your school and press **RETURN.**

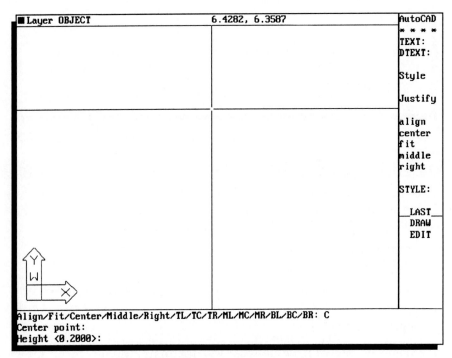

Fig. 9-8 Begin your text at this location.

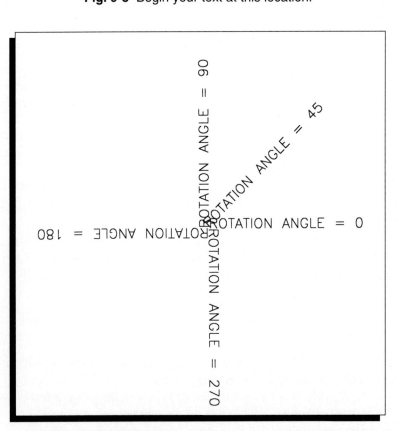

Fig. 9-9 This text has been rotated around its starting point.

Now perform a similar text entry, but use the DTEXT command. Follow these steps.

1. *Command:* **DTEXT** ↵

2. *Justify/Style/<Start point>:* **J** ↵

3. *Align/Fit/Center/Middle/Right/TL/TC/TR/ML/MC/MR/BL/BC/BR:* **C** ↵

4. *Center point:* Pick a point near the middle of the bottom half of your screen.

5. *Height <default>:* ↵

6. *Rotation angle <0>:* ↵

Notice the square that has appeared at the starting point of your text *(Fig. 9-10).* This is the text cursor, and it is completely independent of the crosshairs. The size of the square represents the size of the uppercase characters.

7. *Text:* Once again, enter your full name and press **RETURN.** Watch the screen as you enter your name. The letters appear on the actual drawing, rather than at the prompt line.

Notice that after you press RETURN, the *Text:* prompt still appears at the command line, and the text cursor moves to the center of the line beneath your name.

8. *Text:* Enter the name of your school and press **RETURN.**

9. *Text:* ↵

Pressing RETURN with no text entry indicates the end of the sequence and returns you to the *Command:* prompt.

10. Save your drawing. You will need it for the next "Your Turn."

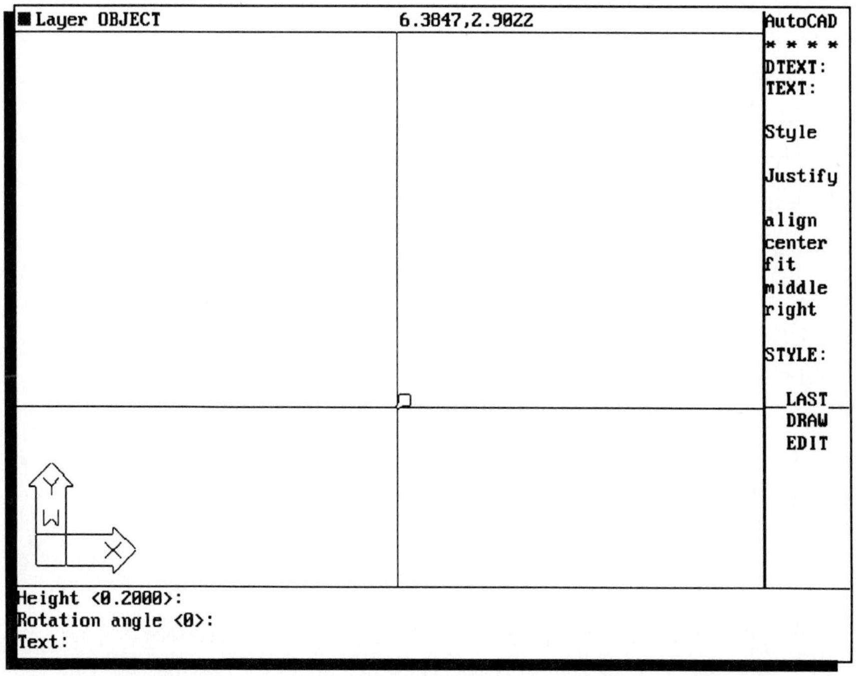

Fig. 9-10 AutoCAD's text cursor is completely independent of the crosshairs.

Modifying Text

"Move that note a bit to the right." "I misspelled WHAT?!" These and other problems require text modifications. You could no doubt solve this problem by erasing the text and beginning again. However, rewriting can be extensive, and it may even create new problems. AutoCAD provides two commands that allow you to make changes to existing text: the DDEDIT and CHANGE commands.

The DDEDIT Command

The DDEDIT command is a text editing command that allows you to insert or delete characters in text you have already created. You cannot change the style or other characteristics of text using this command, however.

ACTIVITY

Open the TEXT drawing you began in previous "Your Turn" sections. Then follow these steps to discover more about how to make changes using the DDEDIT command.

1. Enter the **STYLE** command and enter **STANDARD** to return to the default text style.
2. *Command:* **DTEXT** ⏎
3. *Justify/Style/<Start point>:* Pick a point in any clear space on the left side of the screen.
4. *Height <0.2000>:* ⏎
5. *Rotation Angle (0):* ⏎
6. *Text:* Enter the following text exactly as shown, including the mistakes:
   "This is a text the DdedIT command."

The sentence contains three errors:
- "text" should be "test"
- "of" was omitted between "text" and "the"
- "DdedIT should be all upper case.

Follow the procedure below to correct the errors.

7. *Command:* **DDEDIT** ⏎
8. *<Select a TEXT or ATTDEF object>/Undo:* The crosshairs change to a pick box. Pick the text to be edited.

The Edit Text dialogue box appears. The text appears in the Text box and is highlighted in blue *(Fig. 9-11)*.

If you pressed the BACKSPACE key now, the whole text would disappear and a cursor would appear in the box so that you could type entirely new text. Since you want to change only specific parts of the text, follow the steps below.

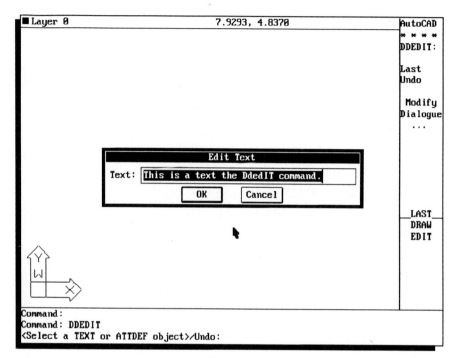

Fig. 9-11

9. Use the mouse to pick the blue block between the "e" and the "x" of "text."

The text becomes white and an edit cursor (a blinking vertical line) appears between the "e" and the "x." If you missed your pick point, use the left or right arrow key to position the edit cursor correctly.

10. Use the DELETE key to remove the "x" and then insert an "s" in its place.

11. Move the edit cursor with the right arrow key and insert a space and the word "of" after "text."

12. Move the edit cursor with the right arrow key to between the "D" and "d" of "DdedIT." Delete "ded" and insert "DED."

13. Pick **OK.**

14. *<Select TEXT or ATTDEF object>/Undo:* ↵

15. Save your drawing. You will need it in the next "Your Turn."

The CHANGE Command

CHANGE permits you to modify various characteristics of existing entities. Note that the CHANGE command is not specific to text; you can use it to change properties of other entities as well. For example, you can use it to change the layer on which a circle or other entity resides. In this section, CHANGE is presented only as it applies to modifying text. For a detailed discussion of the CHANGE command, consult the *AutoCAD Reference Manual.*

ACTIVITY

Open the TEXT drawing that you created in previous "Your Turn" sections. Then follow these steps to change the text location and style.

1. Enter the following text with the **DTEXT** command. Pick any clear space on the screen for a starting point.

    "This is the CHANGE command."

2. Using the mouse, activate the pull-down menus and select **Draw, Text, Set Style...**

3. Select the Roman Simplex style and pick **OK.**

4. *Command: '_style Text style name (or ?) <STANDARD>: romans*

    New style.

    Font file <txt>: romans Height <0.0000>: ⏎

5. *Width factor <1.0000>:* ⏎

6. *Obliquing angle <0>:* ⏎

7. *Backwards? <N>* ⏎

8. *Upside-down? <N>* ⏎

9. *Vertical? <N>* ⏎

10. *ROMANS is now the current text style.*

    Command: **CHANGE** ⏎

11. *Select objects:* Pick the text "This is the CHANGE command."

    The text is selected and *1 found* appears at the command line.

12. *Select objects:* ⏎

13. *Properties/<Change point>:* ⏎

You can make changes to the content of text using only the default Change point option of the CHANGE command. The Properties option can make other changes, such as the layer on which the text resides, but it cannot change the actual text characters.

14. *Enter text insertion point:* Move the text to a new location below the first and pick it.

If you do not want to move the text but want to change its style, you can press RETURN at this prompt to leave the text where it is.

15. *New style or RETURN for no change:* **ROMANS** ⏎

This is the Roman Simplex font which you set up earlier in this sequence.

16. *New height <0.2000):* **0.25** ⏎

17. *New rotation angle <0>:* ⏎

18. *New text <This is the CHANGE command.>:* Retype the text, but add the words **an example of** between the words "is" and "the" so that the new text says "This is an example of the CHANGE command." Then press **RETURN.**

The text now appears in Roman Simplex, and the *Command:* prompt reappears.

You may want to try out other options of the commands presented in this chapter. Try various text fonts and styles. Vary the height and rotation angle. Don't forget, if things get too confused, you can always reset the original text default by using the STYLE command and entering STANDARD. Set the height to 0, width factor to 1, obliquing angle to 0, and reply N (No) to the rest of the prompts. This will return you to AutoCAD's default text font and style.

■ Chapter 9 Review ▬▬▬▬▬▬▬▬

1. What is the difference between a font and a style?
2. Explain why text standards are necessary for working with CAD documents.
3. What is justification? How does it apply to the TEXT and DTEXT commands?
4. Name two advantages the DTEXT command has over the TEXT command.
5. Write out the text you would enter in AutoCAD to create the following text on the screen: "This year's total is <u>23% ±4</u> of last year's total." Underscore only the text "23% ⊥4."
6. What dialogue box appears when you enter the DDEDIT command and select text in a document?
7. Which option of the CHANGE command can be used for text?

Problems

■ Chapter 9 Problems ■

To increase your experience using text in AutoCAD, do the following problems. Save each problem as a separate file for your teacher's approval.

1. Enter this text with a height of 0.125:

   AutoCAD allows drafters to use a variety of text heights and widths to comply with company standards.

   Make each of the changes listed below, using a new copy of the sentence for each change. (HINT: use the COPY command.) When you finish, your screen should have 10 versions of the sentence.

   a. Change its height to 0.5.
   b. Change the height to 0.2.
   c. Change the rotation angle to -45.
   d. Change its style to Roman Simplex and its rotation angle to 0.
   e. Change its width factor to 2.
   f. Change its width factor to 0.7.
   g. Change its obliquing angle to 22.
   h. Change it to appear backwards on the screen.
   i. Change it to appear upside-down.

2. Enter this text with a height of 0.2:

   AutoCAD provides a variety of fonts and styles to suit the needs of most drafters.

   Make each of the changes listed below, using a new copy of the sentence for each change. (HINT: use the COPY command.) When you finish, your screen should have 8 versions of the sentence.

   a. Change the style to English Gothic.
   b. Change the style to Italic Complex.
   c. Change the style to Script Complex.
   d. Change the style to Country Blueprint.
   e. Change the style to Technic.
   f. Change the style to Sansserif Oblique and the width to 1.5.
   g. Change the style to Euroman and the height to 0.3.

3. Create two rectangles (0.5 by 4.5 units) and enter text using the Align and Fit commands, as shown in Fig. 9-12.

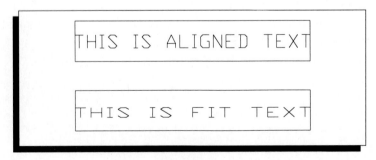

Fig. 9-12

Using CAD
From Burma to the United States

AutoCAD users are found world-wide. One such CAD user is Thet-Win, originally from Burma. (Thet-Win is his entire name.) Thet-Win owned his own architectural firm in Rangoon and was drawing on a board before he discovered CAD. In 1989, he accepted a job with SACON in Adelaide, South Australia, where he trained in AutoCAD. His training included courses at a Technical and Further Education (TAFE) school. TAFE is similar to the community college system in the United States.

After finishing the first two courses at TAFE, Thet-Win joined the South Australian AutoCAD Users' Group. He was elected Editor that year and stayed as a Board Member in 1990. Many communities, including those in the U.S., have AutoCAD users' groups. These groups are a good place to exchange information with other AutoCAD users. Thet-Win feels that his users' group was fun, as well as very productive.

Meanwhile at SACON, Thet-Win worked on a college complex (Tree Gully TAFE Project) that he had been assigned to upon arriving in Adelaide. He later became the CAD manager for the team working on that project. Costing forty million dollars, it was the first large scale project to be done on CAD for SACON. AutoCAD Release 10 and other third party applications were used to create ninety percent of the construction drawings. In 1992, the TTG Project won a Civic Trust Award for "Buildings in their Settings."

Thet-Win was also one of a three-member team whose function was to review, setup, and implement CAD for the Professional Services Division at SACON. He gave presentations to CAD users at the company, as well as to his users' group.

After his second year at SACON, Thet-Win applied for admission and was accepted at UCLA where he is now a Masters of Architecture student. Before leaving for the U.S., Thet-Win was asked to meet with CAD users back in his home country, Burma. He also visited architectural offices in Kuala Lumpur and Singapore and attended a Tropical CAD Camp conference in Kuala Lumpur. By now an experienced CAD user and manager, Thet-Win was able to answer many questions for attendees at the conference.

Thet-Win says AutoCAD was the most significant thing that prompted him to move to the U.S. When he completes his graduate studies, he plans to find a full time job and stay for awhile.

Chapter 10

▶ Key Terms

dimensions
working drawing
dimensioning variables
extension lines
dimension lines
leader
aligned method
unidirectional
   method
prism

cylinder
counterbore
countersink
spotface
datums
limit dimensioning
tolerance
clearance fit
interference fit
transition fit

Dimensioning

▶ ▶ Commands & Variables

DIM	HORIZONTAL
SETVAR	LEADER
DDIM	DIMTOH
STATUS	DIMTIH
DIMSCALE	ALIGN
DIMEXO	DIAMETER
DIMEXE	ANGULAR
EXIT	TEDIT
UPDATE	HOMETEXT
VERTICAL	

▶ ▶ ▶

Objectives

When you have completed this chapter, you will be able to:

- dimension drawings using the DIM command.
- define and apply dimensioning terminology.
- understand and apply the general dimensioning rules.
- recognize and interpret finish symbols.
- apply the theory and concepts of limit dimensioning to orthographic projections.
- understand and apply the concepts of types of fit.
- use the DIM command to dimension drawings with maximum and minimum size values.
- use the DIM command to dimension drawings with tolerance values.

Objects are drawn so that they can be made; and to make an object you need to know two things: 1) what does it look like? and 2) how big is it? The drafter's primary responsibility is to answer those questions by accurately drawing the object and correctly stating its size.

Although drawings need to be precise, they are still only pictures. The person or company that uses the drawing to manufacture the part needs to know the size description, or **dimensions**, of the part.

You cannot estimate actual sizes accurately enough by measuring the drawing features. The dimensions and text you place on the drawing give the manufacturer additional information needed to construct the part. As the drafter, you must present this information clearly and accurately.

To make sure that drawings are understood by the widest possible audience, standard practices, or conventions, have been developed. All drafters should understand and follow conventions that relate to dimensioning as well as those that affect other aspects of drafting. Keep in mind that the conventions and techniques described in this chapter are the ones generally accepted. However, you may need to make changes for one or more reasons. For example, different drafting disciplines such as architectural, electrical, and sheet metal drafting require various specialized conventions and techniques. Be sure to follow your company's policies as well as industrial standards such as ANSI Y14.5M, "Dimensioning and Tolerancing."

Fig. 10-1 shows a **working drawing**, a drawing from which a part is made. The drawing illustrates some of the conventions discussed in this chapter and includes dimensions, notes, and symbols. The dimensions are in hundredths of an inch; however, for greater accuracy, it is not uncommon to express dimensions in thousands or ten-thousandths of an inch.

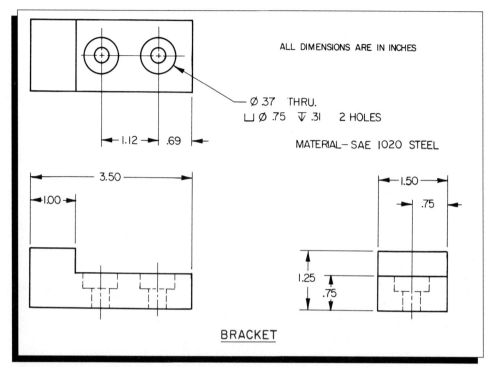

Fig. 10-1 A working drawing contains all of the information needed to manufacture a part.

Configuring the Appearance of Dimensions

Many of the dimensioning conventions are included in AutoCAD's ACAD.DWG prototype drawing. You can change them as necessary using AutoCAD commands, system variables, and dimensioning variables. **Dimensioning variables** are similar to system variables, except that they affect only the dimensioning portion of the AutoCAD drawing. They control the appearance of dimensions on an AutoCAD drawing.

Dimensioning variables always begin with the letters *DIM*. You can access them from the screen menu using the DIM command or from the pull-down menu under Settings and Dimension Style. You can also change them using the SETVAR command. If you know the exact name of the dimensioning variable, you can type it in directly. A list of the dimensioning variables and their meanings is given in Fig. 10-2.

Name	Description	Type	Default
DIMALT	Alternate Units	Switch	Off
DIMALTD	Alternate Units Decimal Places	Integer	2
DIMALTF	Alternate Units Scale Factor	Scale	25.4
DIMAPOST	Alternate Units Text Suffix	String	None
DIMASO	Associative Dimensioning	Switch	On
DIMASZ	Arrow SIze	Distance	0.18
DIMBLK	Arrow Block	String	None
DIMBLK1	Separate Arrow Block 1	String	None
DIMBLK2	Separate Arrow Block2	String	None
DIMCEN	Center Mark Size	Distance	0.09
DIMCLRD	Dimension Line Color	Color Number	BYBLOCK
DIMCLRE	Extension Line Color	Color Number	BYBLOCK
DIMCLRT	Dimension Text Color	Color Number	BYBLOCK
DIMDLE	Dimension Line Extension	Distance	0.0
DIMDLI	Dimension Line Increment	Distance	0.38
DIMEXE	Extension Line Extension	Distance	0.18
DIMEXO	Extension Line Offset	Distance	0.0625
DIMGAP	Dimension Line Gap & Reference Dimensioning	Distance	0.09
DIMLFAC	Length Factor	Scale	1.0
DIMLIM	Limits Dimensioning	Switch	Off
DIMPOST	Dimension Text Prefix, Suffix, or both	String	None
DIMRND	Rounding Value	Scaled distance	0.0
DIMSAH	Separate Arrow Blocks	Switch	Off
DIMSCALE	Dimension Feature Scale Factor	Scale	1.0
DIMSE1	Suppress Extension Line 1	Switch	Off
DIMSE2	Suppress Extension Line 2	Switch	Off
DIMSHO	Show Dragged Dimension	Switch	On
DIMSOXD	Suppress Outside Dimension Lines	Switch	Off
DIMSTYLE	Dimension Style	Name	*UNNAMED
DIMTAD	Text Above Dimension Line	Switch	Off
DIMTFAC	Tolerance Text Scale Factor	Scale	1.0
DIMTIH	Text Inside Horizontal	Switch	On
DIMTIX	Text Inside Extension Lines	Switch	Off
DIMTM	Minus Tolerance Value	Scaled distance	0.0
DIMTP	Plus Tolerance Value	Scaled distance	0.0
DIMTOFL	Text Outside, Force Line Inside	Switch	Off
DIMTOH	Text Outside Horizontal	Switch	On
DIMTOL	Tolerance Dimensioning	Switch	Off
DIMTSZ	Tick Size	Distance	0.0
DIMTVP	Text Vertical Position	Scale	0.0
DIMTXT	Text Size	Distance	0.18
DIMZIN	Zero Suppression	Integer	0

Fig. 10-2 Dimensioning variables control every aspect of dimensions in an AutoCAD drawing. This list explains the use of each variable.

ACTIVITY

When you are beginning a new drawing that will require dimensioning, it is best to set the dimensioning variables when you set up the rest of the drawing. To do this, you could use the DDIM command to bring up the Dimension Styles and Variables dialogue box. This dialogue box allows you to change all of the dimensioning variables without having to know their individual names. However, if you just want to check the current status of the variables, it may be easier to list the variables and their current values. The STATUS dimensioning subcommand allows you to do this. Follow these steps to create a printout of the current settings of the AutoCAD dimensioning variables on the computer you are using.

1. Start AutoCAD.

2. If the printer attached to the computer is not on, turn it on now.

3. Press **CTRL** and **Q** simultaneously to turn on AutoCAD's printer echo feature.

The printer echo feature (CTRL-Q) records every action in the AutoCAD program and feeds this information to the printer or plotter connected to the computer. The information includes keystrokes you make and the resulting AutoCAD text.

4. *Command:* DIM ⏎

The DIM command puts AutoCAD in the dimensioning mode and changes the command prompt to *Dim:*.

In the dimensioning mode, many of the regular AutoCAD commands either do not work or work differently than they do in the command mode.

Note

---

5. *Dim:* **STATUS** ⏎

The STATUS dimensioning subcommand switches AutoCAD to a text display and presents the current value of each dimensioning variable.

6. *Dim:* Press **CTRL-Q** again to turn off AutoCAD's printer echo feature.

Although AutoCAD has recorded the value of each dimensioning subcommand, your printer may or may not have printed the information. The printer echo feature does not issue the "form feed" command that many laser printers need to produce a printed page. Check your printer. If the information has not printed, take the printer off line (press the On Line button to make its light go out) and press the Form Feed button. Ask your instructor if you need additional help.

7. Your printout should look something like the printout in Fig. 10-3, except the settings for your variables may differ. After you have the printout, look at the name and value of each dimensioning variable on your printout. (Many of the individual variables are not discussed in this chapter.) Compare this information with the meaning of the variables, listed in Fig. 10-2. After you have completed this chapter, you may want to refer to this printout and experiment with the settings of individual variables.

```
Dim: status

DIMALT      Off                 Alternate units selected
DIMALTD     2                   Alternate unit decimal places
DIMALTF     25.40               Alternate unit scale factor
DIMAPOST                        Suffix for alternate text
DIMASO      On                  Create associative dimensions
DIMASZ      0.18                Arrow size
DIMBLK                          Arrow block name
DIMBLK1                         First arrow block name
DIMBLK2                         Second arrow block name
DIMCEN      0.09                Center mark size
DIMCLRD     BYBLOCK             Dimension line color
DIMCLRE     BYBLOCK             Extension line & leader color
DIMCLRT     BYBLOCK             Dimension text color
DIMDLE      0.00                Dimension line extension
DIMDLI      0.38                Dimension line increment for continuation
DIMEXE      0.18                Extension above dimension line
DIMEXO      0.06                Extension line origin offset
DIMGAP      0.09                Gap from dimension line to text
DIMLFAC     1.00                Linear unit scale factor
-- Press RETURN for more --
DIMLIM      Off                 Generate dimension limits
DIMPOST                         Default suffix for dimension text
DIMRND      0.00                Rounding value
DIMSAH      Off                 Separate arrow blocks
DIMSCALE    1.00                Overall scale factor
DIMSE1      Off                 Suppress the first extension line
DIMSE2      Off                 Suppress the second extension line
DIMSHO      On                  Update dimensions while dragging
DIMSOXD     Off                 Suppress outside extension dimension
DIMSTYLE    *UNNAMED            Current dimension style (read-only)
DIMTAD      OFF                 Place text above the dimension line
DIMTFAC     1.00                Tolerance text height scaling factor
DIMTIH      On                  Text inside extensions is horizontal
DIMTIX      Off                 Place text inside extensions
DIMTM       0.00                Minus tolerance
DIMTOFL     Off                 Force line inside extension lines
DIMTOH      On                  Text outside extensions is horizontal
DIMTOL      Off                 Generate dimension tolerances
DIMTP       0.00                Plus tolerance
DIMTSZ      0.00                Tick size
-- Press RETURN for more --
DIMTVP      0.00                Text vertical position
DIMTXT      0.18                Text height
DIMZIN      0                   Zero suppression

Dim:
```

Fig. 10-3 The STATUS dimensioning subcommand produces a list of AutoCAD dimensioning variables and their current values.

Controlling Dimension Size

Sometimes a dimension may appear too small or too large for the drawing. You can control the overall size or scale of the dimension entity using the DIMSCALE command. The default is 1; 2 is twice the scale, and 0.5 is half scale (*Fig. 10-4*).

The units you choose to use in the drawing affect the scale you need for the dimensions. For example, you can dimension drawings using either the U.S. customary system or the metric system. However, you should decide on the system you will use at drawing setup. Since AutoCAD drawings that are made for production purposes are usually made full-scale regardless of the units of measure, the limits setting determines the size of the drawing screen.

AutoCAD defaults to the U.S. system. Limits are set at 0,0 and 12,9, with text and dimensions sized accordingly. An A4 sheet, the metric equivalent of an A sheet, measures 210 mm by 297 mm. The upper limits for an A4 sheet, therefore, should be set at 0,0 and 297,210. The drawing dimensions will also be in millimeters, so the DIMSCALE value must be increased to change the physical size of the dimensions, including arrowheads, offsets, and extensions. An A4 sheet at 210 mm by 297 mm is approximately 25 times an A sheet at 9" by 12"; therefore, the DIMSCALE value should be set at about 25.

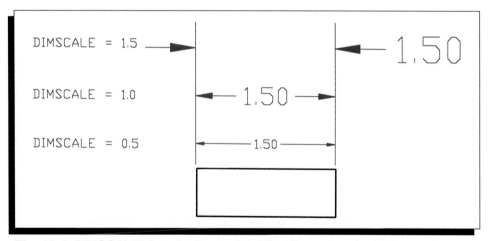

Fig. 10-4 DIMSCALE changes the scale of the dimensions in relation to the scale of the rest of the drawing.

Controlling the Appearance of Lines

The alphabet of lines discussed in Chapter 4 includes the four types of lines that are used for dimensioning: extension lines, dimension lines, center lines, and leaders. In this section, you will learn more about these lines and how to change their appearance.

▶ Extension Lines

The **extension lines** "extend" a feature or surface of an entity and provide an artificial surface to which a dimension can be applied *(Fig. 10-5)*. The extension line never touches the object. Instead, it is separated from the object by a space referred to as the extension line offset. AutoCAD provides a default offset of 0.0625". AutoCAD also extends the extension line 0.180" beyond the outermost dimension line. These values can be changed with the DIMEXO (EXtension Offset) and DIMEXE (EXtension Extension) commands, respectively.

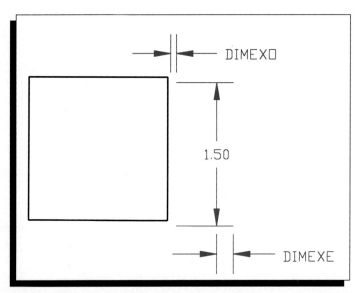

Fig. 10-5 By convention, extension lines are set as shown here. The actual values can be changed using AutoCAD's dimensioning variables.

To see the effect the DIMEXO and DIMEXE variables have on AutoCAD dimensions, follow these steps.

1. Open a new drawing and call it EXTLINES.

For this exercise, you need to know the current value of the DIMEXO and DIMEXE variables. The fastest way to do this is to enter the variable names. AutoCAD shows the current values and allows you to change them if necessary.

2. *Command:* **DIMEXO** ↵

3. Look at the current value. If it is 0.06, press **RETURN** to keep that value. If a different value shows, then enter **0.06** to change the value and then press **RETURN**.

4. *Command:* **DIMEXE** ↵

5. If the current value is 0.18, press **RETURN** to keep that value. If a different value shows, enter **0.18** to change the value and then press **RETURN**.

6. In the upper left corner of your screen, use the **LINE** command to create a rectangle similar to the one in Fig. 10-6. The rectangle should be about 2 units by 3 units.

7. *Command:* **DIM** ↵

8. *Dim:* **VER** ↵

You can enter most of the subcommands by entering the first three letters of the subcommand. For example, VER is the three-letter abbreviation for the VERTICAL dimensioning subcommand, which allows you to create vertical dimensions on a drawing.

9. *First extension line or RETURN to select:* ↵

10. *Select line, arc, or circle:* Pick the right side of the rectangle.

11. *Dimension line location (Text/Angle):* Pick a point about 0.6 units to the right of the rectangle.

12. *Dimension text <measured value>:* ↵

Note the position and spaces between the extension lines on the dimension you just created *(Fig. 10-6)*. Then follow these steps to see how changing the DIMEXO and DIMEXE variables can change the appearance of a drawing.

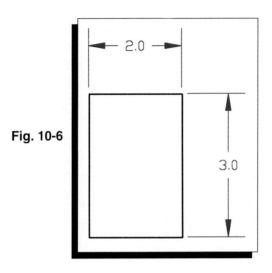

Fig. 10-6

You may need to use the ZOOM command to see the spaces between the extension lines and the object.

13. *Dim:* **E** ⏎

The E stands for "Exit." It exits the dimensioning mode and returns to the *Command:* prompt.

You can also press CTRL-C or pick AutoCAD at the top of the screen menu to exit the dimensioning mode.

14. Copy the rectangle (but not the dimension) to another location on your screen.
15. *Command:* **DIM** ⏎
16. *Dim:* **DIMEXO** ⏎

You may have noticed that we entered the DIMEXO variable from the *Command:* prompt the first time and from the *DIM:* prompt this time. You can change the dimensioning variables from either the *Command:* prompt or the *DIM:* prompt.

17. *New value for DIMEXO <0.06>:* **0.15** ⏎
18. Use the VER subcommand to create a vertical dimension for your copy of the rectangle. Use the same procedure you used in steps 8 through 13.

Notice the difference in the appearance of the dimensions. The difference is due to the change you made in the DIMEXO variable *(Fig. 10-7)*.

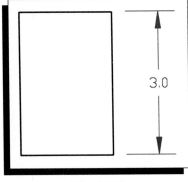

Fig. 10-7

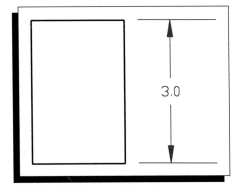

Fig. 10-8

There may be times when you need to change the appearance of an existing dimension. You can do that by changing the value of the appropriate dimensioning variables and then using the UPDATE dimensioning subcommand. The UPDATE subcommand allows you to change all or some of the dimensions to the current dimensioning style. Try it by following these steps.

19. Use the COPY command to make a copy of one of the rectangles *and* its dimension. Place the copy in a clear space on your screen.

20. *Command:* **DIMEXE** ⏎

21. *Command:* **DIM** ⏎

22. *New value for DIMEXE <0.18>:* **0.5** ⏎

23. *Dim:* **UPDATE** ⏎

24. *Select objects:* Pick the dimension (not the line being dimensioned) on the copy you made in step 19.

25. *Select objects:* ⏎

The extension lines now extend farther beyond the dimension line in the dimension you updated *(Fig. 10-8)*.

▶ Dimension Lines

Dimension lines specify the extent of the dimension, from where to where, and the dimension value. By default, AutoCAD places the value in a break at the center of the line. Dimension lines should be placed at least 0.6" from view outlines (outer edge of the views) and 0.3" from other dimension lines *(Fig. 10-9)*.

Because dimension lines point to features, they always end with arrowheads or similar devices. For example, architectural drafters often use tics rather than arrowheads on dimension lines *(Fig. 10-10)*. AutoCAD offers several styles for use on dimension lines.

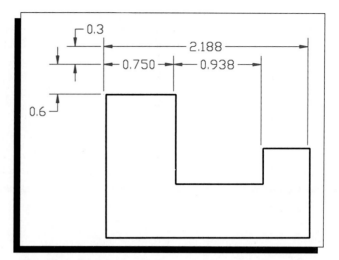

Fig. 10-9 Dimension lines are placed as shown here. The first dimension line is placed 0.6 units from the drawing. Additional dimension lines are placed at 0.3-unit intervals above the first dimension.

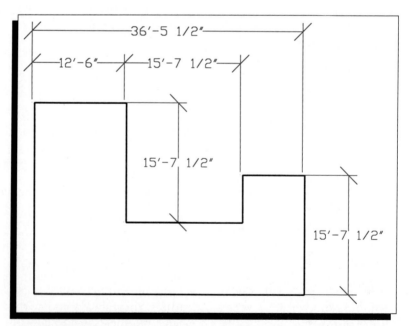

Fig. 10-10 Dimension lines on architectural drawings often have ticks instead of arrows.

ACTIVITY

To experiment with the various styles available for arrowheads and similar devices, follow these steps. When you finish, your drawing should look like the one in Fig. 10-11.

1. Create a new drawing called ARROWS.
2. Draw three rectangles of any size on the screen. Make sure they are far enough apart so that you can dimension them. They should have at least a unit between them.
3. *Command:* **DIM** ⏎
4. *Dim:* **DDIM** ⏎

The Dimension Styles and Variables dialogue box appears.

5. Click on the **Arrows** button.

The Arrows subdialogue box appears. About halfway down the box, you will see small, round "radio" buttons for the following choices: Arrow, Tick, Dot, and User. These are the devices available to put on the ends of dimension lines.

The User selection is user-defined, which means you can define any shape you wish to end the dimension lines. For more information about user-defined devices, refer to the *AutoCAD Reference Manual.*

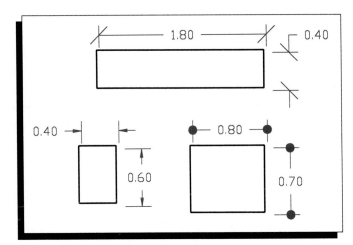

Fig. 10-11

6. Click on the **Arrow** radio button and pick **OK** until you are returned to the drawing.

7. *Dim:* **HOR** ↵

HOR is the three-letter abbreviation for the HORIZONTAL dimensioning sub-command, which allows you to create horizontal dimensions on a drawing.

8. *First extension line origin or RETURN to select:* ↵

9. *Select line, arc, or circle:* Pick the top line of one of the rectangles.

10. *Dimension line location (Text/Angle):* Pick a point about 0.6 units above the top of the rectangle.

11. *Dimension text <measured value>:* ↵

12. Use the VER subcommand to dimension the right side of the same rectangle.

13. *Dim:* **DDIM** ↵

14. When the Dimension Styles and Variables dialogue box appears, pick the **Arrows** button and choose the radio button next to "tic."

15. Repeat steps 7 through 12 to dimension another of the rectangles.

16. Use **DDIM** to select the "dot" arrowhead style.

17. Repeat steps 7 through 12 to dimension the third rectangle.

Note the differences in the appearance of the dimensions for the three rectangles.

18. Save your work. You will need this drawing for the next "Your Turn."

▶ Leaders

A **leader** starts with an arrowhead at a feature and ends at the dimension or note which applies to that feature. Normally, a leader is a single line, drawn at any convenient angle, which ends with a short horizontal shoulder (provided by AutoCAD) *(Fig. 10-12).* Multiple leaders should be drawn paral-lel when possible. If the leader is drawn to a circle or an arc, the leader should point to its center. Keep your leaders as short as possible. Never draw leaders horizontally, vertically, at a shallow angle, or parallel to extension, dimension, or section lines.

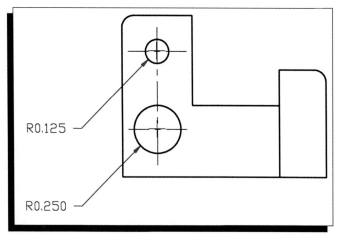

Fig. 10-12 Leader lines are often used to point to circles or holes within an object.

ACTIVITY

Leaders are often used to relate text information to objects on a drawing. Follow these steps to add notes to the drawing you created in the previous "Your Turn." When you finish, your drawing should look similar to the one in Fig. 10-13.

1. Open the ARROWS drawing.
2. *Command:* **DDIM** ⏎
3. Pick the **Arrow** radio button if it is not already selected and pick OK.
4. *Command:* **DIM** ⏎
5. *Dim:* **LEADER** ⏎
6. *Leader start:* Pick a point near one of the arrows on the dimension that contains arrows.

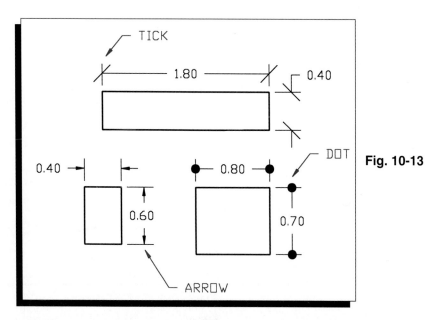

Fig. 10-13

The point you pick in response to this prompt selects the position of the tip of the arrow on the leader. The subsequent *To point:* prompts control the direction and length of the line attached to the arrowhead.

7. *To point:* Select a point at an angle from the first point you chose. Refer to Fig. 10-13 if you need help.

As stated earlier, it is not considered good practice to have leaders that are horizontal or vertical. The lines should slant. The slant and length of the line depends on the amount of space you have and the object you're dimensioning.

8. *To point:* ⏎
9. *Dimension text <measured value>:* **ARROWS** ⏎
10. Repeat steps 5 through 10 to label the dimensions on the other two rectangles **DOTS** and **TICKS**.

▶ Center Lines

Center lines can be used for locating dimensions *(Fig. 10-14)*. Center lines should generally stop 0.25" beyond the feature being centered, and they should never end at another line. If center lines are used as extension lines, they should stop at the same point as other extension lines.

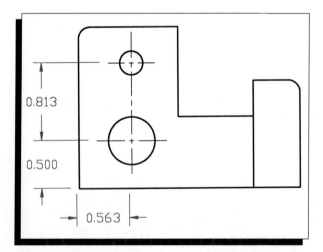

Fig. 10-14 Center lines are generally used to locate holes in objects.

Basic Dimensioning Concepts

Every object, no matter how complex, breaks down into basic geometric shapes. Each of these shapes has a size and a location within the object. Fig. 10-15 shows the breakdown of a bracket into parts that can be dimensioned. To dimension an object completely, you must define the size and location of each of its component shapes. From the simplest part to the most complex, you can dimension objects fully by following an orderly procedure of specifying these features.

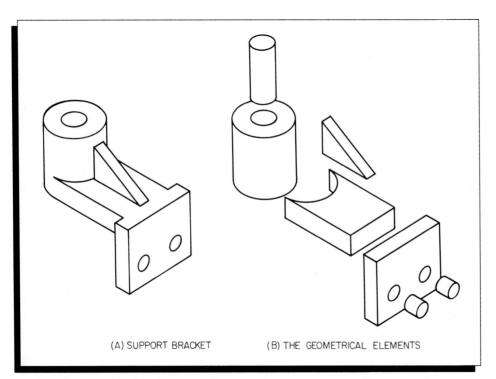

(A) SUPPORT BRACKET   (B) THE GEOMETRICAL ELEMENTS

Fig. 10-15 Every object or part can be broken down into simple geometric shapes. It is necessary to do this, at least in your mind, before you begin to dimension an object. It is the size and relative locations of these geometric shapes that you will dimension.

General Dimensioning Rules

First, you should become familiar with the conventions and other guidelines that govern CAD dimensioning. These rules are summarized in the following list.

1. When dimensioning machine drawings, use two-place decimals for U.S. measuring and one-place decimals for metric, unless greater accuracy is needed. In AutoCAD, use the DDUNITS command to set the drawing precision according to your needs.

2. Omit the units of measure when all measurements on a drawing use the same units.

3. Do not repeat the same dimension in two different views.

4. Give only dimensions that are needed to manufacture the part.

5. Give the diameter of a circle, not the radius. AutoCAD precedes the diameter with the symbol Ø.

6. Give the radius of an arc. AutoCAD precedes an arc radius with R.

7. Place overall dimensions outside smaller dimensions *(Fig. 10-16)*. Do *not* list all smaller dimensions and the overall dimension. This is double dimensioning *(Fig. 10-17)*. Chain smaller dimensions or place them against a datum line *(Figs. 10-18 and 10-19)*.

8. Space dimension lines about 0.6" from the view outline and 0.3" apart. These are visual considerations and you may adjust them as needed; however, always keep the distance from dimension line to view outline greater than the distance from dimension line to dimension line.

9. Give center line to center line dimension of circular end parts, not an overall dimension *(Fig. 10-20)*.

10. Do not use a drawing line or a center line as a dimension line.

11. Do not cross a dimension line with any other line.

12. Extension lines should not cross each other.

13. Dimension from center lines, finished surfaces, or datums.

14. Avoid dimensioning to hidden lines.

15. Avoid placing dimensions inside view lines.

These rules serve as guidelines for dimensioning. All the rules can be broken, but you should do so only if you can justify it.

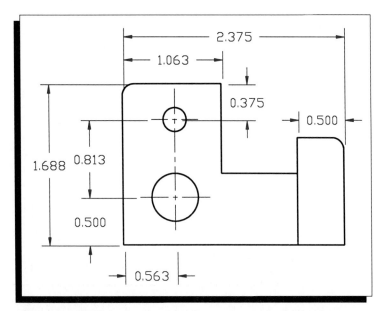

Fig. 10-16 When you must use two or more dimensions along a single object line, always place the longer dimension outside the shorter dimension — in other words, the overall dimension should always be placed outside the partial dimensions.

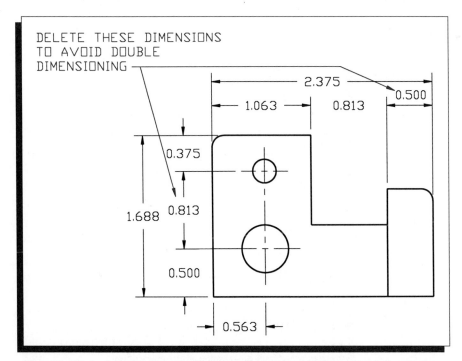

Fig. 10-17 In this drawing, both the horizontal and the vertical dimensions are double dimensioned. Double dimensioning is considered poor drafting technique, and it can actually lead to poorly made parts. Eliminate the least important dimension to compensate for tolerance buildup.

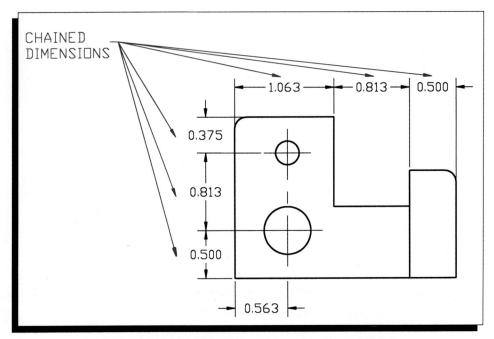

Fig. 10-18 Chain or point-to-point dimensioning.

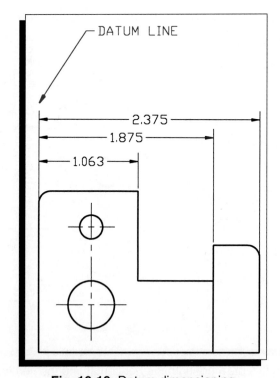

Fig. 10-19 Datum dimensioning.

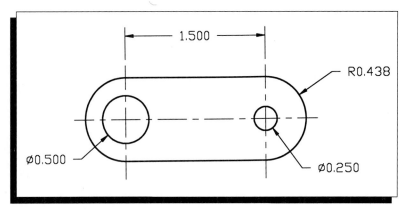

Fig. 10-20 Parts that have circular ends are dimensioned to the center lines, not to the ends of the part.

Placing Dimensions

There are two methods of placing dimensions on a drawing: aligned and unidirectional. With the **aligned method**, dimensions are placed in line with the dimension lines and are read from the bottom of the drawing or from the right side *(Fig. 10-21)*. If possible, dimensions should be kept from within a 45-degree angle of vertical *(Fig. 10-22)*.

The **unidirectional method** places all dimensions horizontally, regardless of the angle of the dimension line. This method allows people to read all the dimensions from the bottom of the drawing *(Fig. 10-23)*.

ANSI specification Y14.5M-1982, "Dimensioning and Tolerancing," follows the unidirectional method. Therefore, AutoCAD defaults to the unidirectional method. However, you can change to the aligned method by using the DIMTIH (Text Inside Horizontal) and DIMTOH (Text Outside Horizontal) dimensioning variables. When you turn these two variables off, AutoCAD uses the aligned method.

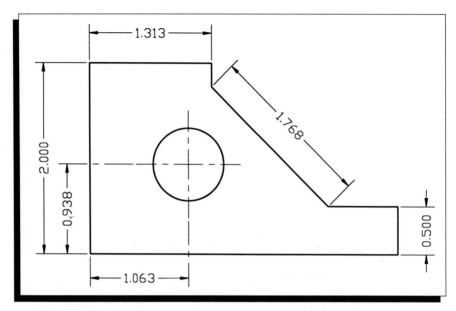

Fig. 10-21 The aligned method of placing dimensions on a drawing.

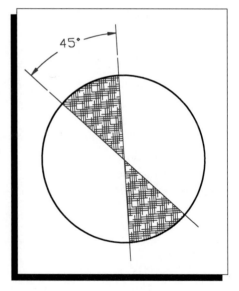

Fig. 10-22 Avoid placing aligned dimensions within the shaded area.

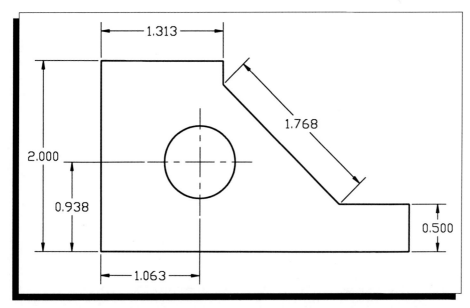

Fig. 10-23 The unidirectional method of placing dimensions on a drawing.

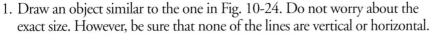

ACTIVITY

Start AutoCAD and begin a new drawing named DIMPLACE. Then follow the steps below to experiment with the unidirectional and aligned methods of placing dimensions. The first procedure uses the unidirectional method. Since AutoCAD defaults to this method, you do not need to change any of the dimensioning variables to perform this procedure.

1. Draw an object similar to the one in Fig. 10-24. Do not worry about the exact size. However, be sure that none of the lines are vertical or horizontal.

2. Create two additional layers: **ALIGNED** and **UNIDIR**. Assign the color **yellow** to ALIGNED and the color **red** to UNIDIR. Make UNIDIR the current layer.

3. *Command:* **DIM** ⏎

4. *Dim:* **ALI** ⏎

ALI is the three-letter abbreviation for the ALIGN subcommand, which allows you to create dimensioning lines that parallel the lines on the object you are dimensioning.

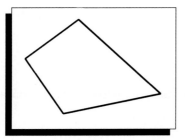

Fig. 10-24

5. *First extension line origin or RETURN to select:* ↵

By pressing RETURN, you allow AutoCAD to find the ends of the line, arc, or circle for you. All you have to do is specify which line you want to dimension.

6. *Select line, arc, or circle:* Pick one of the diagonal lines.

7. *Dimension line location (Text/Angle):* Pick a point about 0.6 units from the object line.

8. *Dimension text <measured value>:* ↵

AutoCAD calculates the measured value and shows it in angle brackets. However, you have the chance to change it. If you wanted to specify a different value, you would type in the desired value before you pressed RETURN.

9. Repeat steps 4 through 8 for each of the other three lines. In each case, accept the measured value that AutoCAD presents.

10. Plot or print your current drawing for reference.

Your drawing should now look like the one in Fig. 10-25.

11. Save your drawing.

Now follow the steps below to dimension the same object using the aligned method.

1. Change the current layer to ALIGNED and freeze layer UNIDIR.

This will make the dimensions you created in the preceding procedure disappear, but they are still in the DIMPLACE file. If you thawed the UNIDIR layer, they would reappear.

2. *Command:* **DIM** ↵

3. *Dim:* **DIMTOH** ↵

4. *Current value <On> New value:* **OFF** ↵

Remember that you have to turn both DIMTOH and DIMTIH *off* to use the aligned placement method.

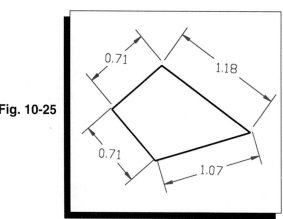

Fig. 10-25

5. *Dim:* **DIMTIH** ⏎

6. *Current value <On> New value:* **OFF** ⏎

7. Now use the **ALIGN** subcommand and the same procedure you used in the previous procedure to dimension the object.

8. Plot or print the drawing and compare it with your copy of the unidirectional placement you did in the previous procedure.

9. Save your drawing.

Your drawing should look like the one in Fig. 10-26. Remember that now you have two sets of dimensions for this object. The dimensioning style that appears depends on the current status of the layers. Freeze the one you do not want to see, and make sure the one you do want to see is not frozen.

AutoCAD will not freeze any layer if it is the current layer. If you try to freeze a layer and it remains on the screen, check to see if that layer is current. If it is, make a different layer current and try again to freeze the layer.

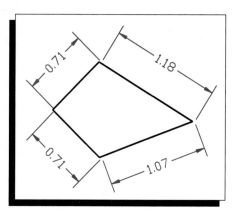

Fig. 10-26

Describing Size

To define the size of an object, you ordinarily need to know its width, depth, and height. For curved or cylindrical objects, you may need to know diameter or radius. One view generally contains the object's outline. Fig. 10-27 shows the outline of an object in the front view. Fig. 10-28 has the outline in the right side view. This section describes in detail how to specify the size of various shapes.

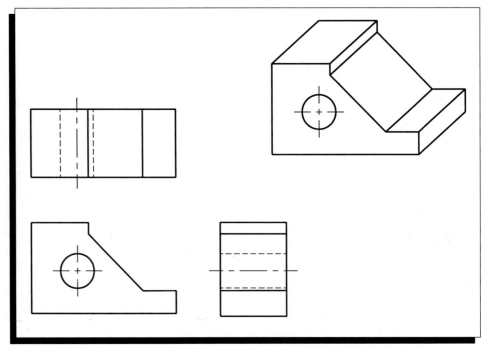

Fig. 10-27 Outline of an object in the front view.

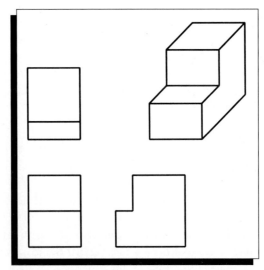

Fig. 10-28 Outline of an object in the right-side view.

▶ Prisms

A **prism** is any cube or other three-dimensional box-like shape. To dimension a prism, place the detail dimensions in the outline view and the overall dimensions in the other views. Dimensions that apply to two adjacent views should be placed between the views. In general, place dimensions where the shapes are shown *(Fig. 10-29)*.

▶ Cylinders

As you may remember from geometry classes, a **cylinder** is any tube-shaped object. Specify the length and diameter of cylinders (Fig. 10-30). When you use AutoCAD's DIAMETER dimensioning subcommand, the Ø symbol preceding the dimension indicates a diameter.

You can also use the LEADER subcommand to show the diameter of a cylinder. When you use the LEADER subcommand, however, the Ø symbol does not appear automatically. You can enter it manually by entering the code %%c before the diameter.

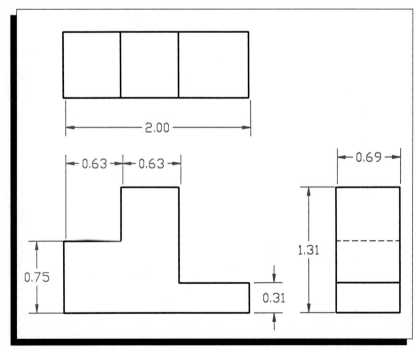

Fig. 10-29 Prisms should be dimensioned as shown here.

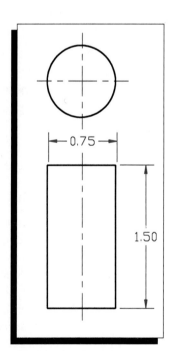

Fig. 10-30 Dimension a cylinder as shown here.

▶ Other shapes

Other basic geometric shapes that you may need to dimension are cones, pyramids, frustums, and spheres. The methods for dimensioning these shapes are shown in Fig. 10-31.

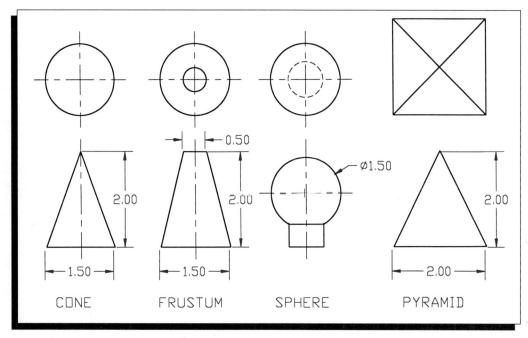

Fig. 10-31 Dimension cones, frustums, spheres, and pyramids as shown here.

▶ Holes

Information about holes and holes with counterbores, countersinks, and spotfaces is generally given in notes at the feature's location *(Fig. 10-32)*. When possible, place this information in the outline view, especially when the method of forming or finishing the hole is specified.

A **counterbore** is an enlargement of a hole to a specific depth *(Fig. 10-33)*. Give the diameter (or drill size) and depth when dimensioning. Counterbores are usually used to recess the heads of screws and bolts.

A **countersink** is a conical entrance to a hole *(Fig. 10-34)*. It is made to fit a flathead screw so that the head of the screw is flush with the surface of the part. Give the diameter and angle when dimensioning.

A **spotface** smooths an uneven surface around a hole so that the bolt head, washer, or nut will seat properly. *(Fig. 10-35)*. A spotface is a clean-up cut less than 0.06" deep and is called out (labeled) but not dimensioned.

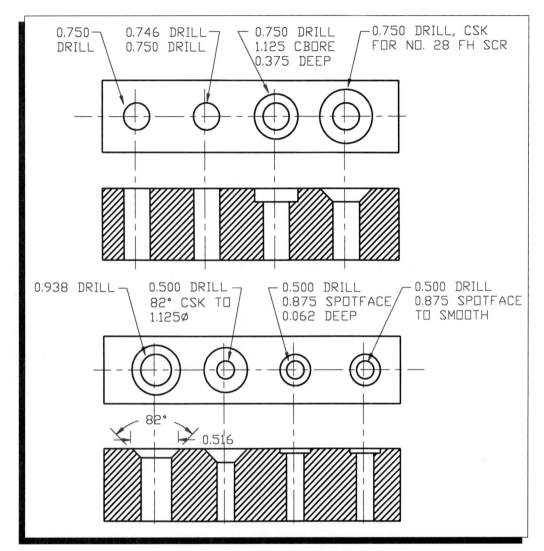

Fig. 10-32 Dimensioning holes.

Fig. 10-33 A counterbored hole is dimensioned with diameter and depth. Counterbored holes provide a recessed area for bolts and other fasteners so that they do not interrupt the smoothness of an object's surface.

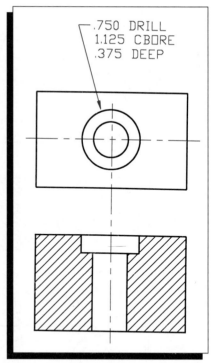

.750 DRILL
1.125 CBORE
.375 DEEP

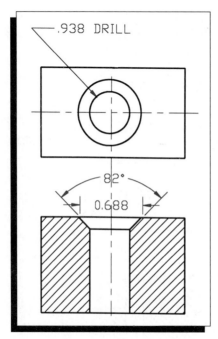

.938 DRILL

82°

0.688

Fig. 10-34 A counterbored hole is dimensioned with diameter and angle. Countersunk holes are slanted to allow screw heads to fit flush with an object's surface.

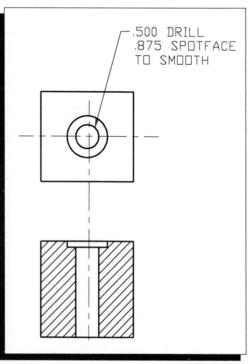

.500 DRILL
.875 SPOTFACE
TO SMOOTH

Fig. 10-35 Some rough castings, such as engine blocks, require smoothing before the head of a fastener will fit snugly against their surfaces. The smoothed areas are known as spotfaces.

▶ Angles

Angle measure is determined by the UNITS or DDUNITS command at drawing setup. Fig. 10-36 illustrates several methods of dimensioning angles.

Chamfers are variations of angular measurement. Fig. 10-37 shows correct ways of measuring each variation.

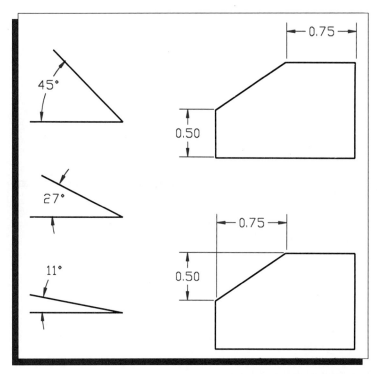

Fig. 10-36 Dimensioning angles.

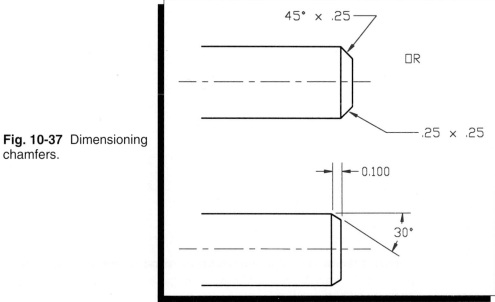

Fig. 10-37 Dimensioning chamfers.

▶ Curves

Fillets and radii are dimensioned where the curves occur *(Fig. 10-38)*. However, not all curves are uniform arcs. Some curved surfaces are composed of compound arcs, or numerous arcs of different sizes blended together, whereas others are ellipses, parabolas, and so on. The method of dimensioning the curve depends on the type of curves present. Figs. 10-39 through 10-42 show examples of appropriate dimensioning methods.

Fig. 10-38 Dimensioning fillets and radii.

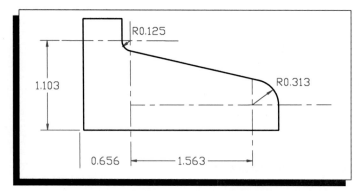

Fig. 10-39 When curves are not composed of circular arcs, you can dimension them as shown here.

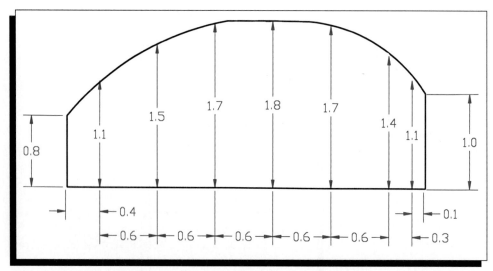

Fig. 10-40 When curves are not composed of circular arcs, the method of dimensioning depends on the curve itself.

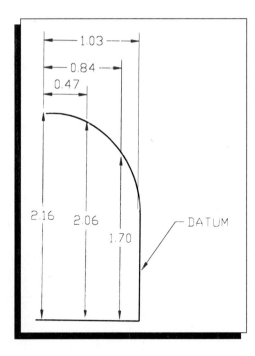

Fig. 10-41 Another way to dimension curves is to use datum lines.

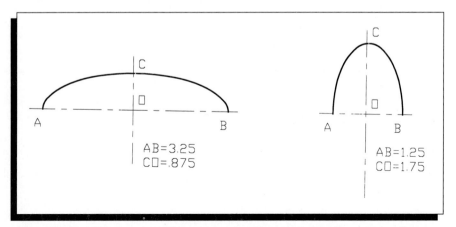

Fig. 10-42 A third way to dimension noncircular curves is to name them and show their component dimensions.

ACTIVITY

To become more familiar with dimensioning angles and curves, follow these steps.

1. Create a new drawing called ANGLE.

2. Create an angle and a polyline similar to those in Fig. 10-43.

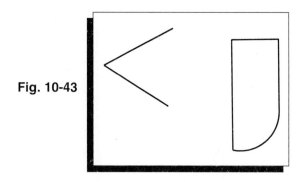

Fig. 10-43

Use the LINE command to create the angle. Use the Line and Arc options of the PLINE command to create the polyline.

3. *Command:* **DIM** ↵
4. *Dim:* **ANG** ↵

ANG is the abbreviated form of the dimensioning subcommand ANGULAR. You can also select ANGULAR from the screen menu.

5. *Select arc, circle, or line, or RETURN:* Pick the top line of the angle.
6. *Second line:* Pick the bottom line of the angle.

After you pick the second line, a rubberbanding dimension appears on the cursor. Move the cursor around the outside of the angle to see the various options. You will see that AutoCAD will dimension any of the four possible angles created by your two lines.

7. *Dimension arc location (Text/Angle):* Position the cursor so that the dimension shown in Fig. 10-44 is visible and click the mouse button.
8. *Dimension text <measured angle>:* ↵
9. *Enter text location:* ↵

By entering RETURN, you are allowing AutoCAD to center the text automatically on the dimension.

10. Now repeat this procedure to dimension the arc on the polyline. Use the **ANGULAR** dimensioning subcommand, and place the dimension as shown in Fig. 10-44.
11. Dimension the rest of the polyline using the **HORIZONTAL** and **VERTICAL** commands.

Your finished drawing should look like the one in Fig. 10-44.

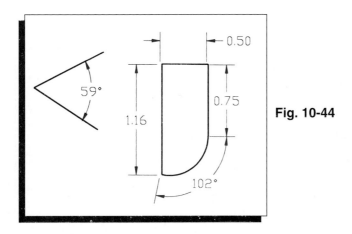

Fig. 10-44

Describing Location

Location dimensions show the position of features relative to surfaces or other features. Generally, three dimensions are needed to locate features in both prisms and cylinders: up and down, left and right, and front and back *(Fig. 10-45)*. Locations are fixed by the axes of center lines and finished surfaces, as shown in Figs. 10-46 and 10-47.

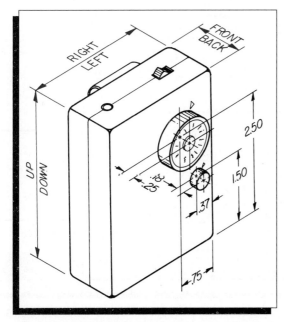

Fig. 10-45 To locate features on an object, you must consider location from three points of view: front-back, up-down, and right-left.

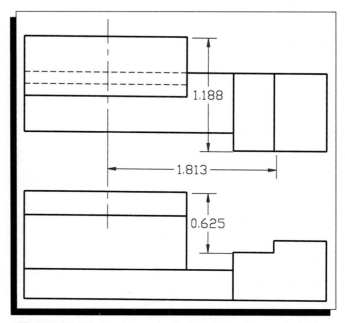

Fig. 10-46 Dimensions can be located using center lines.

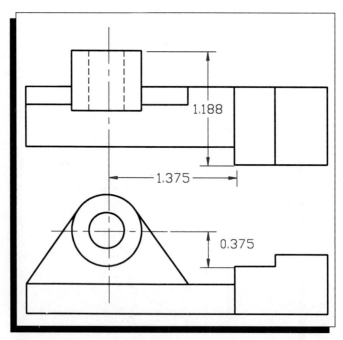

Fig. 10-47 You can also locate dimensions using finished surfaces.

▶ Datums

Datums are points, lines, or surfaces whose dimensions are assumed to be exact. They are used as references for other measurements during the dimensioning process. For example, the length of a true edge of a part can become a datum against which measurements for other dimensions are made. Datums are always used in pairs because in any given view it takes two dimensions to locate an object. Typical datums include two center lines, two surfaces, or a center line and a surface *(Fig. 10-48)*.

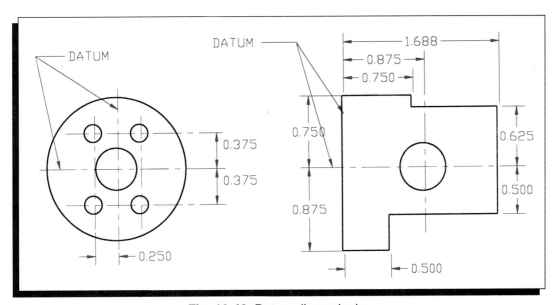

Fig. 10-48 Datum dimensioning.

▶ Reference Dimensions

Reference dimensions are for information only. They must not be used for construction because they may not be accurate. Reference dimensions are marked "Ref" or are enclosed in parentheses *(Fig. 10-49)*.

The inclusion of a dimension which should be marked "reference" may result in double dimensioning. Double dimensioning occurs when too many dimensions are given and they cannot all be met.

For example, in Fig. 10-50, the overall part is dimensioned in addition to each of the steps. If the tolerance is +/- 0.05, each step can be acceptable on the high side or low side, but the part may still be rejected for the overall dimension. If all (or most) of the steps were on the high side of the tolerance, for example, the overall size would be higher than the overall dimension. The overall dimension should be marked as a reference dimension so that the height of the individual steps can be controlled. In this case, the reference dimension means, "This is the nominal height. You don't have to go to the trouble of adding the steps to find your approximate clearance."

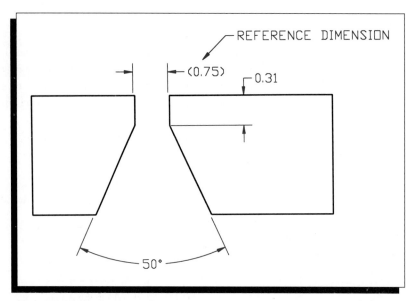

Fig. 10-49 Reference dimensions are marked "Ref" or are enclosed in parentheses.

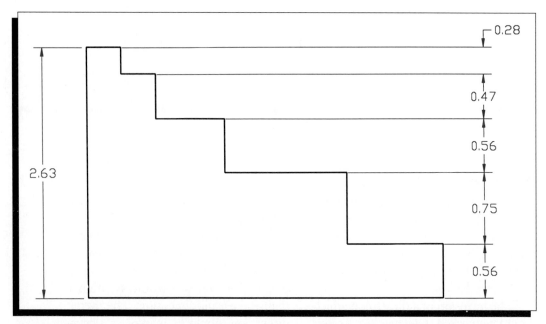

Fig. 10-50 The addition of the overall dimension creates a condition of double dimensioning. The overall dimension should be marked as a reference dimension so that the piece can be manufactured to these standards.

▶ Location of Holes

The locations of holes may be specified in several different ways. Fig. 10-51 shows examples of holes dimensioned using vertical and horizontal dimensions, polar coordinates, and datum lines, as well as chained dimensions and dimensions with notes. These methods can also be applied to other features. Drafters frequently use notes to specify methods of finishing, spotfacing, counterboring, and countersinking holes *(Fig. 10-32)*.

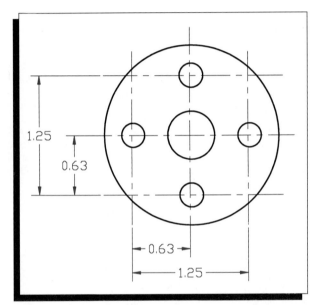

Fig. 10-51A Holes dimensioned with rectangular coordinates against horizontal and vertical center lines.

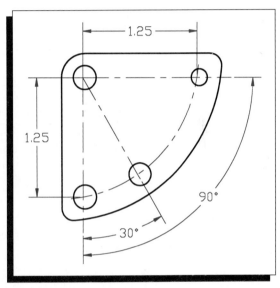

Fig. 10-51B Holes dimensioned with polar coordinates giving distance and angle.

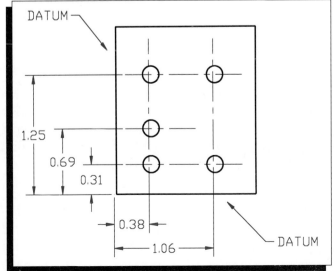

Fig. 10-51C Holes dimensioned against datum lines.

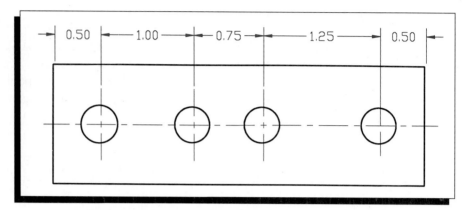

Fig. 10-51D Holes dimensioned in chains.

Fig. 10-51E Holes dimensioned with notes.

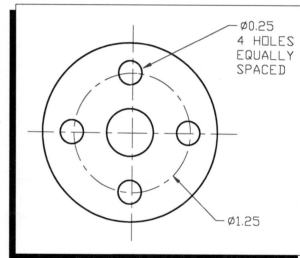

Surface Finish

Surfaces of objects can be left as formed, or they can be machined after forming. If the condition of a surface is important beyond the normal forming or machining process, note this with a surface finish symbol. Give the finish requirements at the symbol or in a general drawing note. Drawing notes consist of text that appears with the dimensions. They tell the manufacturer more about the specifications for the part.

▶ Types of Finish Requirements

Surface finish, sometimes referred to as surface texture, includes roughness, waviness, lay, and flaws *(Fig. 10-52)*.

Roughness is the general condition of the surface after forming or machining. It is measured in microinches.

A microinch is one millionth of an inch (0.000001 in.) and is abbreviated *µin.*

Waviness is a measure of the flatness of a surface. For example, imagine that you are looking at a 4' × 8' piece of plywood. Ideally, it should be perfectly flat. However, if you were to place it on a perfectly flat surface, you might notice that one part does not conform to the perfectly flat surface. It might miss the surface by ¹⁄₁₆" or more. Waviness is the degree of imperfection in the flatness of the plywood.

Roughness is similar to waviness, but it is a more closely spaced component. You might think of roughness as the "bumps" and topical imperfections in that same piece of plywood.

Lay is the direction of the predominant surface pattern. Flaws are surface defects.

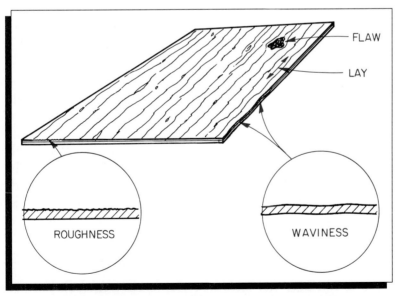

Fig. 10-52 Surface roughness, waviness, flaw, and lay.

▶ Finish Symbols

As you work with drafted documents, you will encounter three basic finish symbols. Fig. 10-53. The italicized "*f*" appears mostly on older drawings, but both the "V" and the "check" are currently in use. With the check symbol, the finish information appears in the drawing note. Finish information may be a number indicating roughness in microinches, a symbol or letter indicating lay, a waviness height and width measured in inches, a description of the finish process, or a combination of these. AutoCAD does not provide a surface finish symbol. However, you can insert a capital "V" as needed using the TEXT command.

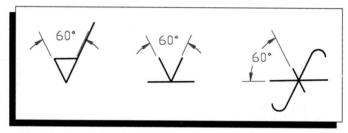

Fig. 10-53 Finish symbols. Construct the symbols with angles as shown.

Changing Dimensions

At times, you may need to move the text associated with a dimension, so that two dimensions do not overlap, for example. AutoCAD's TEDIT command allows you to do this. When you enter the TEDIT command, AutoCAD prompts you to select the dimension to change. After you select the dimension, the prompt changes to *Enter text location (Left/Right/Home/Angle)*. You can choose a new location for the text by moving the mouse (or other pointing device) and picking the new location. You can also choose one of the options listed on the command line to place the text at the extreme right or left of the dimension line or at an angle to the dimension line. The Home option moves the text back to its original position.

You can also use the HOMETEXT command to move dimension text back to its original location.

Limit Dimensioning

With the theory in place and the rules set, limit dimensioning actually puts the numbers on the drawing. **Limit dimensioning** is a dimensioning method that includes a **tolerance**, or range of acceptable values, for each dimension.

To understand limit dimensioning, you must first understand a few key terms.

- *Nominal size* is the general identification of a size. Examples: ½" bolt, ¾" pipe.
- *Basic size* is the dimension to which the allowances and tolerances are applied.
- *Design size* is the dimension to which the tolerances are applied. When there are no allowances, the design size equals the basic size.
- *Actual size* is the measured size.
- *Allowance* is the intentional maximum interference (negative allowance) or minimum clearance (positive allowance) between mating parts.
- *Tolerance* is the maximum amount that a dimension can vary. It should be expressed in the same form as its basic or design size.
- *Limits of size* are the maximum and minimum tolerance values.
- *Unilateral tolerance* is a tolerance that occurs on one side of the design size. For example, the dimension 1.500 +0.001/-0.000 permits the actual size to vary from 1.500 to 1.501. This method is often used for close-fitting shafts and holes.
- *Bilateral tolerance* is a tolerance that occurs on both sides of the design size. For example, the dimension 1.500 +/- 0.001 permits the actual size to vary from 1.499 to 1.501. This method is usually used with locating dimensions.

Remember, someone will build your part. Be accurate, be complete, and be realistic.

Interchangeability

Mass production permits the interchangeability of parts. When a part for your vacuum cleaner, saw, or car breaks, you can purchase and install a new one. The new part may be made by the original manufacturer or by a third party. If the replacement part is made to the original specifications, your vacuum cleaner, saw, or car will run again. Interchangeable manufacturing means that replacement parts will fit and run like the original ones.

Tolerance

Parts cannot be made to exact sizes. To be truly interchangeable, parts must be made to fit within a size range, or tolerance. How close the tolerance or size range is depends on the part and its use. A part for a high-power microscope must be made to a much closer tolerance than a part for a screen door latch.

The tolerance range is one of the factors that determines the price of a part. Generally, the closer the tolerance, the more expensive the part. Limit dimensioning is dimensioning with a tolerance.

Dimension Limits

There are three methods of expressing dimension limits *(Fig. 10-54)*. In Fig. 10-54A, the limits are expressed as maximum and minimum values. The first or top value is the largest. In B, the design size is given with positive and/or negative values to be added to and/or subtracted from the design size. In C, a note specifies a tolerance for all drawing dimensions unless otherwise specified. Only the design size appears at the dimension. When MAX or MIN appears at a dimension, the other limit is not important.

Types of Fit

Shafts can be fitted to holes with three classifications of tightness: 1) clearance, 2) interference, and 3) transition. With a **clearance fit**, the maximum diameter of the shaft is smaller than the minimum diameter of the hole. A clearance will always occur when the two parts are mated *(Fig. 10-55)*.

An **interference fit** will always result in interference between the shaft and the hole because the minimum shaft size is larger than the maximum hole size *(Fig. 10-56)*. The **transition fit** combines the other two fits. When the shaft is at maximum and the hole is at minimum there will be interference, but when the shaft is at minimum and the hole is at maximum, there will be clearance *(Fig. 10-57)*.

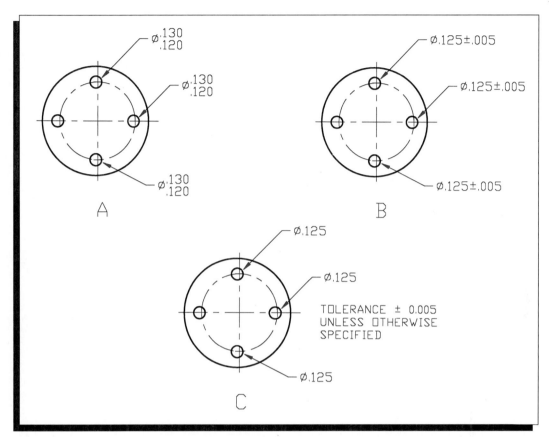

Fig. 10-54 Dimension limits can be shown as A) minimum and maximum values;
B) variance (+/–) from the actual design size; or C) a note that applies to the entire drawing.

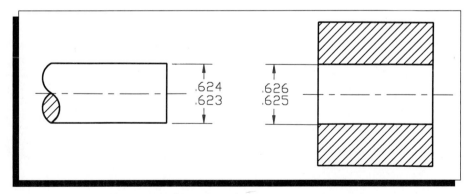

Fig. 10-55 Clearance fit between mating parts.

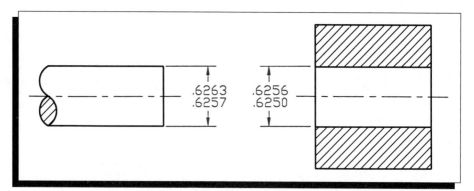

Fig. 10-56 Interference fit between mating parts.

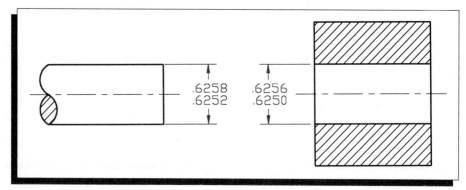

Fig. 10-57 Transition fit between mating parts.

▶ Basic Hole System

When designing the type of fit for a given situation, you must decide which of the two parts is the controlling part: the shaft or the hole. With the basic hole system, the hole is considered the controlling part. The minimum hole size is regarded as the basic size and the allowance is applied to the shaft. First, select the minimum hole size. Then determine the shaft size for clearance or interference fit. Finally, adjust the tolerances of the hole and shaft for the minimum interference or maximum clearance. This is the preferable method because standard tools can be used for machining.

► Basic Shaft System

With the basic shaft system, a similar process is used, except that the shaft is the controlling part. First, you select the maximum shaft diameter. Then you determine the minimum hole size for interference or clearance fit. Finally, adjust the tolerances for minimum interference or maximum clearance. Since the basic shaft system requires special tooling, you should use it only if there is a good reason.

ACTIVITY

Your Turn

Performing the following procedures will help you understand the concepts presented in this section. You will use all three types of limit dimensioning: using notes, using maximum and minimum sizes, and using tolerance values.

AutoCAD dimensioning defaults to notes. When you dimension an object using AutoCAD's defaults, only the design size is given and you must use notes to identify the tolerances. Open a new drawing and name it DIM1. Then follow these steps to dimension an object using notes.

1. Draw the bracket in Fig. 10-58. Use a grid of **1** and a snap of **0.125**, and use the **DDUNITS** command to select 2-place decimal numbers.

2. *Command:* **DIM** ↵

3. *Dim:* **VER** ↵

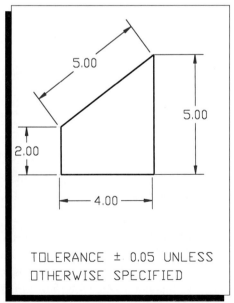

5.00

5.00

2.00

4.00

Fig. 10-58

TOLERANCE ± 0.05 UNLESS
OTHERWISE SPECIFIED

4. *First extension line origin or RETURN to select:* ⏎

5. *Select line, arc, or circle:* Pick the 5-unit vertical line.

6. *Dimension line location (Text/Angle):* Pick a point about 0.6 units from the object line.

7. *Dimension text <5.00>:* ⏎

8. *Dim:* ⏎

9. *First extension line origin or RETURN to select:* ⏎

10. *Select line, arc, or circle:* Pick the 2-unit vertical line.

11. *Dimension line location (Text/Angle):* Pick a point about 0.6 units from the object line.

12. *Dimension text <2.00>:* ⏎

13. *Dim:* **HOR** ⏎

14. *First extension line origin or RETURN to select:* ⏎

15. *Select line, arc, or circle:* Pick the horizontal line.

16. *Dimension line location (Text/Angle):* Pick a point about 0.6 units from the object line.

17. *Dimension text <4.00>:* ⏎

18. *Dim:* **ALI** ⏎

19. *First extension line origin or RETURN to select:* ⏎

20. *Select line, arc, or circle:* Pick the diagonal line.

21. *Dimension line location (Text/Angle):* Pick a point about 0.6 units from the object line.

22. *Dimension text <5.00>:* ⏎

23. *Dim:* **E** ⏎

24. *Command:* **TEXT** ⏎

25. *Justify/Style/<Start point>:* Pick a point below the bottom left of the horizontal dimension.

26. *Height <0.20>:* **0.15** ⏎

27. *Rotation angle <0>:* ⏎

28. *Text:* **TOLERANCE +/- 0.05 UNLESS OTHERWISE SPECIFIED.** ⏎

This completes the first example of limits dimensioning. Note that tolerance text might also be placed in the drawing's Notes section or in the title block. Remember to save your drawing.

In the following procedure, you will dimension a similar object, but this time you will show the tolerance of each dimension. Tolerance values are expressed as variations from the design size, either bilaterally (plus *and* minus) or unilaterally (plus *or* minus). This example uses bilateral tolerancing. For unilateral tolerancing, set the appropriate value to 0.00.

Since AutoCAD defaults to the design size dimension, you will need to change the defaults. Open a new drawing and name it DIM2. Then follow the steps below to dimension an object using variances.

1. Set the units for two-place decimal numbers. Then draw the object shown in Fig. 10-59. Do not include the dimensions at this time.

2. *Command:* **DDIM** ⏎

The Dimension Styles and Variables dialogue box appears *(Fig. 10-60)*.

3. Using the mouse, doubleclick on the Text Format text box.

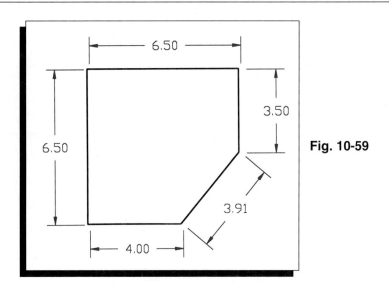

Fig. 10-59

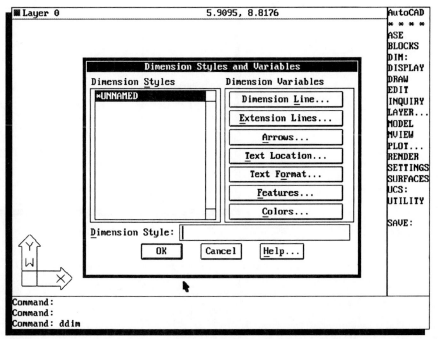

Fig. 10-60 The Dimension Styles and Variables dialogue box can help you set several dimensioning variables at one time. You do not need to know the names of the individual variables to use this method.

A subdialogue box appears. In the upper right corner is a section called Tolerances. In this area, click on the check box next to the word *Variance.*

4. Position the cursor in the text box next to Upper Value and enter **0.05.**

5. Position the cursor in the text box next to Lower Value and enter **0.05.**

6. Pick the **OK** box to return to the main dialogue box, and pick **OK** again to exit the dialogue box and save your changes.

AutoCAD is now set up to dimension using tolerances, or variances.

7. *Command:* **DIM** ⌐

8. *Dim:* Use the same dimensioning subcommands you used in the first procedure (**VERTICAL, HORIZONTAL, ALIGNED**) to dimension the object.

Your drawing should now appear similar to the one in Fig. 10-61. Notice that the dimensions appear with the tolerances you specified. This concludes the procedure for dimensioning using tolerance values. Remember to save your drawing.

The last procedure demonstrates dimensioning using maximum and minimum values only. When you use this method, you do not give the design size. Open a new drawing and name it DIM3. Then follow the steps below to dimension an object using maximum and minimum values.

Make sure Grid and Snap are on before you draw the object for this exercise.

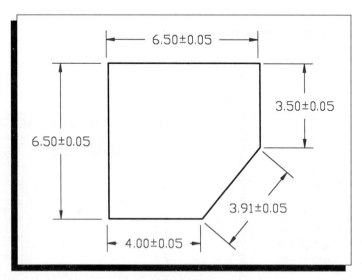

Fig. 10-61

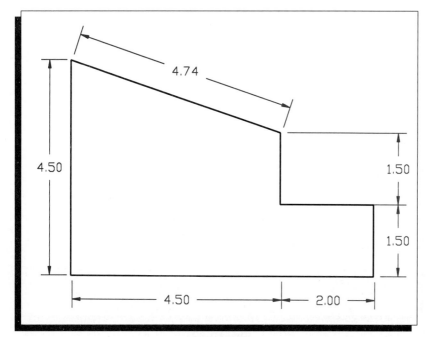

Fig. 10-62

---

1. Draw the object shown in Fig. 10-62. Set the units for two-place decimals. Do not include the dimensions at this time.

2. *Command:* **DDIM** ⏎

The Dimension Styles and Variables dialogue box appears.

3. Using the mouse, doubleclick on the Text Format text box.

A subdialogue box appears. In the Tolerance section in the upper right corner, click on the check box next to the word *Limits*.

4. Position the cursor in the text box next to Upper Value and enter **0.03**.

5. Position the cursor in the text box next to Lower Value and enter **0.03**.

6. Pick the **OK** box to return to the main dialogue box, and pick **OK** again to exit the dialogue box and save your changes.

AutoCAD is now set up to dimension using maximum and minimum values for each dimension.

7. *Command:* **DIM** ⏎

8. *Dim:* **HOR** ⏎

9. *First extension line or RETURN to select:* **END** ⏎

   of: Use the mouse to pick the left end of the bottom line.

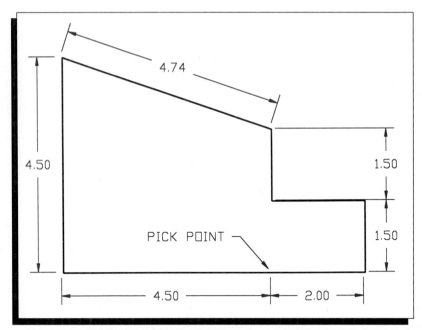

Fig. 10-63

10. *Second extension line origin:* Using the Snap, align the cursor on the bottom line directly beneath the vertical line on the top surface of the object *(Fig. 10-63).* Click the mouse button.

11. *Dimension line location (Text/Angle):* Pick a point about 0.6 units below the bottom line of the object.

12. *Dimension text <measured value>:* ⏎

13. *Dim:* ⏎

    HOR

    First extension line or RETURN to select: Pick the same point you picked in step 10.

14. *Second extension line origin:* **END** ⏎

    of: Pick the right end of the bottom line.

15. *Dimension line location (Text/Angle):* Align this dimension with the first one you did.

16. *Dimension text <measured value>:* ⏎

17. *Dim:* Use the same dimensioning subcommands you used in the first procedure (**VERTICAL, HORIZONTAL, ALIGNED**) to dimension the object.

Your drawing should now appear similar to the one in Fig. 10-64. Notice that this time each dimension appears as an upper and lower limit only. The design size is not specified.

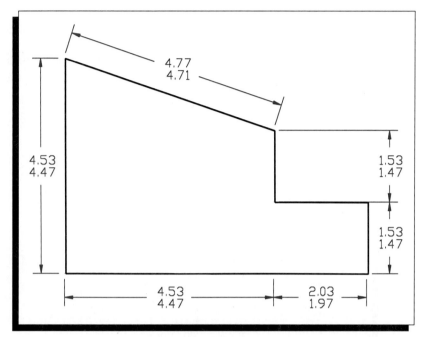

Fig. 10-64

Chapter 10 Review

1. What are dimensioning variables? What is their purpose?
2. What dimensioning subcommand would you use to change the size, or scale, of dimensions on a drawing?
3. What are the two basic methods for placing dimensions on a drawing? How do they differ?
4. What two dimensions do you need to specify to dimension a basic cylinder correctly?
5. What is a datum line? What is its purpose?
6. What three symbols are used to denote a surface finish?
7. What is limit dimensioning? Why is it necessary for machined parts?
8. Describe the three basic types of fit.

■ Chapter 10 Problems

To practice the skills you learned in this chapter, create and dimension each of the following drawings using AutoCAD. For each drawing, one square on the grid represents 0.25 inch.

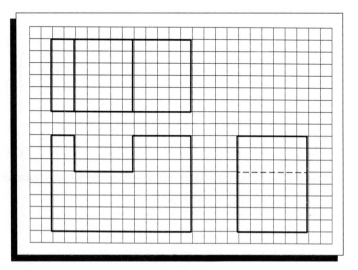

Problem 10-1

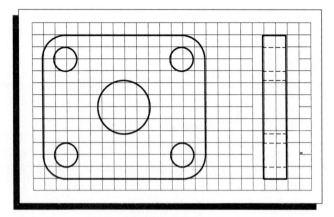

Problem 10-2

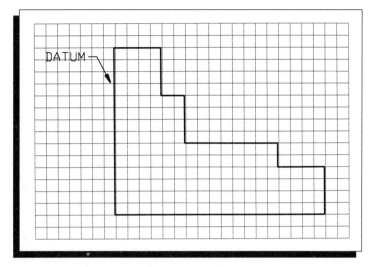

Problem 10-3

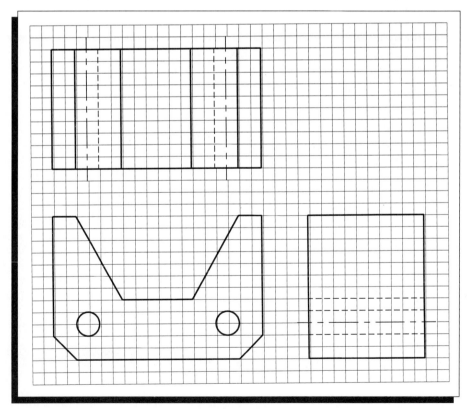

Problem 10-4

317

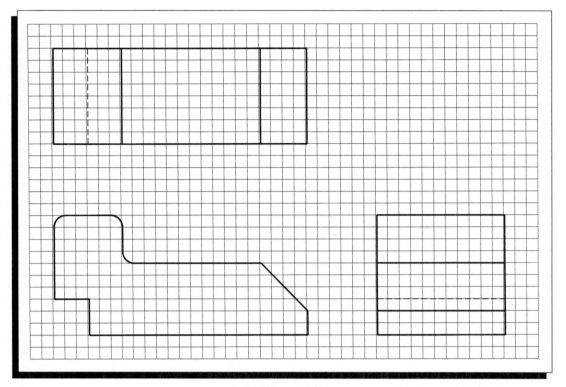

Problem 10-5

Using CAD

Problems Solved with AutoCAD and CADPIPE

How can drafters keep up with the largest expansion of the pipeline and compression facilities in the history of a company? By choosing AutoCAD as the new CAD standard.

TransCanada PipeLines Ltd. (TCPL) provides natural gas from Western Canada to markets throughout Canada and the United States. In 1989, it launched the largest expansion in the history of the company. Expanding its mainline transmission to accommodate the increasing demand for natural gas provided a dilemma for the drafting department. It would be too expensive to upgrade and expand its existing CAD system, so management decided to "start again from scratch."

AutoCAD was chosen as TCPL's new CAD standard, along with CADPIPE ISO, ORTHO, and ELECTRICAL software programs. CAD-PIPE'S customizing features were able to maintain TCPL standards. The CADPIPE database was easily modified to accommodate the 48" diameter pipe used by the company. Using this new program, many specific dimensions found only in the pipeline industry were easily added.

AutoCAD and CADPIPE were able to meet the specific applications for pipeline alignment sheets and crossing drawings. TCPL also developed a layering and symbology standard and a drawing indexing system.

In 1991, a new problem developed. The construction schedule was doubled with virtually no increase in staff. The staff on board had little CAD experience using the new CAD system. TCPL solved the problem by providing designers and draftspeople with basic AutoCAD training. The training, along with the intuitive nature of CADPIPE, allowed the dedicated staff of TCPL to meet all drawing requirements for 1991.

Larry Haines, CAD Systems Developer for TCPL, is pleased with the productivity gains of the new CAD system. "Since implementing the AutoCAD/CADPIPE system, the productivity gains have been almost two-fold. Through further staff training and further development on this system, the productivity can only continue to increase."

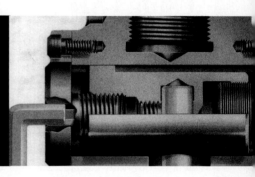

Chapter

11

▶

Key Terms

sectional view   broken-out section

cutting plane   revolved section

section lines   removed section

full section   offset section

half section

▶ ▶

Commands & Variables

HATCH   BHATCH

EXPLODE

Sectional Views

Objectives

When you have completed this chapter, you will be able to:

- demonstrate the need for sectional views.
- identify various types of sectional views.
- use the HATCH command to crosshatch objects.
- use the BHATCH command to crosshatch objects.

Hidden lines may be confusing, and they do not always explain hidden features completely. Sometimes it is necessary to "cut" an object apart on paper to show what is really inside. Creating a sectional view allows you to look inside an object without actually cutting it up.

Using Sectional Views

A **sectional view** is a drawing view which passes an imaginary cutting plane through an object and removes the material on one side of the plane so you can see what is inside. It is like sawing through an object and removing a piece of it *(Fig. 11-1)*. The **cutting plane** is the plane in which the cut is made. It is shown with a dashed line or a phantom line with arrowheads at 90 degrees to the plane. The arrowheads show the viewing direction of the section *(Fig. 11-2)*.

When you have more than one section on a drawing, it is necessary to match each section with its cutting plane. When this is necessary, place capital letters near the arrowheads and label the sectional view with the same letters *(Fig. 11-2)*.

Section lines are lines drawn to indicate material which has been cut away.

Some companies and drafters refer to section lines as *crosshatching*. AutoCAD refers to section lines as *crosshatching* or simply *hatching* and calls the individual lines *hatch lines*.

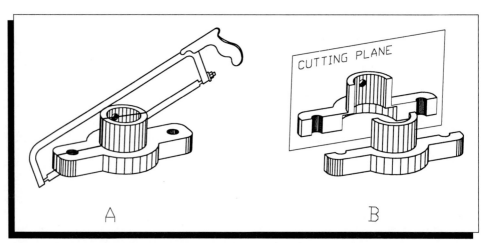

Fig. 11-1 A sectional view removes part of an object so that the inside can be seen.

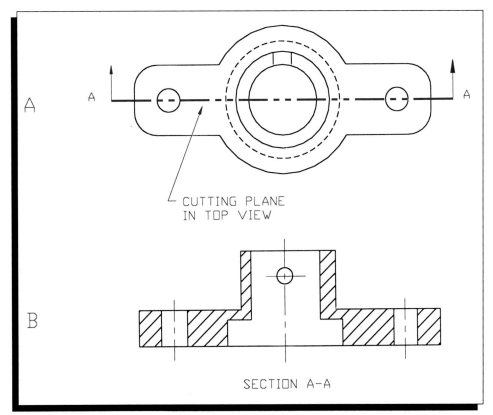

CUTTING PLANE
IN TOP VIEW

SECTION A-A

Fig. 11-2 When a drawing has more than one section, the cutting plane and the sectional view are identified alphabetically as section A-A, B-B, and so on. A) In this case, the top view of the drawing shows the cutting plane A-A. B) The section is also labeled A-A.

The style of section lines depends upon the nature of the material *(Fig. 11-3)*. General-purpose section lines are evenly spaced parallel lines. They are usually drawn at a 45-degree angle running from lower left to upper right. In practice, the angle and direction of the section lines will vary with the shape of the area sectioned. Section lines should not run parallel or perpendicular to visible lines.

In sectional views, pieces in cross section are shown using visible lines. Any hidden lines present in the plan drawing are not shown. This practice simplifies the drawing and makes the section easier to understand. However, center lines are often necessary and may be included in sectional views *(Fig. 11-4)*.

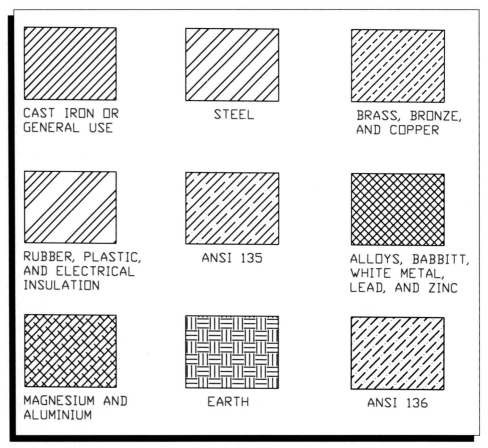

Fig. 11-3 Symbols for indicating section materials.

Types of Sections

Several types of sections are commonly used in drafting, depending on the characteristics of the object being drafted. The various types of sections are described here.

You will learn about another type of section, the auxiliary section, in Chapter 12.

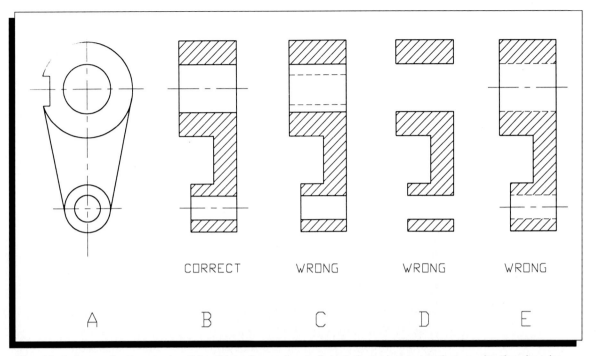

CORRECT   WRONG   WRONG   WRONG

A   B   C   D   E

Fig. 11-4 Some linetypes should not be shown in sectional views because they make the drawing more complicated and may confuse the viewer. A) The object to be sectioned; B) the correct section; C) incorrect section (hidden lines should not be included); D) incorrect section (visible lines are missing); E) incorrect section [all lines except center lines should be shown as visible (continuous) lines].

▶ Full Section

The way an object is cut depends on the complexity of the object and the information you need about the inside. The simplest method is to use the full section *(Fig. 11-5)*. With a **full section**, the cutting plane passes completely through the object and half of the object is removed. Material that has been cut is shown with section lines.

In Fig. 11-5, the cutting plane line is in the top view and the sectional view appears in the front view. Notice that neither the cutting plane nor the sectional view contains a section label. If there is only one section and its identification is obvious, labeling is not necessary.

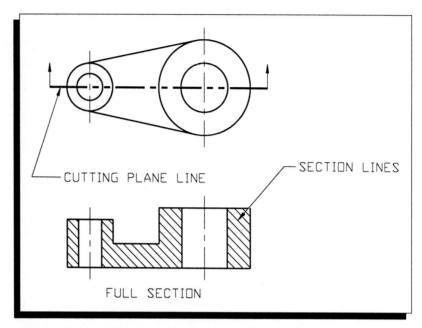

Fig. 11-5 In a full sectional view, the cutting plane passes completely through the object. The resulting section contains section lines to indicate where solid materials have been cut.

▶ Half Sections

A full section may not be the best way to look at the inside of an object. If the object is symmetrical, a half section may be enough to describe the interior fully. While the full section slices through the entire object and removes one half of it, the **half section** cuts half of the object and removes the quarter in front of the cutting plane *(Fig. 11-6)*. The half section can show both the inside and outside of an object in a single view because only half of the view is cut away.

Only one cutting plane arrow shows the direction of sight in a half section. The section line separating the cut half from the uncut half should be a center line when appropriate. Hidden lines are generally omitted in both parts of a half section *(Fig. 11-6)*.

▶ Broken-Out Sections

The **broken-out section** cuts away a small portion of the object with a cutting plane and a break line *(Fig. 11-7)*. The cutting plane is not shown because it is obvious where the cut takes place. Draw the break line with a series of short lines. You can accomplish this in AutoCAD using the SKETCH command.

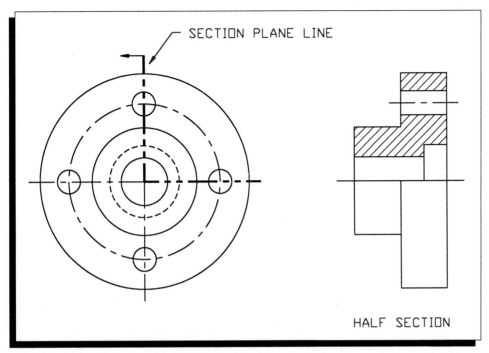

Fig. 11-6 The half section removes a quarter of the object so that you can see both the inside and the outside of a symmetrical object.

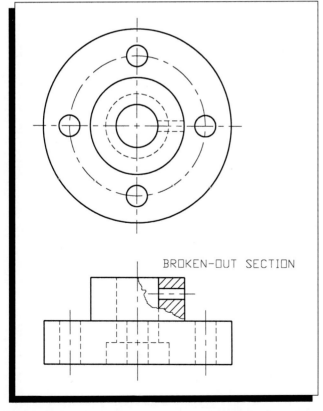

Fig. 11-7 A broken-out section allows you to see the inside of an object in an area not bound by formal cutting planes.

▶ Revolved Sections

A **revolved section** shows the cross section of a spoke, rib, airfoil, or some other object which needs to be described at a specific location *(Fig. 11-8)*. The cutting plane passes perpendicular to the object's length at the point where the section is to be shown. The section is then revolved and drawn parallel to the plane of projection. The view may be broken on both sides of the revolved section, or the revolved section may be drawn at the point of cutting *(Fig. 11-9)*. Remember, the revolved section presents a view at a specific point. Draw it accurately to represent that point.

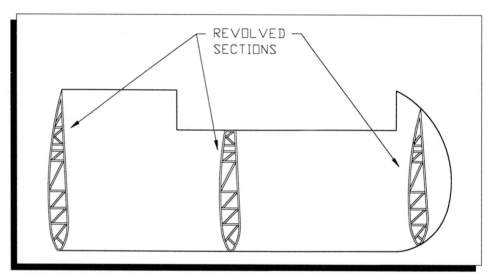

Fig. 11-8 An aircraft wing as a revolved section.

▶ Removed Section

A **removed section** is a section which is moved from its normal place on a drawing and put in a more convenient location *(Fig. 11-10)*. A removed section is a detail of an object and is often drawn to a larger scale. The visible parts behind the section are usually omitted for clarity.

It is important to label the cutting plane lines at each end and at the section because the removed section has been relocated. Label them with successive capital letters: A-A and SECTION A-A; B-B and SECTION B-B, and so on, as shown in Fig. 11-10.

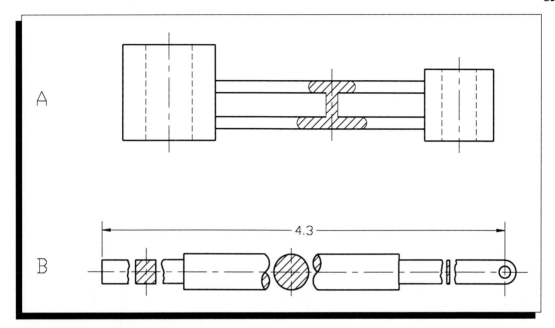

Fig. 11-9 A) A view can be broken on both sides of the revolved section. B) The section can be revolved at the point of cutting.

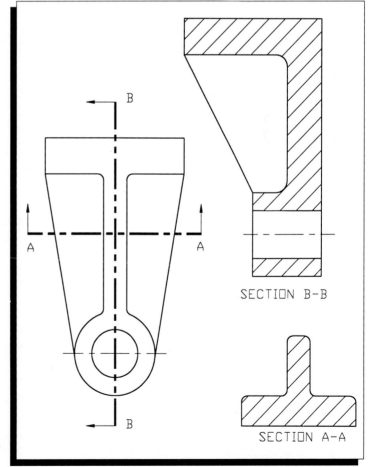

SECTION B-B

SECTION A-A

Fig. 11-10 Removed sections must always be labeled.

▶ Offset Section

An **offset section** staggers the cutting plane so that features which are not aligned can be shown in the same sectional view *(Fig. 11-11)*. The cutting plane line always changes direction by 90 degrees. The offset is shown only in the cutting plane view, never in the sectional view. The offset section is usually drawn as a full section.

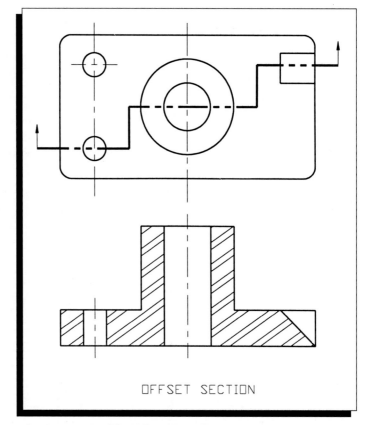

OFFSET SECTION

Fig. 11-11 An offset section.

▶ Revolved Features

A sectional view may show a feature more clearly if the feature is revolved. The cutting plane may be straight and the feature revolved to the cutting plane, *(Fig. 11-12A)*, or the cutting plane may be bent to include the feature which is then revolved in the sectional view *(Fig. 11-12B)*. A wheel with an odd number of spokes will be easier to understand if the sectional view has one of the spokes revolved and others omitted *(Fig. 11-12C)*.

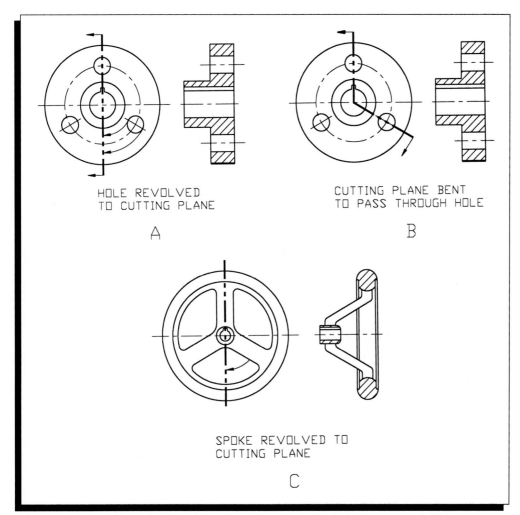

HOLE REVOLVED
TO CUTTING PLANE

A

CUTTING PLANE BENT
TO PASS THROUGH HOLE

B

SPOKE REVOLVED TO
CUTTING PLANE

C

Fig. 11-12 Revolved sections: A) straight cutting plane with feature revolved; B) cutting plane bent to include feature; C) odd number of spokes on wheel revolved for clarity.

Practice recognizing the various types of sectional views by identifying each of the following as a full, half, broken-out, revolved, removed, or offset section. Refer to the descriptions and illustrations in the first part of this chapter if you need help.

1.

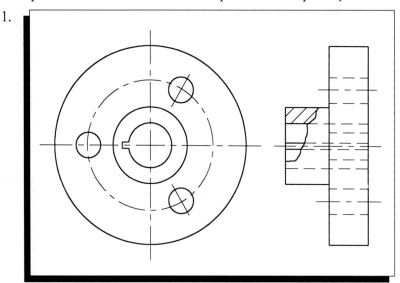

Fig. 11-13

2.

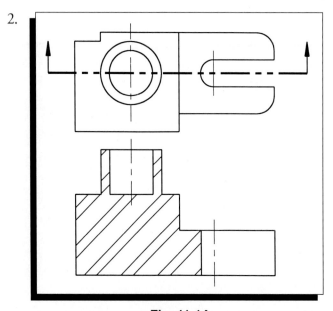

Fig. 11-14

3.

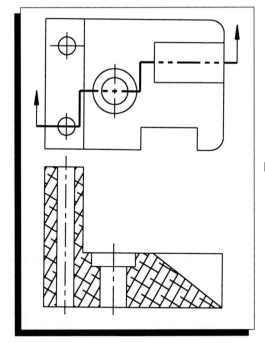

Fig. 11-15

4.

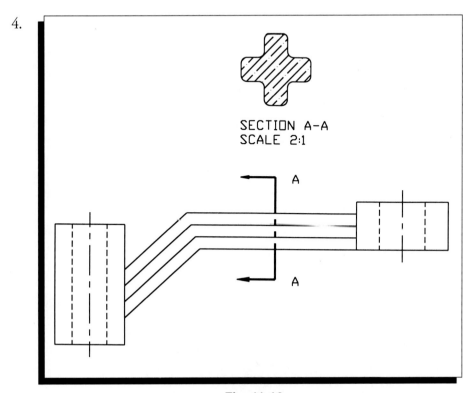

SECTION A-A
SCALE 2:1

Fig. 11-16

5.

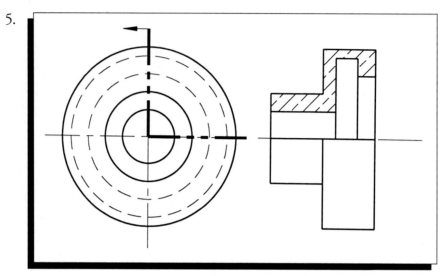

Fig. 11-17

Sections of Support Structures

Support structures are items such as the spokes on a wheel — they connect and support other parts of an object.

Drafters show support structures in sectional views according to accepted conventions.

▶ Spokes and Webs

Spokes, ribs, and webs are support structures. Spokes permit a feature such as a wheel to be tied to a hub without having a solid circular connection, or web. By convention, if a wheel is connected to a hub with spokes, the sectional view does not cut the spokes, even if the cut-

ting plane does *(Fig. 11-18)*. However, if the wheel is connected to a hub with a solid web, the sectional view must include the web. In other words, a web is always marked with section lines *(Fig. 11-19)*. Spokes are not marked with section lines.

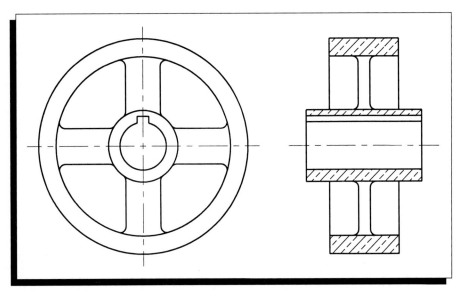

Fig. 11-18 A wheel with spokes in sectional view. Note that the spokes are not sectioned (hatched).

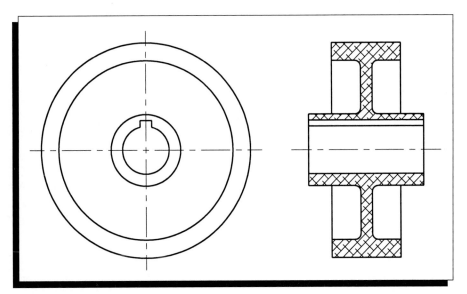

Fig. 11-19 A wheel with a solid web in sectional view. Note that the web is sectioned (hatched).

▶ Ribs

Ribs are used for bracing. Fig. 11-20A shows the front view of a rib with two possible cutting planes for sections. Cutting plane A cuts across the rib, so the rib should contain section lines, as shown in Fig. 11-20B. Cutting plane B cuts lengthwise with the flat of the rib, so the rib is not marked with section lines *(Fig. 11-20C)*.

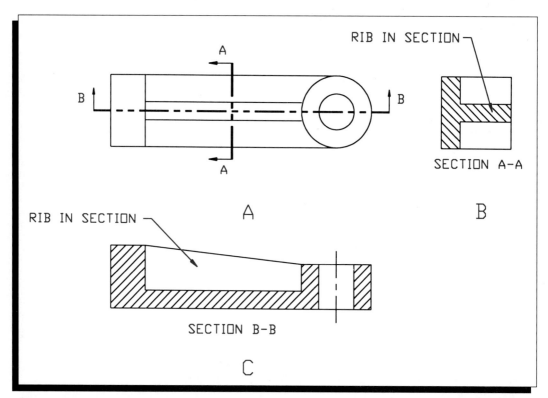

Fig. 11-20 A rib section should be hatched only if the cutting plane cuts across the rib. A) The front view showing cutting planes A and B. B) A rib sectioned along cutting plane A should be hatched. C) A rib sectioned along cutting plane B should not be hatched.

Assemblies in Section

Sectional drawings may be used to show how parts are assembled and to show the relationships between the parts *(Fig. 11-21)*. When two separate parts of the same material are sectioned, you must hatch them differently. You should use the same pattern, but run the hatching symmetrically in a different direction.

The same part in different locations has the same hatch. You can show different materials with different hatch patterns, but be sure you use patterns appropriate to the materials. Do not hatch fasteners, bolts, rivets, or screws, even if the cutting plane goes through them.

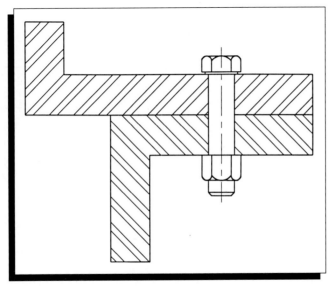

Fig. 11-21 Assemblies in section. Notice that the fasteners are not sectioned.

Using AutoCAD to Create a Sectional View

Sectional views are relatively easy to create using AutoCAD because Auto-CAD can create the section lines for you automatically. First, you draw the sectional views in AutoCAD in the same way that you would any other views. Then you can use AutoCAD's HATCH or BHATCH command to create the appropriate section lines, or crosshatching.

The HATCH Command

AutoCAD's HATCH command is a generic command that includes standard section lines as well as other patterns available for crosshatching. The HATCH command requires keyboard entry. With the HATCH command, you can select the pattern from a list of available patterns, control the scale of the pattern, and set the angle of the pattern.

When you use the HATCH command, you must tell the computer which area to hatch. First, AutoCAD asks you to select objects. This selection set creates the boundaries within which AutoCAD places the pattern. If you do not properly enclose the area, or if your area is not *completely* closed, your hatching will probably not look the way you expect or want it to *(Fig. 11-22)*.

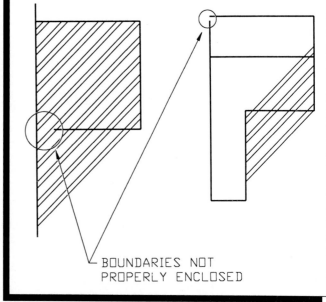

Fig. 11-22 Poor hatching is caused by incomplete boundaries.

BOUNDARIES NOT
PROPERLY ENCLOSED

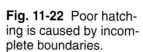

Use the following procedure to view the hatch patterns available and to hatch a simple object.

1. *Command:* **HATCH** ⏎

2. *Pattern (? or name/U,style):* ? ⏎

3. *Pattern(s) to list <*>:* ⏎

AutoCAD displays 53 different patterns on three screens. The standard section line, or crosshatching, pattern is ANSI31.

4. When you have viewed all three screens, press **F1** at the *Command:* prompt to return to the graphics screen.

5. *Command:* Draw a circle completely inside a rectangle *(Fig. 11-23)*.

6. *Command:* HATCH ⏎

7. *Pattern (? or name/U,style):* ANSI31 ⏎

8. *Scale for pattern <1.0000>:* ⏎

9. *Angle for pattern <0>:* ⏎

10. *Select objects:* Select the rectangle with the pick box. AutoCAD responds with "1 found." Make sure all sides of the rectangle have been selected.

11. *Select objects:* Select the circle with the pick box. AutoCAD again responds with "1 found."

12. *Select objects:* ⏎

You hatched the area inside the rectangle with the general pattern, but you did not hatch the circle *(Fig. 11-24)*. HATCH put the pattern between the two boundaries which you selected.

The hatch pattern is treated as one entity. For example, to erase the entire hatching, pick any point within the hatch pattern. If you want to edit small pieces of the hatch pattern with TRIM or EXTEND or if you want to erase only a few lines, insert an asterisk (*) in front of the hatch pattern name when you create the hatch. The asterisk breaks the pattern into individual entities.

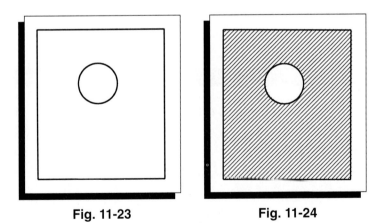

Fig. 11-23   Fig. 11-24

If you forget to use the asterisk when you create the hatch, or if you decide later that you want to edit portions of the hatch, you can use the EXPLODE command to break the pattern into individual entities.

The BHATCH Command

BHATCH, which stands for "Boundary Hatch," is an expanded version of the HATCH command. It allows you to use dialogue boxes to select the characteristics of your hatching pattern. You can activate this command from the keyboard by typing BHATCH, from the Draw pull-down menu, or from the screen menu under DRAW.

On the pull-down menu, this command is called Hatch, but when you select it, BHATCH appears on the command line and the dialogue box appears. The screen menu gives you the option of selecting either HATCH or BHATCH.

When you use BHATCH, you can select your own boundaries or have AutoCAD select them for you. The Select Objects option allows you to create your own selection set to be hatched. Whichever method you choose, make sure the area to be hatched is fully enclosed. The Boundary Hatch dialogue box has two ways to define the hatch area: Pick Points and Select Objects. Pick Points automatically constructs a boundary. When you use this option, be sure to pick a point near the boundary to avoid a Boundary Definition Error.

A Boundary Definition Error is an internal error that occurs in AutoCAD if you do not close the boundary completely or if you choose a point that is not inside a boundary. When such an error occurs, a dialogue box appears informing you of the specific error.

ACTIVITY

Create a drawing, including text, similar to the one in Fig. 11-25. Then follow these steps.

1. *Command:* **BHATCH** ⏎

The Boundary Hatch dialogue box appears *(Fig. 11-26)*.

2. Pick **Hatch Options**...

The Hatch Options subdialogue box appears.

3. Check to be sure that the **Stored Hatch Pattern** button is selected. If it is not, select it.

4. Pick **Pattern**...

The first of five screens of hatch patterns appears. Each hatch pattern is identified graphically and by name.

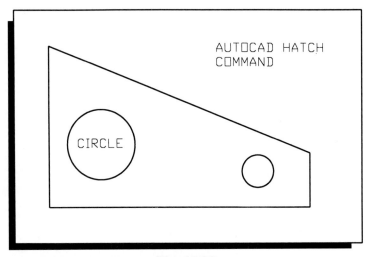

Fig. 11-25

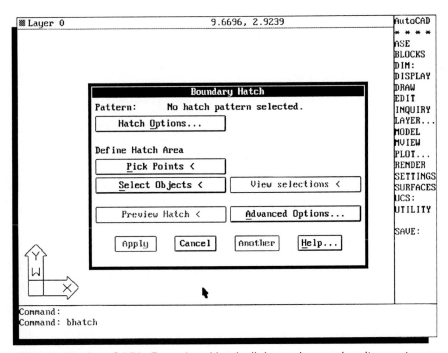

Fig. 11-26 AutoCAD's Boundary Hatch dialogue box makes it easy to
select a pattern visually.

5. Pick ANSI31 on the first page (screen). Pick the pattern, not the name.

You are returned to the Hatch Options subdialogue box. The pattern type is identified as ANSI31.

6. Set the following:

Scale: **1**

Angle: **0**

Hatching Style: **Normal**

Exploded Hatch: Be sure this box is not checked. "Checked" is the same as placing an asterisk before the pattern type using the HATCH command.

7. Pick **OK.**

AutoCAD returns you to the Boundary Hatch dialogue box.

8. Pick the **Pick Points** < button to construct a boundary automatically.

9. *Select internal point:* Pick inside near the edge of the large circle with the word "Circle" in it.

10. *Select internal point:* ↵

You are returned to the Boundary Hatch dialogue box.

11. Pick the **Select Objects** < button.

12. *Select objects:* Pick the word "Circle" with the pick box.

This tells BHATCH not to run the hatch lines through the word "Circle." The text becomes the inside boundary.

13. *Select objects:* ↵

You are returned to the Boundary Hatch dialogue box.

14. Pick **Apply**.

Your drawing should look like the one in Fig. 11-27.

AutoCAD allows you to see what the applied hatch will look like *before* the hatch is actually applied. This option can be helpful for beginning CAD drafters. To preview the hatch, pick the Preview Hatch button before you pick Apply.

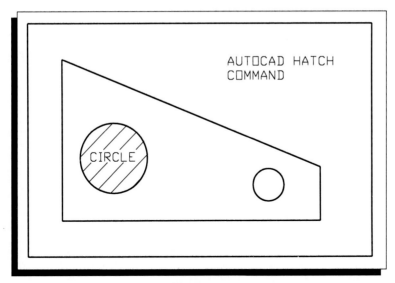

Fig. 11-27

Chapter 11 Review

1. Explain why or under what circumstances sectional views may be needed.
2. In what way do drafters indicate material that has been cut away from an object, as on a sectional view?
3. How does a full section differ from a half section?
4. When might an offset section be useful?
5. How can you tell whether a support structure in a specific section is a web or a spoke?
6. What two AutoCAD commands are available for inserting section lines into a drawing?
7. Why is it important to enclose a hatch boundary completely?

Chapter 11 Problems

Do the following problems to increase your understanding and skill in working with sectional views. For each problem, reconstruct the views shown and then create the section as indicated by the cutting plane, if shown. Use the scale provided.

1. Create a half section based on the information provided in Problem 11-1. All holes are symmetrically placed.
2. Create a full section based on the information provided in Problem 11-2.
3. Create two full sections as shown by the cutting planes in Problem 11-3.
4. Create an offset section as indicated in Problem 11-4.

5. Create a breakaway section as indicated by the break line in Problem 11-5.

6. Create at least two sections to describe the part in Problem 11-6: one horizontal and the other vertical. Place the cutting planes appropriately and show the cutting planes on your drawing.

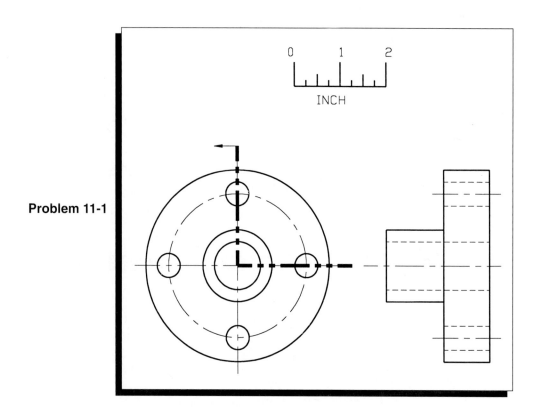

Problem 11-1

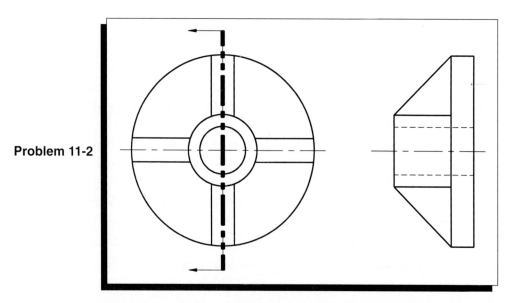

Problem 11-2

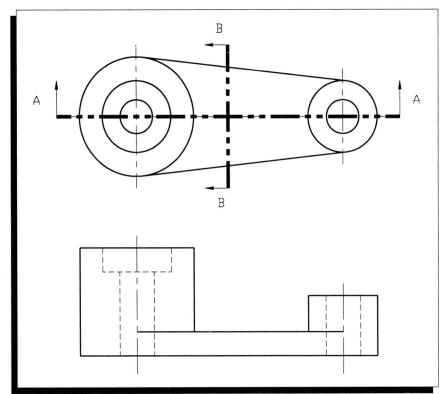

Problem 11-3

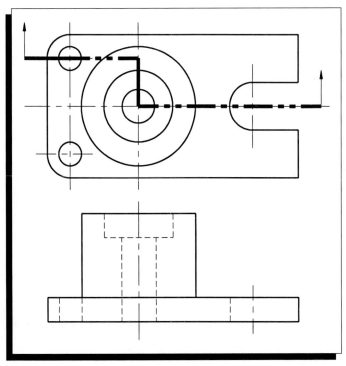

Problem 11-4

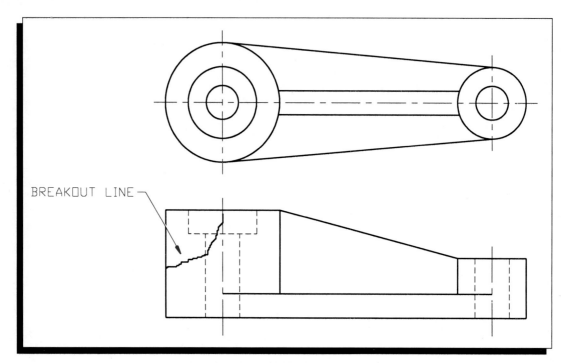

BREAKOUT LINE

Problem 11-5

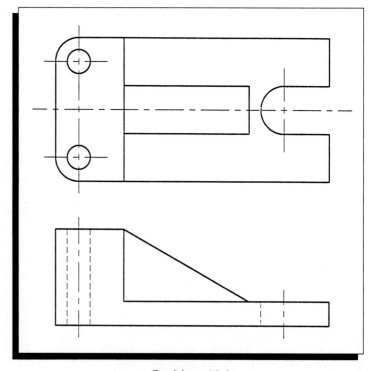

Problem 11-6

Using CAD
Structural Engineer Increases Productivity with AutoCAD

Tony Aimer, Johannesburg, South Africa, is a structural engineer who specializes in the use of steel and concrete, mainly for buildings. Aimer owns his own firm, Tony Aimer and Associates.

When Aimer purchased his first CAD system in September, 1987, he was spending about half of his time working on a drawing board and half of his time with design and client liaison. He also had three employees drawing on boards. At various times during 1987, all three employees left, leaving Aimer to face 1988 with only himself, his computer, and AutoCAD Release 2.62. Because there were no training courses available, he spent a two-week period over the Christmas holiday teaching himself AutoCAD with the help of *Applying AutoCAD* by Terry Wohlers.

Soon after, on January 4, 1988, he was called by an architect and was told "by the way, we appointed you as engineer on a 70,000 square foot shopping center in the Katlehong Township, and we start on site on February 1st!" As a rookie AutoCAD user, Aimer worked many hours into the night to get five structural foundation drawings ready and on site for the contractor. This included all of the conceptual design for the steel work and preliminary stress calculations. Says Aimer, "I could not have achieved this but for AutoCAD."

In spite of the loss of his staff, that was the beginning of greatly increased productivity for Aimer's company. During that year, he managed to increase the company's output by 200% over what they did with four people the year before. In 1989, that figure jumped to 300%.

Aimer's work consists primarily of creating and editing geometric shapes in 2D, along with sectional work. For example, a structural floor plan consists of a plan view of the shape of the building with numerous small sectional details showing how the concrete beams are formed.

Aimer believes that customizing AutoCAD to the type of work you do is vital to overall productivity. One way he customizes AutoCAD is by creating applications that automatically generate commonly used entities. For example, to create a cross section of a staircase is time consuming, even with AutoCAD. Aimer has written a program that will create the staircase in about 30 seconds. All staircases are not exactly alike, so you enter the parameters, such as length, width, etc., that define the staircase.

Parametric programs such as the ones Aimer creates are also available for purchase from software companies. Because AutoCAD can be customized in this way, Aimer thinks that in the future CAD will stand for computer-*automated* drawing. Use of the keyboard and pointing device will be minimal. One thing is sure: AutoCAD will remain as the center of Aimer's business.

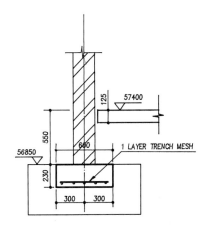

TYPICAL SECTION
THROUGH STRIP FOOTING

Chapter 12

Key Terms

front auxiliary view

top auxiliary view

right-side auxiliary
  view

reference plane

reference line

auxiliary section

secondary auxiliary
  views

revolution

frontal plane
  revolution

top plane revolution

side plane revolution

Commands & Variables

ROTATE

Auxiliary Views and Revolutions

▶ ▶ ▶

Objectives

When you have completed this chapter, you will be able to:

- determine the need for auxiliary views.
- draw auxiliary views, partial auxiliary views, auxiliary sections, and secondary auxiliary views.
- determine the need for revolutions.
- define and draw revolutions in the frontal plane, top plane, and side plane.
- determine the true length of an oblique plane.

Drafters are responsible for selecting enough views to describe an object fully; three views may not be sufficient. The glass box discussed in Chapter 8, "Views and Techniques of Drawing," shows the six orthographic planes of projection, but the glass box does not take into consideration inclined surfaces.

Auxiliary Views

To show inclined surfaces in their true size and shape, a new glass box must be developed. The new box has a vertical plane (front view), horizontal plane (top view), and a profile plane (side view) with an additional auxiliary plane parallel to the inclined surface *(Fig. 12-1)*. An auxiliary view is a projection on an auxiliary plane that is parallel to an inclined surface. In an auxiliary view, you look at the inclined surface in a direction perpendicular to it.

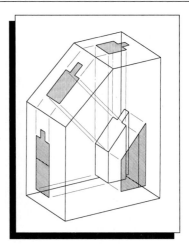

Fig. 12-1 The auxiliary view glass box allows you to project inclined surfaces onto the box in their true size and shape.

Primary Auxiliary Views

Drafting involves the visual description of an object in three primary planes: vertical, horizontal, and profile. Auxiliary views can be described relative to each of these planes. These views represent the primary auxiliary views.

In the vertical or **front auxiliary view**, the glass box hinges to the front view. In the horizontal or **top auxiliary view**, the glass box hinges to the top view. In the profile or **right-side auxiliary view**, the glass box hinges to the right side.

In all cases, the auxiliary view lies at right angles to the inclined surface. To find the true size and shape of the inclined surface, one dimension (height, width, or depth) is projected directly from the primary view. However, one dimension must be referred from another view that is hinged to the primary view. This dimension is the primary reference for the auxiliary view.

For example, in the front auxiliary view, the glass box hinges to the front view. The height for the auxiliary view is taken directly from the front view, as shown with straight lines. The front view does not show the depth, so it must be referred from the top view, as shown by the curved lines. Therefore, the primary reference for this view is the depth. *(Fig. 12-2)*. For the top auxiliary view, the primary reference is the height *(Fig. 12-3)*. For the right-side auxiliary view, the primary reference is the width *(Fig. 12-4)*.

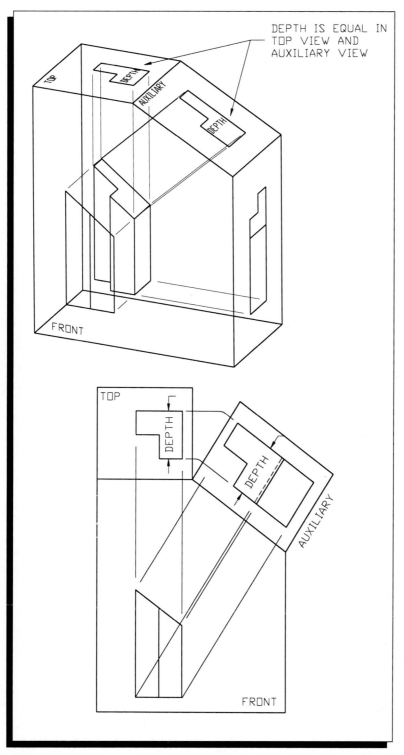

Fig. 12-2 The front auxiliary view is also known as the depth auxiliary view.

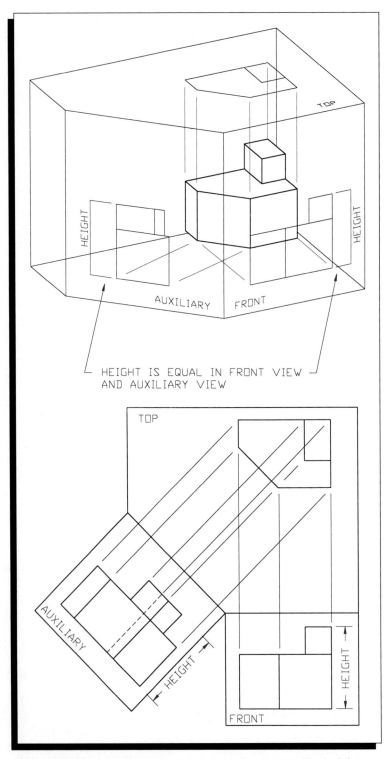

Fig. 12-3 The top auxiliary view is also known as the height auxiliary view.

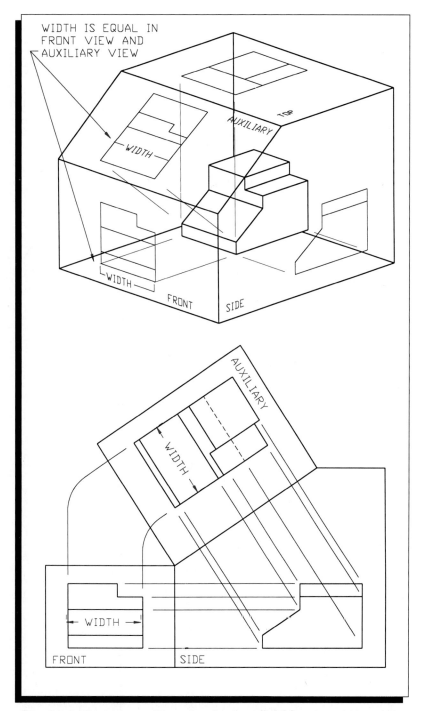

Fig. 12-4 The side auxiliary view is also known as the width auxiliary view.

ACTIVITY

AutoCAD provides several techniques and commands that are helpful in the construction of auxiliary views. The grid can be rotated to aid with view alignment, and the OFFSET command and object snap tools assure the accuracy of feature positioning. The following procedure describes a method for drawing a primary vertical (front auxiliary) view. The result of this exercise is shown in Fig. 12-5.

1. Draw the front and top views of the H-block, as shown in Fig. 12-6. Use a minimum of 1.5 units of space between views. Do not dimension.

Use polar coordinates to draw the 60° line in the front view.

2. Examine the views and determine the edge of the inclined surface.
3. If Grid is not on, turn it on using the **F7** key.

The following steps will rotate the grid to the angle of the inclined surface.

4. *Command:* **SNAP** ↵
5. *Snap spacing or ON/OFF/Aspect/Rotate/Style <default>:* **R** ↵
6. *Base point <0.0000,0.0000>:* ↵
7. *Rotation angle <0>:* **30** ↵

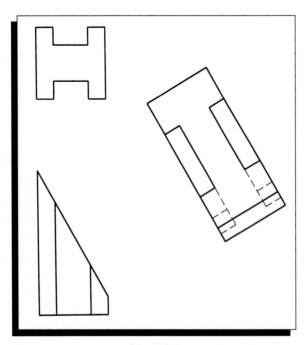

Fig. 12-5

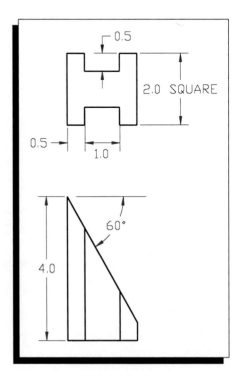

Fig. 12-6

The angle of the inclined surface is 60 degrees from horizontal, measured in a counterclockwise direction, as shown in Fig. 12-6. However, grid rotation is always done in a clockwise direction. Therefore, the angle of grid rotation will be the complement of 60 degrees: $90° - 60° = 30°$.

The grid and the crosshairs have now been rotated to a 30 degree angle counterclockwise from horizontal *(Fig. 12-7)*.

Now you need a **reference plane**, which is a plane parallel to the incline on the object. In this exercise, the reference plane will become the back side of the object in the auxiliary view. Since the reference plane is parallel to the incline, you cannot draw the plane in this view. Instead, you will draw a **reference line** to represent the edge of the plane.

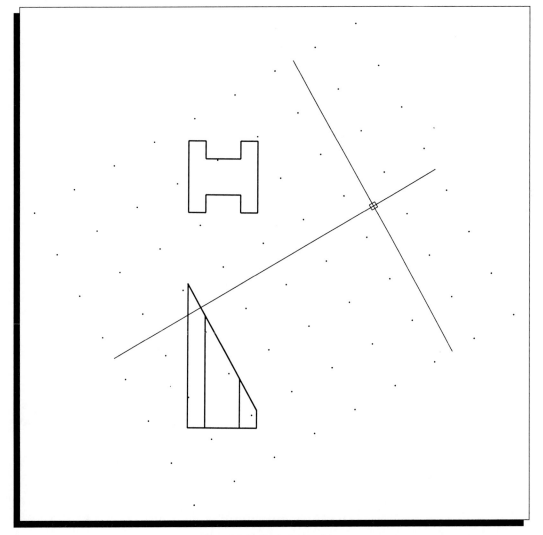

Fig. 12-7 Rotated grid.

8. Draw a vertical reference line parallel to the incline to represent the back surface of the H-block *(Fig. 12-8)*.

Note

The word "vertical," in this case, refers to the orientation of the object. The reference line itself will *not* necessarily be vertical.

The placement of the reference plane depends on the type of auxiliary view and the nature of the object. A front auxiliary view, as in this example, requires a vertical reference plane; however, a top auxiliary view requires a horizontal reference plane, and a right-side auxiliary view requires a profile reference plane. A center reference plane may be more practical for a symmetrical object *(Fig. 12-9)*.

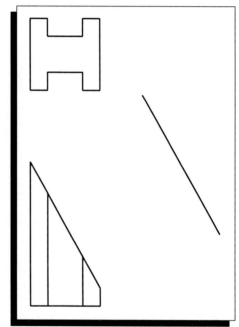

Fig. 12-8 Vertical reference plane.

9. Set **OSNAP** for a running INTersection and draw projection lines perpendicular to the inclined plane from all intersecting points on the front view *(Fig. 12-10)*.

When you enter the LINE command, the OSNAP box appears at the intersection of the crosshairs. The box does not rotate with the grid. Remember to place the intersection to be picked within the box.

10. Use the **OFFSET** command to offset the back surface the correct depth from the front surface *(Fig. 12-11, dimension D)*.

11. Offset the cut-outs *(Fig. 12-12, dimension C)*.

12. Use the **TRIM, ERASE,** and **REDRAW** commands as needed *(Fig. 12-13)*.

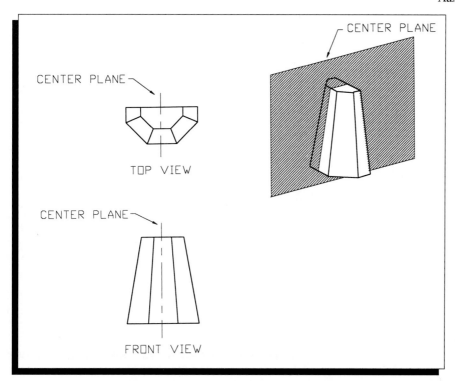

Fig. 12-9 Center reference plane for a symmetrical object.

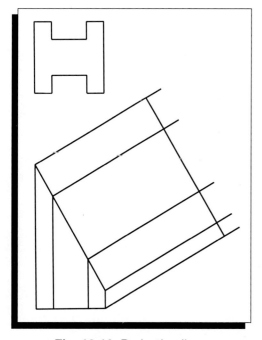

Fig. 12-10 Projection lines.

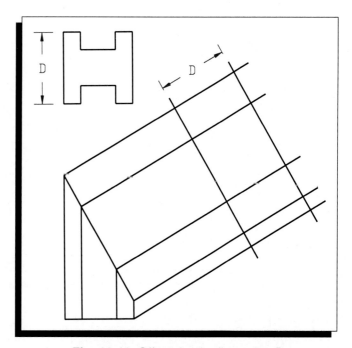

Fig. 12-11 Offset depth, dimension D.

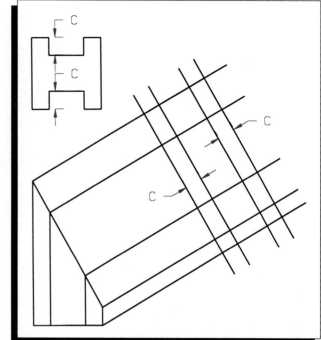

Fig. 12-12 Cut-outs, dimension C.

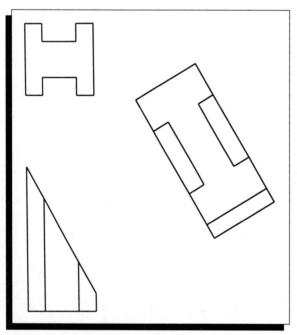

Fig. 12-13 Auxiliary view without hidden lines.

13. Load the hidden linetype using the **LINETYPE** command, and place the hidden linetype on a separate layer.

14. Draw the two hidden-line projection lines and extend the cut-outs with hidden lines *(Fig. 12-14)*.

15. Use the **TRIM**, **ERASE**, and **REDRAW** commands as needed *(Fig. 12-5)*.

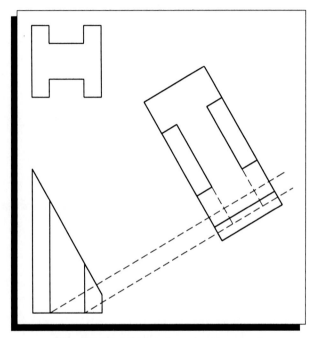

Fig. 12-14 Hidden line development.

Curves on Auxiliary Views

Regular curved surfaces such as circles and arcs transfer to auxiliary views as ellipses. Irregular curves must be transferred point by point, and the points must be joined with continuing arcs (CONTIN from the ARC command).

The following procedure draws an ellipse as an auxiliary view of a cylinder *(Fig. 12-15)*.

1. Draw the front and top views of the cylinder *(Fig. 12-16)*.

2. Rotate the grid as in step two in the preceding "Your Turn" *(Fig. 12-7)*.

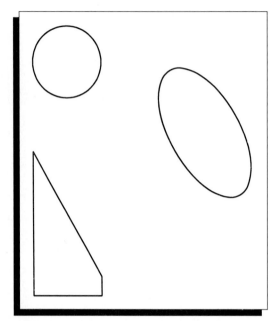

Fig. 12-15 Finished auxiliary drawing of a cylinder.

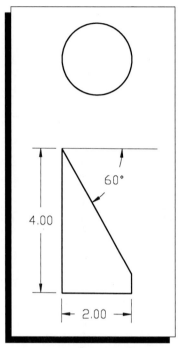

Fig. 12-16 The cylinder.

3. Draw a center reference plane parallel to the incline to serve as the major axis for the ellipse *(Fig. 12-17)*.
4. Draw a projection line perpendicular to the reference plane from the midpoint of the inclined plane on the cylinder to serve as the minor axis of the ellipse *(Fig. 12-18)*.
5. Draw the projection lines from each end of the inclined plane *(Fig. 12-18)*.
6. Form the ellipse with the **ELLIPSE** command *(Fig. 12-19)*.

First enter the major axis length. This length is determined by the intersection of the reference line and the projection lines from the ends of the inclined plane on the cylinder. The minor axis of the ellipse is equal to the diameter of the cylinder in the top view. AutoCAD's *Other axis distance:* prompt asks for the distance away from the major axis, which is half the total axis length. Therefore, enter the *radius* of the cylinder at that prompt.

7. Use the **ERASE** and **REDRAW** commands to finish the drawing *(Fig. 12-15)*.

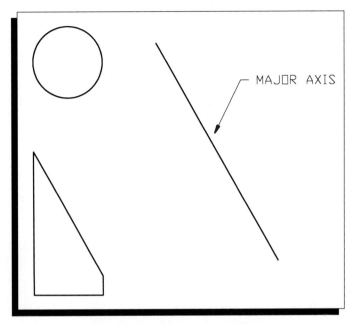

Fig. 12-17 Major axis of ellipse.

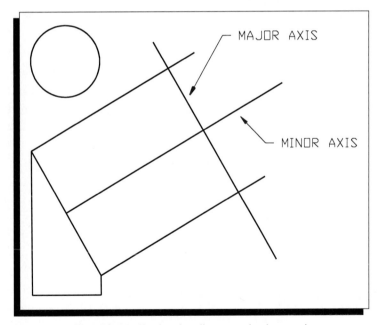

Fig. 12-18 Projection lines and minor axis.

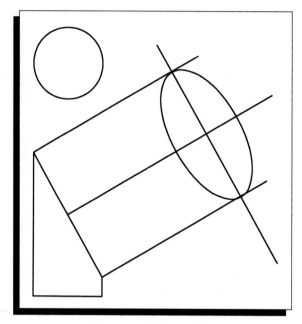

Fig. 12-19 The ellipse.

Partial Auxiliary Views

The auxiliary view of the cylinder represented by the ellipse in Fig. 12-15 did not include all the features normal to an auxiliary view. For clarity and simplicity, only the inclined surface was shown. A partial auxiliary view may show just the inclined surface, or it may incorporate other features of the object. With the proper use of center lines and break lines, an appropriate shape description of an object can be given in a partial view *(Fig. 12-20)*.

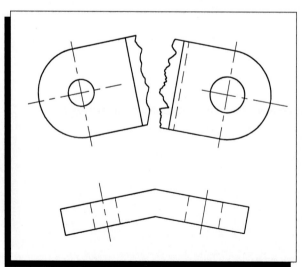

Fig. 12-20 Partial auxiliary views are often clearer than full auxiliary views, providing the proper projection lines are shown.

Auxiliary Sections

An **auxiliary section** is a section viewed at an angle. The inclined surface is made by a plane that cuts through the object. In Fig. 12-21, cutting plane A-A locates the sectioned auxiliary view to be crosshatched. For more information about sectional views, refer to Chapter 11, "Sectional Views."

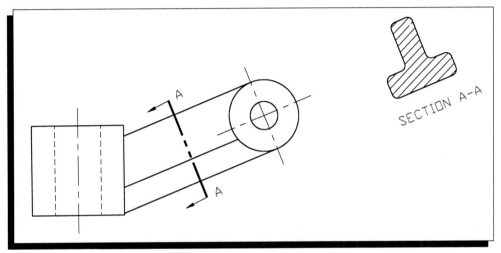

Fig. 12-21 Auxiliary section.

Secondary Auxiliary Views

Secondary auxiliary views describe oblique surfaces, or surfaces that are inclined to all three primary planes. A secondary auxiliary view is a view projected from a primary auxiliary view. In Fig. 12-22, neither of the two primary auxiliary views shows the true shape and location of either the holes or the surface containing them. The secondary auxiliary view projected from primary auxiliary view A accurately describes both the holes and the surface.

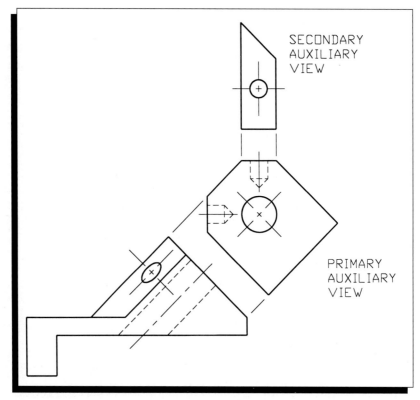

Fig. 12-22 A secondary auxiliary view is taken from a primary auxiliary view.

Revolutions

Drafters may also use a revolution to obtain the true size of an inclined surface. A **revolution** is a view in which the object is revolved around an axis so that a standard plane of projection (usually the front, top, or side view) can show the true size of the inclined surface. An auxiliary view is no longer necessary *(Fig. 12-23)*. The inclined surface is parallel to the plane of projection, which in Fig. 12-23 is the side view. In the front view, the axis appears as a point because it is perpendicular to the front plane. In the top view, the axis appears as a line because it is parallel to the top plane.

Note

The difference between a revolution and an auxiliary drawing is that in a revolution the entire object is revolved and the basic three views are drawn in their revolved positions. In other words, the object revolves in relation to the viewer. When an auxiliary view is used, the object is not revolved in the three basic views. The auxiliary is an additional view that gives the viewer a chance to see the object from a different angle.

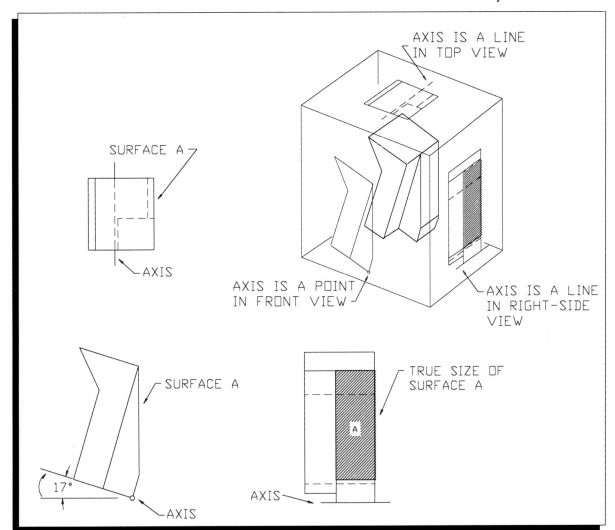

Fig. 12-23 A revolution is another way to view the true size of an inclined surface. Instead of creating a separate (auxiliary) view, the drafter revolves the object to show the inclined surface at its true size. Note that when one view is revolved, the remaining views change also.

The location of the inclined surface determines:
- which view is rotated (front, top, or side).
- the direction of revolution (clockwise or counterclockwise).
- which view shows the true size of the surface.

When you revolve objects, be sure to follow these rules *(Fig. 12-24).*

1. In the view where the axis shows as a point, the view is revolved. This view is not changed in either size or shape.
2. In the views where the axis shows as a line, the dimensions parallel to the axis remain unchanged. All other dimensions will be changed.

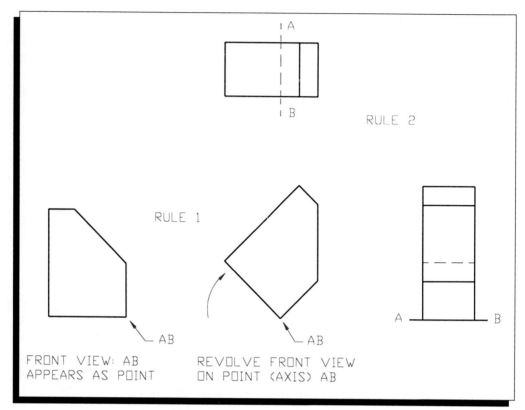

Fig. 12-24 The rules of revolution.

Frontal Plane Revolution

In a **frontal plane revolution**, axis AB appears as a line in the top and side views. Since the axis of revolution, AB, is a point in the front view, Fig. 12-25 represents a frontal plane revolution. The incline is on the top at a 45-degree angle; therefore, the object is revolved 45 degrees in the counterclockwise direction. The top view shows the true size of the surface. If the incline is on the top of the object, the top view shows the true size of the surface. The front view of the object retains the same size and shape, but it is now in a revolved position (Rule 1). The depth dimension shown in the top and right-side views remains unchanged, but the height and width have changed (Rule 2).

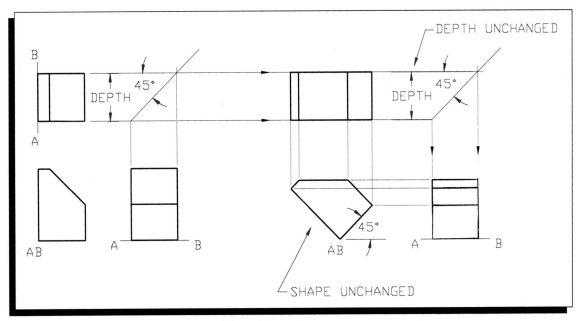

Fig. 12-25 Frontal plane revolution.

ACTIVITY

Fig. 12-26 shows the front and top views of a block in its normal position. The following procedure revolves the frontal plane 25 degrees in the clockwise direction and draws the front, top, and right-side views. The result of this exercise is shown in Fig. 12-27.

1. Set the following drawing parameters:

   Units: two-place decimals

   Grid: **1** and **ON**

   Snap: **0.25** and **ON**

   Ortho: **ON**

2. Draw the front view in its normal, unrevolved position *(Fig. 12-26)*. Do not dimension, and do not draw the top and side views yet.

   Now use the following procedure to revolve the front view 25 degrees clockwise *(Fig. 12-27)*.

The angle of the inclined surface is 25 degrees from vertical measured in a counterclockwise direction. To show this surface at its true size in both the top and right-side views, the surface must be vertical in the front view. Rotate the front view 25 degrees in a clockwise direction. The inclined surface will now be shown true side in the top and right-side views. All other surfaces will be foreshortened in the top and right-side views.

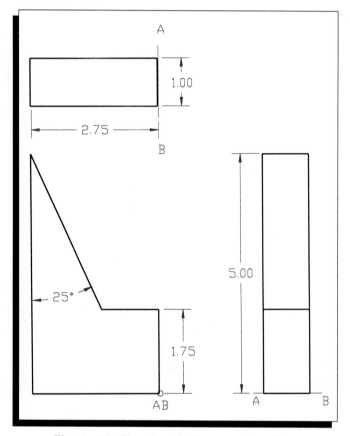

Fig. 12-26 The block in its normal position.

3. *Command:* **ROTATE** ↵

4. *Select objects:* Enclose the entire front view within a window.

5. *Select objects:* ↵

6. *Base point:* Pick the lower right corner where the axis of revolution, AB, is a point.

7. *<Rotation angle>/Reference: –25* ↵

Be sure to include the "–" because the default for measuring angles is counterclockwise.

8. Draw the top and side views by projecting points from the revolved front view. The 1-unit depth is parallel to the axis of revolution and remains constant in both views. Load the hidden linetype with the **LINETYPE** command, and place hidden lines on a separate layer.

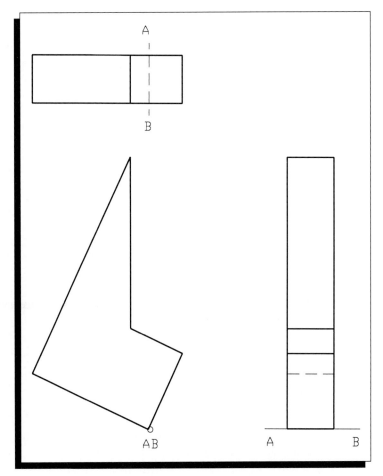

Fig. 12-27 The block in its revolved position.

Top Plane Revolution

In a **top plane revolution**, the axis appears as a line in the front and side planes. Since the axis of revolution, AB, is a point in the top view, Fig. 12-28 represents a top plane revolution. The top view has been revolved 45 degrees in the clockwise direction, but it retains its same shape and size. The front view shows the true size of the surface. In the front and right-side views, the width and depth change, but the height is parallel to the axis and remains unchanged.

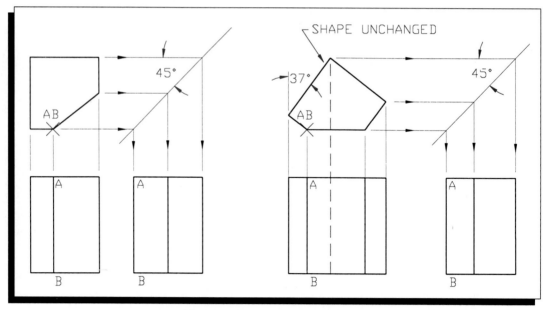

Fig. 12-28 Top plane revolution.

Side Plane Revolution

In a **side plane revolution**, the axis appears as a line in the top and front views. Since the axis of revolution, AB, is a point in the side view, Fig. 12-29 represents a side plane revolution. The side view has been revolved 27 degrees in the counterclockwise direction, but it retains its same shape and size. The front view shows the true size of the surface. In the top and front views, the height and depth change, but the width is parallel to the axis and remains unchanged.

True Length of a Line

Use revolutions to find the true length of any oblique line. An oblique line is one that appears as an angle in all views. For example, to determine the true length of the oblique edge of the roof shown in Fig. 12-30, revolve line AB to position AC and transfer point C to the front view. The true length of the oblique edge is represented by line AC in the front view.

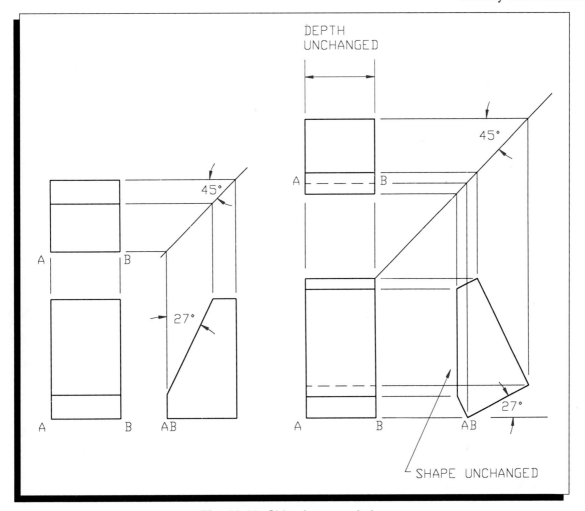

Fig. 12-29 Side plane revolution.

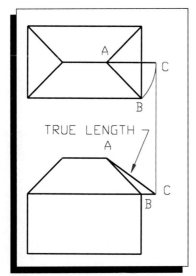

Fig. 12-30 Revolve line AB to create the true length AC.

Practical Uses for Revolutions

Ribs and spokes frequently need to be revolved to clarify views and dimensions *(Fig. 12-31)*. Some parts are bent after machining, and a revolution may give a simpler perspective of the part than an auxiliary view *(Fig. 12-32)*. Accuracy, clarity, and simplicity should be the governing criteria when you decide how to present an object for viewing.

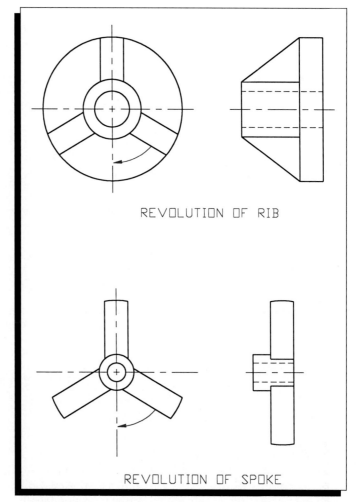

REVOLUTION OF RIB

REVOLUTION OF SPOKE

Fig. 12-31 Revolutions applied to ribs and spokes.

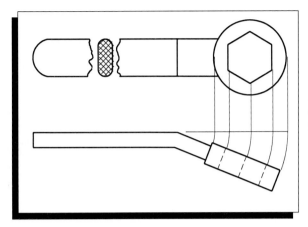

Fig. 12-32 Revolutions applied to objects bent after machining.

Chapter 12 Review

1. What is the purpose of an auxiliary view?
2. What are the three primary auxiliary views?
3. If an object has a circular hole in the front view, what shape will the hole take in an auxiliary view?
4. Why are many of the normal features of a view often omitted in an auxiliary view, so that the auxiliary view shows only the inclined surface?
5. What is an auxiliary section?
6. What is a secondary auxiliary view? What is its purpose?
7. What is the difference between an auxiliary view and a revolution?
8. In which view should a revolution occur?

Chapter 12 Problems

Each of the following drawings contains a pictorial view and either a front, a top, or a side view. Based on the information in each drawing, create a complete CAD drawing, using auxiliary views as necessary to define the object completely. Include dimensions, but do not include instructions or the pictorial view in your drawing.

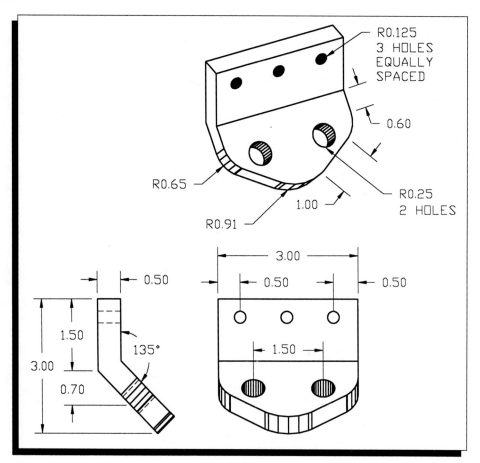

R0.125
3 HOLES
EQUALLY
SPACED

0.60

R0.65

R0.91

1.00

R0.25
2 HOLES

0.50

3.00

0.50

0.50

1.50

0.50

3.00

1.50

135°

0.70

Problem 12-1

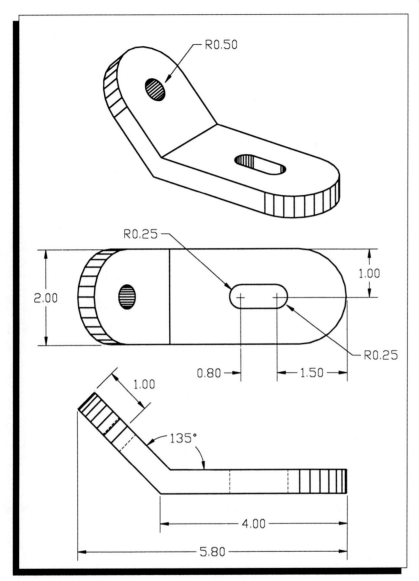

R0.50

R0.25

R0.25

2.00

1.00

0.80

1.50

1.00

135°

4.00

5.80

Problem 12-2

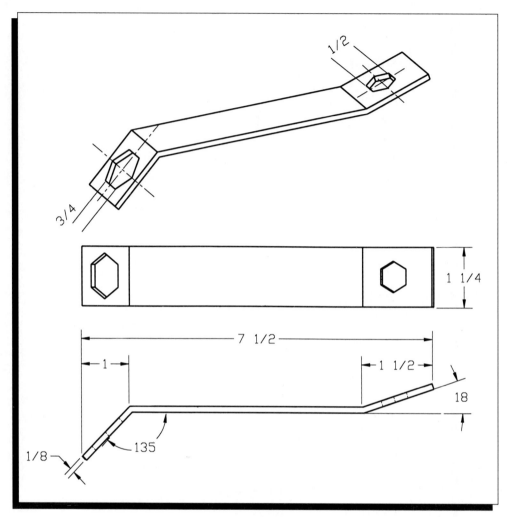

Problem 12-3

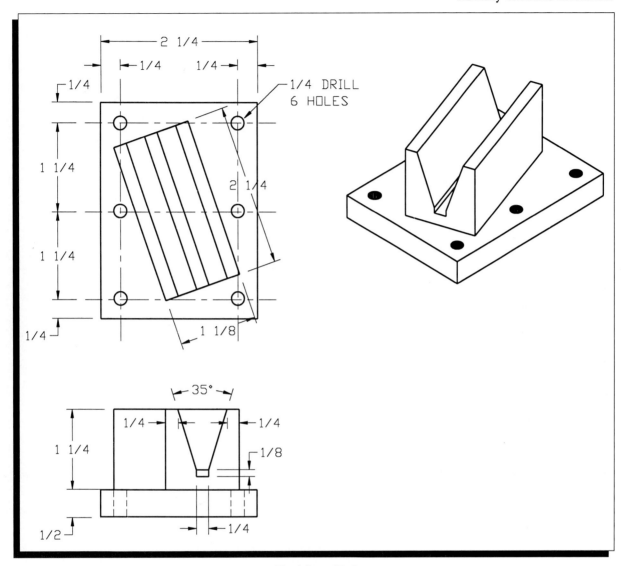

Problem 12-4

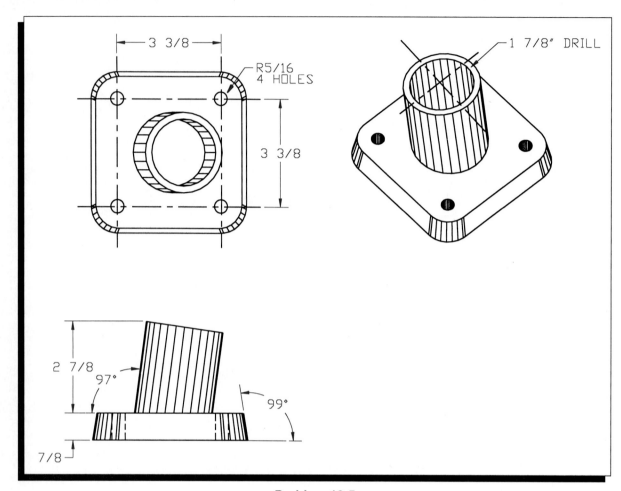

Problem 12-5

Using CAD

Art to Parts

Greg Haminger works in Product Engineering for Monroe, Inc., Grand Rapids, Michigan. His company uses AutoCAD Release 12 and AME for the design and manufacture of an automotive component called the Pointer. Pointers are needles inside the gages that tell a driver vital information such as speed, water temperature, and so on. Each individual pointer rotates on a center shaft driven by an electronic coil, which is mounted within the instrument cluster.

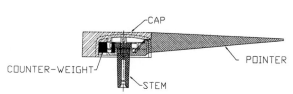

The pointers built by Monroe, Inc. must be built and balanced to very stringent requirements. This ensures that the gage in the vehicle will calibrate properly and give the driver accurate information. Each pointer consists of three or four individual components including a counterweight that balances the pointer.

Haminger uses AutoCAD and the Advanced Modeling Extension (AME) to calculate the weight and balance of the part before a mold is built. AME is solid modeling software produced by Autodesk, Inc., makers of AutoCAD. It is used within AutoCAD for creating 3D objects that are solid or, in other words, have volume. Without these tools, many hours would have to be spent creating a prototype of the part. Many hours of reworking the prototype are also saved when changes must be made. Monroe, Inc. also uses AutoCAD for plant layout, as well as in the custom design of equipment used to paint and assemble their pointers.

CAD/CAM (Computer-Aided Design/Computer-Aided Manufacturing) systems also play a part in automating the manufacture of the pointers. CAD/CAM systems take designs from CAD systems directly into computer-controlled machinery such as CNC (Computer Numerically Controlled) milling machines.

The CNC milling machines at Haminger's company have in the past been programmed manually. The company is currently implementing a new product to further automate the manufacturing process — AutoSurf with AutoMill, another software package produced by the makers of AutoCAD. AutoSurf with AutoMill is a 3D CAD/CAM system for the design and machining of complex mechanical parts and assemblies. According to Haminger, this is a very powerful product, and the productivity of their CNC milling machines is expected to be very high. He says, "We are seeing more and more product designs with freeform and NURB-type surfacing than we have in the past, which are a programming nightmare for even the most experienced CNC programmers." (NURBS are non-uniform rational b-splines. A spline is a smooth line that passes through a set of control points.) CNC programming is one of the many things AutoSurf with AutoMill will take care of automatically.

Haminger says they are also working to set up file translation standards with their customers so the efficiency of their "Art to Part" CAD/CAM capabilities will provide them with the fastest product delivery possible. "The ability to read true surface data directly from customer CAD files, to create these types of surfaces, and to do true 3D modeling at the PC level should keep us competitive — a definite requirement for sustaining any type of U.S. automotive business in the 90s."

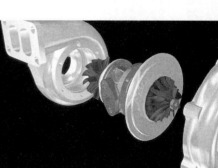

Chapter
13

───── ▶ ─────

Key Terms

working drawing

detail drawing

assembly drawing

reference assembly
   drawing

exploded pictorial
   assembly drawings

cutaway pictorial
   assembly drawing

normal views

bill of materials

attribute

block

symbol library

───── ▶ ▶ ─────

Commands & Variables

BLOCK

WBLOCK

INSERT

ATTDEF

ATTDISP

ATTEXT

ATTEDIT

Production Drawings

Objectives

When you have completed this chapter, you will be able to:

- recognize the various types of production drawings.
- select the best views to describe an object.
- describe criteria for checking drawings.
- create blocks.
- create a symbol library.
- generate a bill of materials.

Drafters create drawings so that products can be manufactured. Sketches and layouts may never be used for production, but ultimately they should lead to finished drawings that will. Production drawings come in many sizes and configurations. There are detail drawings and assembly drawings, multiview drawings and pictorial drawings, and exploded drawings and cutaway drawings. All of these have something in common: someone will use them for manufacturing, assembly, or repair. Someone will use them for production.

Working Drawings

A **working drawing** provides all the information necessary to build an object. It tells the shape and size of the object, the materials that should be used, and the type of finish it should have. In short, it tells everything you need to know about the object in order to produce it.

A working drawing can be a detail drawing or an assembly drawing. A **detail drawing** is an enlarged, finely detailed version of a small portion of the overall drawing. An **assembly drawing** is a drawing that shows how two or more objects are put together, or assembled. An assembly drawing can contain details, but a detail drawing cannot contain assemblies.

Detail Drawings

Detail drawings usually describe individual objects *(Fig. 13-1)*. Each detail drawing may be drawn on a separate sheet, or more than one object may be drawn on the same sheet. Some detail drawings are made for specific assemblies, and others have a general application. A complex machine such as a car or an airplane contains many parts described by detail drawings. One part may be common to several models of airplanes; in fact, a part used on an airplane may also be used on a refrigerator. The detail drawing is the same for both applications.

Sometimes a detail drawing is made for each operation required to make the product *(Fig. 13-2)*. For example, an object may have one detail drawing for casting, another for machining, and a third for welding. Each drawing contains all the information needed to complete the specific operation.

One detail drawing may be made for several parts if their configuration is the same, even if the dimensions differ. A tabular drawing gives the shape of the part, and the dimensions are identified by letters *(Fig. 13-3)*. The actual sizes for the corresponding letters are given in a table on the drawing. The views may not be to scale, and the relationships between the dimensions may not be the same for all configurations.

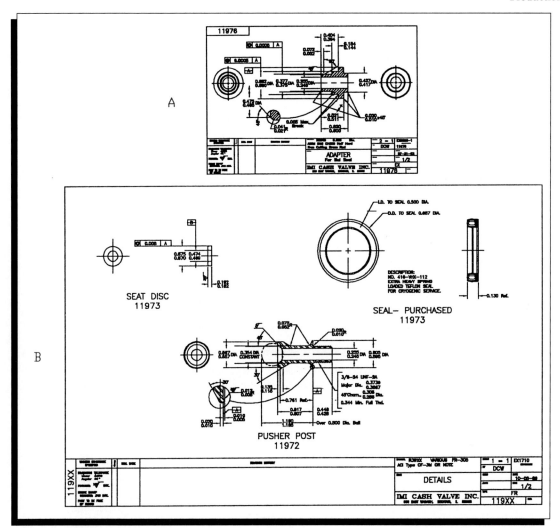

Fig. 13-1 Each detail drawing may be on a single sheet (A), or several details may be shown on one sheet (B).

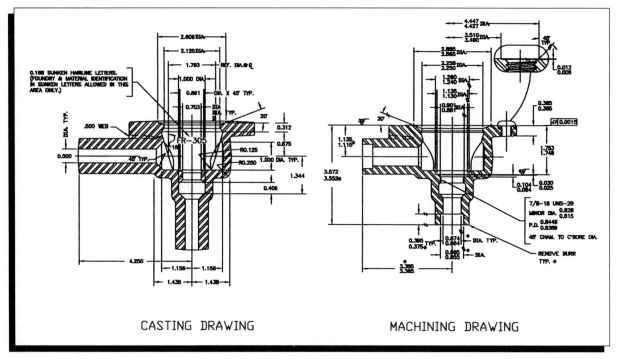

Fig. 13-2 Sometimes a different detail drawing is needed for each operation required to make the product. For example, a casting drawing (left) and a machining drawing (right) may be needed.

Fig. 13-3 In a tabular detail drawing, the parts are identified by letters on the illustration. To find the part number and ordering information, you must find the corresponding letter in the accompanying table, or tabulation.

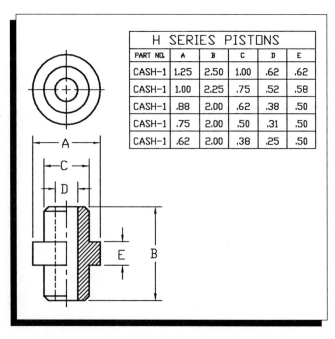

H SERIES PISTONS					
PART NO.	A	B	C	D	E
CASH-1	1.25	2.50	1.00	.62	.62
CASH-1	1.00	2.25	.75	.52	.58
CASH-1	.88	2.00	.62	.38	.50
CASH-1	.75	2.00	.50	.31	.50
CASH-1	.62	2.00	.38	.25	.50

Assembly Drawings

Assembly drawings show how objects go together *(Fig. 13-4)*. Although some include detail drawings, assembly drawings are primarily concerned with how parts fit and how they are fastened. An assembly drawing that gives all the details of the parts and the assembly is called a working assembly drawing *(Fig. 13-5)*.

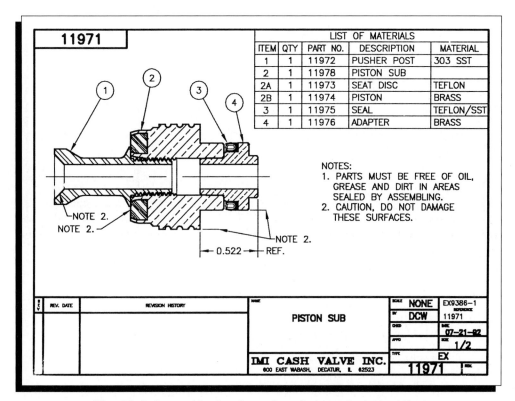

ITEM	QTY	PART NO.	DESCRIPTION	MATERIAL
1	1	11972	PUSHER POST	303 SST
2	1	11978	PISTON SUB	
2A	1	11973	SEAT DISC	TEFLON
2B	1	11974	PISTON	BRASS
3	1	11975	SEAL	TEFLON/SST
4	1	11976	ADAPTER	BRASS

LIST OF MATERIALS

11971

NOTES:
1. PARTS MUST BE FREE OF OIL, GREASE AND DIRT IN AREAS SEALED BY ASSEMBLING.
2. CAUTION, DO NOT DAMAGE THESE SURFACES.

NOTE 2.
NOTE 2.
NOTE 2.
0.522
REF.

PISTON SUB

IMI CASH VALVE INC.
600 EAST WABASH, DECATUR, IL 62523

SCALE NONE EX9386-1
BY DCW REFERENCE 11971
DATE 07-21-92
SIZE 1/2
TYPE EX
11971

Fig. 13-4 Assembly drawings show how parts go together.

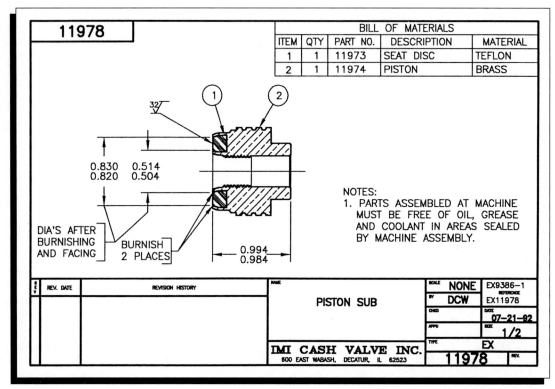

Fig. 13-5 A working assembly drawing shows details as well as assembly information.

▶ Reference Assembly Drawing

A **reference assembly drawing** shows how parts go together and how they work *(Fig. 13-6)*. It has no dimensions. Each part on the drawing is identified and described in a table. Reference assembly drawings are used by assemblers and inspectors who must be sure that all the parts are there and that the parts are put together correctly. Many purchased items — from barbecue grills to bicycles — require some assembly. Their instructions make extensive use of reference assembly drawings.

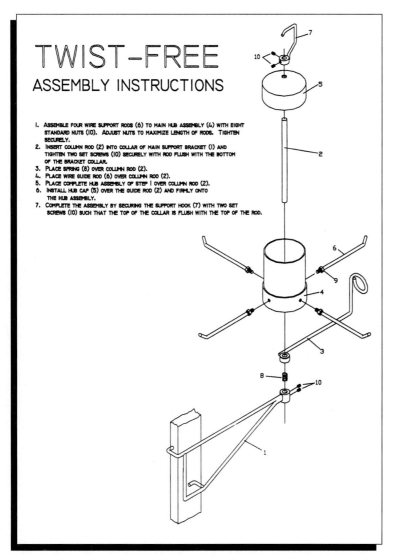

Fig. 13-6 Reference assembly drawings are often used to show consumers how to put together a product that needs assembly.

▶ Exploded Pictorial Assembly Drawings

Exploded pictorial assembly drawings show products disassembled, but with the parts adjacent to their assembly positions *(Fig. 13-7)*. Illustrated parts lists and parts catalogs frequently use exploded assembly drawings to identify parts for replacement. The parts are numbered in the drawing, and each number corresponds with a list that gives the part's description and ordering information.

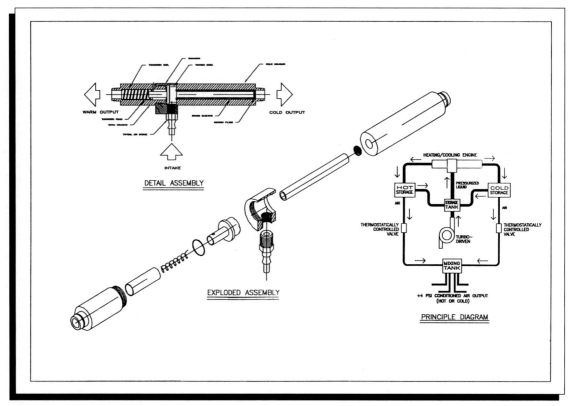

Fig. 13-7 This exploded pictorial assembly drawing shows the assembly of an air heat exchange valve.

▶ Cutaway Pictorial Assembly Drawing

The **cutaway pictorial assembly drawing** shows a product as if a section had been cut out of it so that details of the inner parts can be seen *(Fig. 13-8)*. Sales literature frequently uses cutaway drawings to emphasize attributes of products. Cutaways are also used to show the inside workings of objects for educational and training purposes.

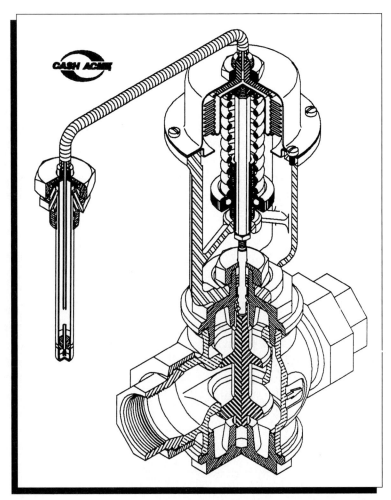

Fig. 13-8 This cutaway pictorial assembly drawing of a valve shows the internal functions of the valve.

ACTIVITY

Visit a local major appliance dealer and ask the salesperson to show you some working drawings that come with the appliances. Discuss the drawings. Ask the salesperson the following questions and any others you think are important or interesting.

- Does the salesperson find them helpful?
- Which ones are used most often? Why?
- How would the salesperson change the drawings if he or she were the drafter?
- How would his or her job be different if the appliance manufacturers no longer supplied working drawings?

Write a short paper or make a class presentation based on your visit.

Selecting Views

We all know that one picture is worth 10,000 words. For this reason, select your pictures — in this case, views — carefully. In working drawings, as in all other drawings, you should choose views that clearly describe the object and are easy to read. Make sure that all features are understood, but do not be redundant. Each view must play a role in the overall description of the part. If it does not, it should not be drawn.

Simplicity is very important. It may be more practical in some cases to use notes instead of drawing additional views. Auxiliary and sectional views may help simplify a complex object. When in doubt about which views to include, think about why you are making the drawing and how the drawing will be used. If necessary, ask a more experienced drafter.

Normal Views

The front view, top view, and right-side view are considered the **normal views**, or those used most often to describe an object *(Fig. 13-9)*. Think back to the discussion of views in Chapter 8. Remember that, in some cases, only one view is needed. A stencil can be described in one view. Its thickness can be specified in a note. Two views will suffice if nothing can be added by drawing a third. For example, a symmetrically round object can usually be shown in two views. Other objects may need more than three views to be described correctly. The additional view or views may be another side, an auxiliary, or a sectional.

Sectional Views

As you may recall from Chapter 11, sectional views open objects so that you can see complex hidden features *(Fig. 13-10)*. Sectionals may be part of detail drawings or assembly drawings. To create a sectional, you pass a cutting plane through the object and remove the material on one side of the cutting plane. The cutting plane is the viewing plane for the sectional drawing.

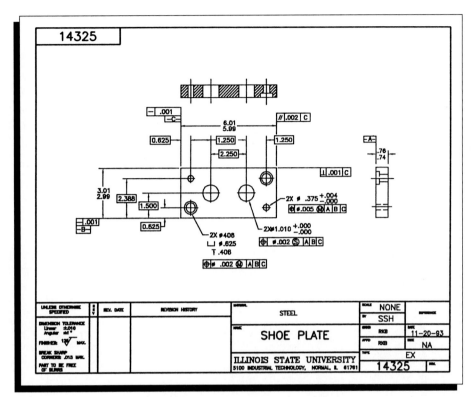

Fig. 13-9 The three normal views.

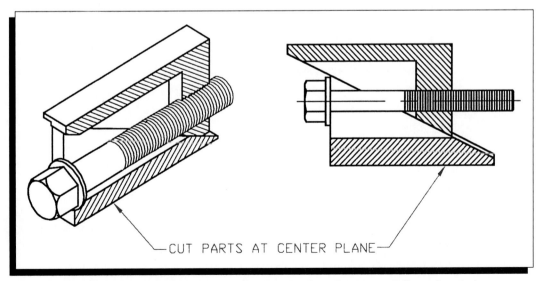

Fig. 13-10 The cutting plane is also the viewing plane on a full sectional view.

391

Auxiliary Views

Auxiliary views are drawn so that inclined features can be seen in their true size and shape *(Fig. 13-11)*. The viewing plane of an auxiliary view is perpendicular to the angle of the inclined feature and parallel to the surface. The location of the auxiliary on the drawing depends on the position of the inclined feature and the best position from which to view that surface. See Chapter 12 for more information about auxiliary views.

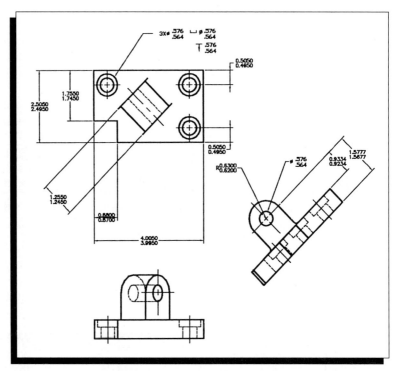

Fig. 13-11 The top portion of the object shows an auxiliary view to show its true size and shape.

Pictorial Drawings

Pictorial drawings show objects in three dimensions and bring a level of realism to them *(Fig. 13-12)*. Pictorials display several faces at one time, and they often supplement multiview draw-ings. Typical uses of pictorials include pictures in catalogs, sales brochures, and technical literature. Refer to Chapter 2 for more information about pictorials.

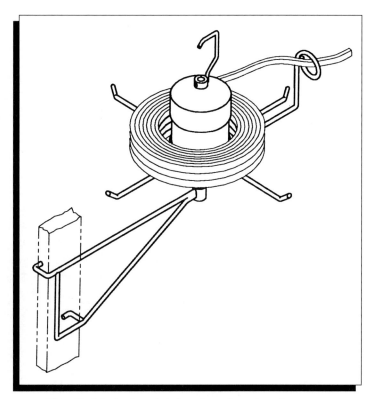

Fig. 13-12 A pictorial drawing of a wire dispenser.

ACTIVITY

Select a simple object in your classroom, such as your instructor's desk, and do the following.

1. Sketch a pictorial view.
2. Sketch the three normal views.
3. Sketch a sectional view.
4. Sketch an auxiliary view.

Analyze your sketches based on the information you have read in this section. Which views best describe the object? Which views are not needed? Do you need more views or other information to describe the object completely? Save this information for use in the next "Your Turn."

Finishing a Drawing

A production drawing is not necessarily just a combination of shapes and objects. On some drawings, you may need to add notes. If any individual shapes seem too large or out of proportion, you need to fix them. After you have finalized the shapes and objects, you must edit them to create finished production drawings.

Adding Notes

Notes explain drawings. They contain information that cannot be made clear by a drawing alone *(Fig. 13-13)*. General notes such as tolerances, heat treating, and overall finish are placed in a Notes section on the drawing and are numbered sequentially. Notes specific to a feature — such as hole data, chamfers, and threads — are placed at the feature.

When drawings must comply with military or industrial standards, notes are used to specify standard practices and procedures. Drawing compliance with company procedures is understood and is not included as a note. For more information about notes, see Chapter 9.

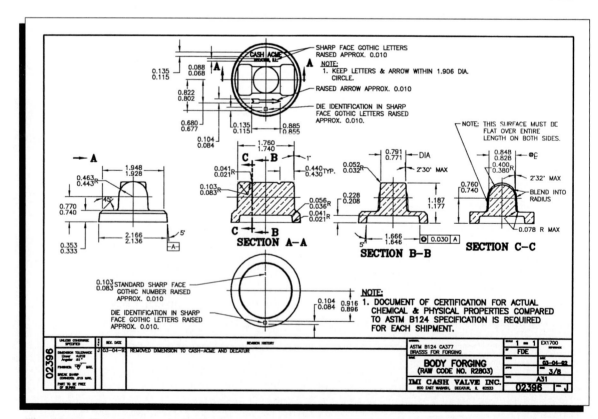

Fig. 13-13 The placement of notes on a drawing depends on the nature of the notes. General notes are placed in a notes section; notes specific to a feature are placed near the feature they describe.

Scaling

When you draw with AutoCAD, you draw objects in their full size because you can set the AutoCAD units to mean anything you want. The units can be inches, feet, decimeters, angstroms, or anything you specify. You can set the units to feet, for example, and set the limits to 64 feet by 52 feet. Then you can draw a building using real units, eliminating the possibility of scaling errors.

If you are drawing a large object such as a house, you can use AutoCAD's ZOOM command to enlarge a small detail for easier drawing. Since you are using real units, the zoomed-in area stays in proportion to the full-size house. When you have completed your drawing, you can plot it in any scale you want so that it fits on the drawing sheet.

Checking Drawings

After drawings have been completed and before they can be released for production, they must be checked. The person who checks the drawing must usually sign off on the drawing, approving it. Checking is very important work. Although drafters inspect their own drawings, they never officially check them. Someone who has not worked on the drawing will see any errors much more easily than the person who created the drawing.

Some companies employ drafters to be full-time checkers. The American Design Drafting Association lists Checker as a professional group; refer to Chapter 1, "Introduction to Drafting." Small companies and companies that have small drafting departments usually ask drafters to double as checkers.

Checkers usually follow a very exacting procedure to ensure that all errors are found. Checks should include, but not be limited to, the following:

- Views should describe the shape of the drawing.
- There should be enough views, but not too many.
- The views must be to the proper scale.
- Dimensions must describe the object completely.
- Dimensioning and tolerancing must be done correctly.
- Drawing must comply with company, industrial, and military specifications when necessary.
- Proper notes must be given for finish, material, processing, etc.
- Standard parts for fasteners, handles, knobs, etc., should be specified when possible.

ACTIVITY

Your Turn

From your analysis in the previous "Your Turn," make a complete working drawing of the object using AutoCAD. Add dimensions and notes. Then plot the drawing, making sure the scale is correct. Submit your drawing to another student for checking and have that student sign off on it before handing it to your instructor.

Bill of Materials

Most large, complex objects are made from many parts. The parts have identification numbers, descriptions, and other information necessary for assembly. A **bill of materials** is a list of the parts and the information necessary to make the product *(Fig. 13-14)*. Sometimes this list appears on the assembly drawing. However, if the assembly is complex, the bill of materials may be a separate list.

AutoCAD provides for the creation of a bill of materials with a series of attribute commands. An **attribute** is text information that you assign to a block using the ATTDEF command. This and other attribute commands manipulate text information that is stored in AutoCAD blocks. A **block** allows you to combine several entities into one and store them for later use. For example, you can draw a casement window (using several lines and rectangles) and store it as a block. You can add attributes to the block to describe the window. Then, when you draw a house, you can insert the block wherever you need a casement window. Create the block using AutoCAD's BLOCK command. Then you can insert the block into a drawing as many times as necessary. Later, you can extract the attributes from all the blocks and print them as part of the bill of materials for the house.

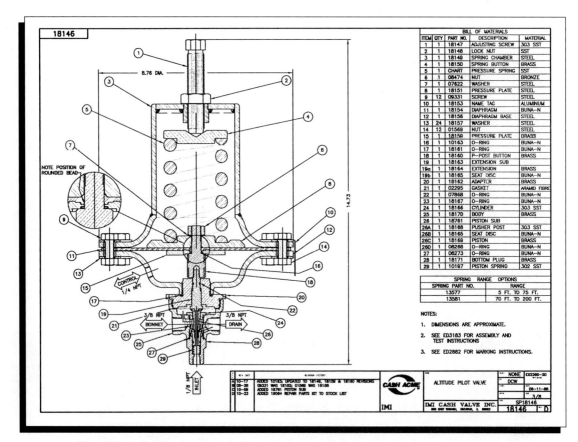

Fig. 13-14 A bill of materials lists the materials needed to produce a product.

ACTIVITY

In the following procedure, you will create a symbol library for the components of a computer system: console, monitor, keyboard, and printer. Then you will add the attributes for each component's description, manufacturer, and model. Finally, you will generate a bill of materials directly from the symbol library.

All the commands in this procedure will be given for keyboard entry. However, you may prefix many of the commands with "DD" to use a dialogue box instead of keyboard entry. Vary some of the repetitive steps below by trying the dialogue boxes. If you get lost, cancel the dialogue box and return to keyboard entry.

1. Begin a new drawing called **LIBCOM**.
2. Create the simplified representations of a computer console, monitor, keyboard, and printer shown in Fig. 13-15. Do not include the component labels yet.

Since you do not want the labels to be part of the blocks, it is good practice to place them on a separate layer. Be careful, however. Blocks can include different layers. To exclude a layer from a block, you must freeze the layer before creating the block.

3. Create a new layer called **TEXT**, make it the current layer, and label each of your components.
4. Make layer 0 the current layer and freeze layer TEXT.

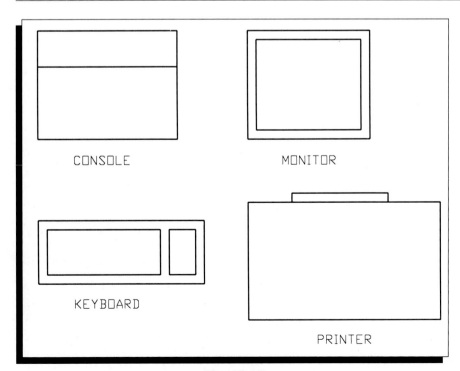

Fig. 13-15

5. *Command:* **BLOCK** ↵

6. *Block name (or ?):* **CONSOLE** ↵

7. *Insertion base point:* **INT** ↵

8. *of* Pick the lower left corner.

9. *Select objects:* Select the console representation.

AutoCAD responds with "2 found."

The number of entities AutoCAD "finds" may vary depending on how you constructed the console. For example, if you used a rectangle and a line, AutoCAD would find two entities. If you used five lines, however, AutoCAD would find five entities.

---

10. Select objects: ↵

The console representation disappears.

11. Repeat steps 5 through 10 for blocks named **MONITOR**, **KEYBOARD**, and **PRINTER.**

You now have a blank screen.

12. Thaw layer TEXT.

Now use the INSERT command to place the blocks you have created in your drawing.

13. *Command:* **INSERT** ↵

14. *Block name (or ?):* **CONSOLE** ↵

15. *Insertion point:* Notice that a movable console is attached to the crosshairs at the insertion point you selected in step 7. Pick a point to center the console above its name.

16. Press **RETURN** three times to accept the insertion point defaults.

17. Repeat steps 13 through 16 to insert the monitor, keyboard, and printer.

18. *Command:* **QSAVE** ↵

QSAVE is another command that allows you to save your drawing. It saves the current drawing without asking for a filename, so you can use it only to save files that you have named. If you have not yet named the file, you must use the SAVE command to name the file before you can use QSAVE.

At this point, you have created a drawing named LIBCOM which is made up of four separate blocks. As you will see beginning in step 45, you can insert the LIBCOM drawing file into any other drawing. By doing so, you can insert all of the blocks you have stored in the LIBCOM drawing into your current drawing. For this reason, the LIBCOM file is considered a **symbol library** — a place to store blocks that you want to use in other drawings.

If you want to insert just one block rather than the entire drawing, you can use the WBLOCK command to create the block. WBLOCK takes one block and converts it into a drawing file that you can then insert into another drawing.

---

19. Explode each block.

The EXPLODE command unblocks the entities. Now that you have exploded the blocks, you can add attributes. Then you will redefine the blocks to include the attributes.

20. Zoom in on the console.

21. *Command:* **ATTDEF** ↵

The ATTDEF (ATTribute DEFinition) command defines the attributes to be assigned to each block. First, AutoCAD prompts you for the attribute modes. The four modes are Invisible, Constant, Verify, and Preset. By default, they are all off. For this activity, the attributes should be invisible and constant. In other words, you want to set the values of the attributes permanently so that they are always the same. Finally, you want AutoCAD to ask for verification.

If you set Constant to No, or off, AutoCAD asks you for a prompt that will appear each time you insert the block, asking for the value of the attribute. This allows you to use one block, for example, for a whole roomful of computers, even if the room contains various kinds of computers. As you store each block representing a computer, you can store its individual attributes.

---

22. *Attribute modes — Invisible:N Constant:N Verify:N Preset:N*
    Enter (ICVP) to change, RETURN when done: **I** ↵

23. *Attribute modes — Invisible:Y Constant:N Verify:N Preset:N*
    Enter (ICVP) to change, RETURN when done: **C** ↵

24. *Attribute modes — Invisible:Y Constant:Y Verify:N Preset:N*
    Enter (ICVP) to change, RETURN when done: **V** ↵

25. *Attribute modes — Invisible:Y Constant:Y Verify:Y Preset:N*
    Enter (ICVP) to change, RETURN when done: ↵

26. *Attribute tag:* **DESCRIPTION** ↵

27. *Attribute value:* **Console** ↵

Use upper and lower case letters as shown.

28. *Justify/Style/<Start point>:* Pick a point inside the console near the top. You will be placing four lines of text, so judge your start point accordingly.

29. *Height <current value>:* **0.1** ↵

You may have to adjust the height so that it is appropriate to the size of your console.

30. *Rotation angle <0>:* ⏎

The word DESCRIPTION should appear. Don't worry if it runs off the right edge.

31. Press **RETURN** to repeat the ATTDEF command. Change the Attribute tag to **MFGR** (for "manufacturer"). Change the Attribute value to the name of your computer.

Press RETURN at the *Justify/Style/<Start point>:* prompt to place this attribute directly below the first one.

32. Press **RETURN** to repeat the ATTDEF command. This time, change the Attribute tag to **MODEL** and the attribute value to the model of your computer.

If you don't know your computer console manufacturer's name or the computer model number, ask your instructor.

33. Press **RETURN** to repeat the ATTDEF command. This time, change the Attribute tag to **COST** and the attribute value to the approximate cost of your computer.

Your console should contain attributes as shown in Fig. 13-16.

34. Repeat steps 20 through 32 for the monitor, keyboard, and printer. Use **DESCRIPTION, MFGR, MODEL,** and **COST** for attribute tags and appropriate terms for the attribute values.

You may want to try the DDATTDEF command to create the attributes for one or more of these blocks.

Fig. 13-16

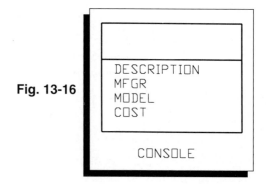

When you complete step 34, redefine your blocks to include the four attributes.

35. *Command:* **BLOCK** ⏎

36. *Block name (or ?):* **CONSOLE** ⏎

37. *Block CONSOLE already exists.*
    Redefine it? <N>: **Y** ⏎

38. *Insertion base point:* **INT** ⏎

39. *of* Pick the lower left corner.

40. *Select objects:* Use a window to select the console and the four attributes.

AutoCAD responds with "5 found."

The number of entities AutoCAD finds depends on how you constructed your drawing.

---

41. *Select objects:* ⏎

AutoCAD responds with "Block CONSOLE redefined. Regenerating drawing." The console disappears.

42. Repeat steps 35 through 41 for the monitor, keyboard, and printer. Your blocks should all disappear.

43. Insert each of your blocks. If you don't remember how, refer to Steps 13 through 16.

All four blocks are in place, but the attributes are missing! Not really — you just can't see them. Let's check them using the ATTDISP (ATTribute DISPlay) command.

44. *Command:* **ATTDISP** ⏎

45. *Normal/ON/OFF/<Normal>:* **ON** ⏎

Your blocks should now display the attributes you entered, as shown in Fig. 13-17. Remember that your attributes will be different than those shown here, depending on the values you entered for the attributes. Everything okay? If not, review your procedure before going on to the next part of the sequence: creating the bill of materials. If you need to make changes, explode the block, erase the offending attribute, and replace it using the ATTDEF command. Don't forget to save the block again. You may also use the ATTEDIT (ATTribute EDIT) command. This is a more powerful way to edit attributes, but it is also more difficult. Try it if you have time. When everything is set, turn the ATTDISP back to "Normal" and save your drawing with QSAVE.

46. Begin a new drawing named **EXT**.

47. *Command:* **INSERT** ⏎

48. *Block name (or ?):* **LIBCOM** ⏎

49. *Insertion point:* Pick a location so that the drawing appears centered on your screen. Press **ENTER** three times to accept the defaults.

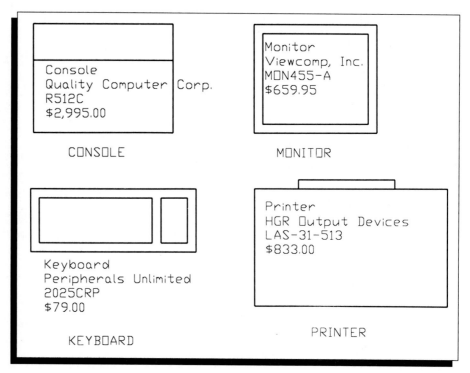

Fig. 13-17

50. *Command:* **ATTEXT** ⌐

The ATTEXT (ATTribute EXTraction) command creates a separate file of the attributes that can be read using another computer program, such as BASIC.

51. *CDF, SDF, or DXF Attribute extract (or Entities)? <C>:* Enter **D** for DXF.

The Create extract file dialogue box appears. Notice that the Pattern box shows a file extension of "dxx." AutoCAD uses this variation of the Data Interchange File (DXF) format to import and export extraction files among CAD systems. The other two options, CDF (Comma Delimited File) and SDF (Space Delimited File) are designed for importing the file into a database program such as dBASE IV or into a text editor. AutoCAD uses the "dxx" extension to distinguish text attribute extraction from DXF drawing files.

52. Pick **OK**.

AutoCAD responds with the number of entities in the extracted file.

 **Note**

You also may want to try using the DDATTEXT command to extract the attributes into a file.

Generating the Bill of Materials (Optional)

The bill of materials file now exists; however, to see the bill of materials, you must run the file. AutoCAD cannot do this. A special file has been written in the computer language BASIC for generating the bill of materials from a .DXX file. This file, called ATTEXT.BAS, is available on the disk package sold separately from this text. See the Introduction for ordering information.

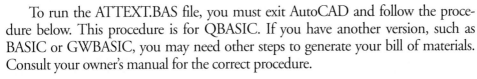

ACTIVITY

Your Turn

To run the ATTEXT.BAS file, you must exit AutoCAD and follow the procedure below. This procedure is for QBASIC. If you have another version, such as BASIC or GWBASIC, you may need other steps to generate your bill of materials. Consult your owner's manual for the correct procedure.

1. *Command:* **END** ⏎
2. At the C:> prompt, copy **QBASIC.EXE** (from your DOS directory), **ATTEXT.BAS** (from the data package), and **EXT.DXX** (from the directory where you store your AutoCAD drawings) into a single directory. This step is not absolutely necessary, but it will certainly make the generation of the bill of materials a lot easier.
3. Enter your new directory and start QBASIC.
4. From the File pull-down menu, pick **Open**.
5. From the list of files, doubleclick on **ATTEXT.BAS**.
6. Press **F5** or pick **Start** from the Run pull-down menu to run the program.
7. *Enter extract file name:* **EXT** ⏎

You should see a list similar to the one in Fig. 13-18, except your list will display the attributes you entered for your computer system.

DESCRIPTION	MFGR	MODEL	COST
Console	Quality Computer Corp	R512C	$2,995.00
Monitor	Viewcomp, Inc.	MON455-A	$ 659.95
Keyboard	Peripherals Unlimited	2025CRP	$ 79.00
Printer	HRG Output Devices	LAS-31-513	$ 833.00

Fig. 13-18

■ Chapter 13 Review

1. What information does a working drawing contain?
2. What is the difference between an exploded pictorial assembly drawing and a cutaway pictorial assembly drawing?
3. Give an example of a use for a cutaway pictorial assembly drawing.
4. How do you decide which views of an object to draw?
5. Why is a checker's job important?
6. What are some of the things that a checker looks for?
7. What is a bill of materials?
8. What are attributes in AutoCAD, and what role do they play in the generation of a bill of materials?

■ Chapter 13 Problems

1. Create working drawings of the objects shown in Problem 1. Notice that no dimensions are given for these objects. Use the approximate proportions shown. Decide which views are necessary, and create any detail drawings, sectional views, or auxiliary views that are appropriate. Add dimensions, and include notes if necessary.
2. Plot the drawings you created in Problem 1. Make sure your drawings are plotted to the correct scale.
3. Create a symbol library for office equipment. Give your drawing a descriptive name such as OFCEQUIP. Use the drawings shown in Problem 3. Define the following attributes for each object: DESCRIPTION, MFGR, MODEL, DATE PURCHASED, and COST. You may use actual data for the attribute values, or you may make up the manufacturer, model, date, and cost.
4. Extract the attributes from Problem 3 into a file called OFFICE.
5. (Optional) Generate the bill of materials for the office equipment.

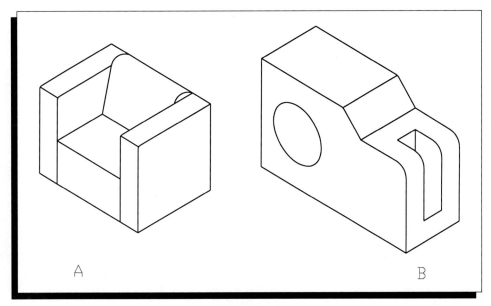

Problem 1

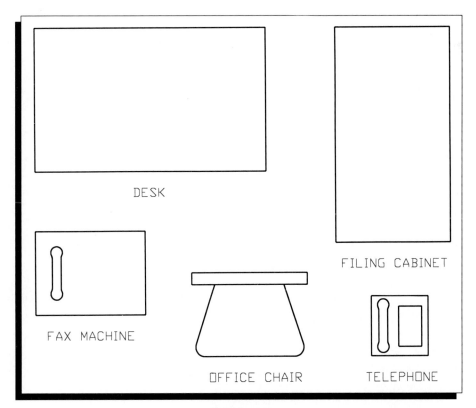

Problem 3

Using CAD

Behind the Scenes with AutoCAD

The next time you attend a theatre production, keep in mind that AutoCAD may have been used to make it happen. Dave Tosti-Lane, Chair of the Performance Production Department at Cornish College of the Arts in Seattle, Washington, is a theatrical technical director who uses AutoCAD to translate the ideas of a scenic designer into real stage scenery.

A designer arrives at the theatre with a set of drawings, often a painted sketch of the various scenes. The designer also provides a ground plan that shows the outlines of the various levels and walls. One of the technical director's many functions is to use these drawings to figure out how to build everything from platforms to walls.

Many problems must be solved, such as how to deal with heavy three-dimensional scenery or situations where things have to be flown into the air or dropped into a trap. The technical director then produces a set of working drawings from which each piece will be built. The drawings include all mechanical details, materials, etc., much like a contractor's set of drawings in a construction setting.

Every show requires a set of precise drawings. Shows run typically for about a month, so as soon as one begins Tosti-Lane starts creating drawings for the next one. In the educational setting, the runs are even shorter—one often has two or three shows to do in a single semester term. Theatre groups tend to recycle many components such as platforms and flats (walls) with some modifications for each production. This requires an extensive and detailed inventory.

Prior to CAD, Tosti-Lane had been drafting on a board and was tired of the drudgery of repetitive detail. He was determined to make the conversion to CAD from about 1983 on. He began with Generic CADD, which is now produced by Autodesk, producers of AutoCAD. In 1988, he bought a new computer and AutoCAD Release 9. The day it was delivered, he cut up his drafting board to use the wood to build a computer desk!

AutoCAD makes the inventory of existing components much simpler and eases work that is repetitive. With AutoCAD, Tosti-Lane finds he can afford the time to provide 3D perspective sketches of a piece that must be built, including enough detail to help the students who will build the scene. He includes precise notations about radius points and information on types of fasteners, glue, and other materials. Because the students don't have to ask questions, this information speeds up operations and saves them from making costly errors.

According to Tosti-Lane, "Most technical directors have some engineering background, a good grounding in math and science, an inveterate tendency to tinker, and a passion to know something about everything there is to know." They also share an unwillingness to sit in one place for very long and a fondness for things like computers — and AutoCAD.

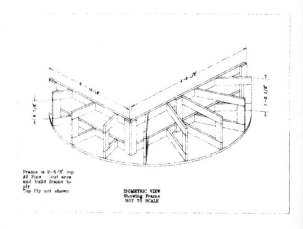

Frame is 2-5/8" rip
#2 Pine cut arcs
and build frame to
ply
Top Ply not shown

ISOMETRIC VIEW
Showing Frame
NOT TO SCALE

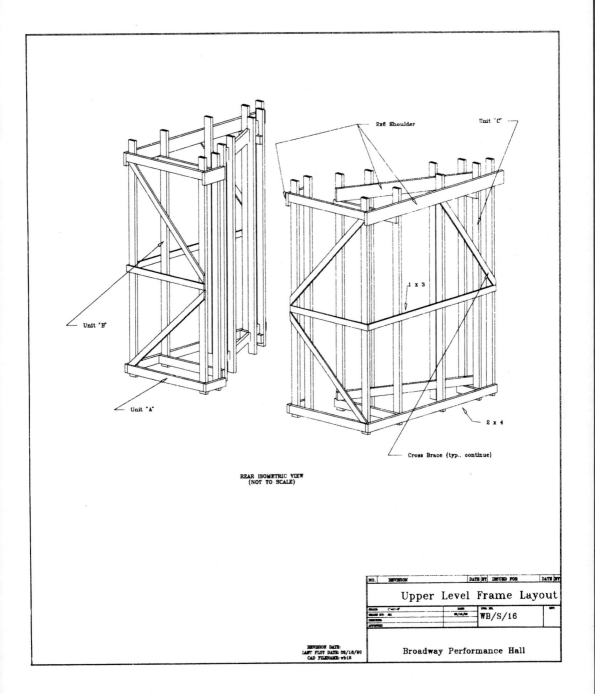

2x6 Shoulder

Unit "C"

Unit "B"

1 x 3

Unit "A"

2 x 4

Cross Brace (typ., continue)

REAR ISOMETRIC VIEW
(NOT TO SCALE)

REVISION DATE:
LAST PLOT DATE: 08/18/90
CAD FILENAME: wb16

NO.	REVISION	DATE	BY	ISSUED FOR	DATE	BY

Upper Level Frame Layout

SCALE: 1"=1'-0"	DATE: 08/18/90	DWG. NO. WB/S/16	REV.

Broadway Performance Hall

Chapter
14

▶

Key Terms

patterns

developments

stretchout line

parallel line
  developments

right prism

truncated prism

oblique prism

right cylinder

truncated cylinder

elbows

radial line
  developments

right rectangular
  pyramid

truncated right
  pyramid

right cone

truncated right cone

triangulation
  developments

oblique cone

transition pieces

intersections

piercing point

▶ ▶

Commands &
Variables

AREA

Developments and Intersections

Objectives

When you have completed this chapter, you will be able to:

- create parallel line developments.
- create radial line developments.
- create triangulation developments.
- find the intersections of various solids.

Many objects start as flat pieces and then are shaped, formed, and joined to create familiar objects we use every day. The hood of your car, the case for your glasses, a teapot, and the shirt or blouse you are wearing are common and obvious examples. Other examples that may not be quite so obvious include heating and air handling ducts, electrical outlet boxes, and metal chimneys. In their flat form, these objects are called **patterns**.

Examine a cereal box closely, and you will find an overlapping seam on one edge. Open the seam carefully with a knife. The cereal box unfolds into a flat form. (You may have to open the top and bottom seams, too) *(Fig. 14-1)*. This flat form, which ultimately becomes a cereal box, is the pattern.

Patterns can be made from almost any material, including cloth, paper, wood, and metal. They include the tabs and seam allowances that are necessary to put the object or product together.

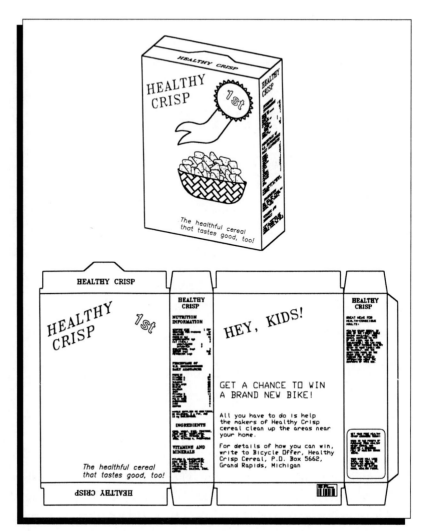

Fig. 14-1 A cereal box is made from a flat pattern such as the one shown here.

Developments

Developments are the individual shapes that make up a pattern. An object is made from a pattern, and a pattern is made up of one or more developments. Sometimes the development can be used as the pattern without any changes. Sometimes tabs must be added to the development to create the pattern. In some cases, developments must be combined to create the pattern. The sheet metal elbow in Fig. 14-19 is a pattern that consists of three developments. Developments are made around a **stretchout line** — a line that represents the opened-up length of the pattern *(Fig. 14-2)*. Examples of developments of the four basic solids are shown in Fig. 14-3.

Most objects fall into one of three types of developments: parallel line, radial line, or triangulation. Triangulations (shown later in the chapter) are used for developments that cannot be created by either parallel line or radial line developments.

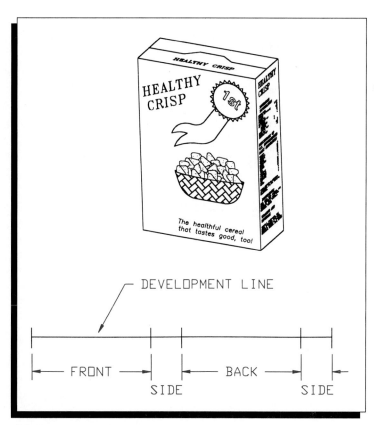

Fig. 14-2 The stretchout line represents the opened-up length of the pattern.

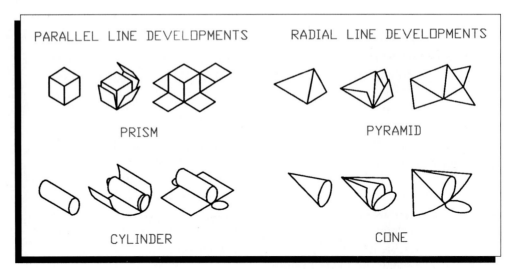

Fig. 14-3 Development of the four basic solid forms.

Sheet Metal Work

Developments are frequently associated with the sheet metal trade. Examples range from heating and air conditioning ducts and vents to automobile and aircraft parts and fabrications *(Fig. 14-4)*. Sheet metal can be formed with a break, shear, seamer, press, or even a mallet and vise *(Fig. 14-5)*. An important part of sheet metal pattern-making is determining what type of joints to make, where to place them, and how to bind raw edges *(Fig. 14-6)*.

Parallel Line Developments

Parallel line developments have parallel features. That means that when you roll them out, they make rectangular patterns. Parallel line developments include prisms and cylinders *(Fig. 14-3)*.

Fig. 14-4 Examples of sheet metal parts and fabrications.

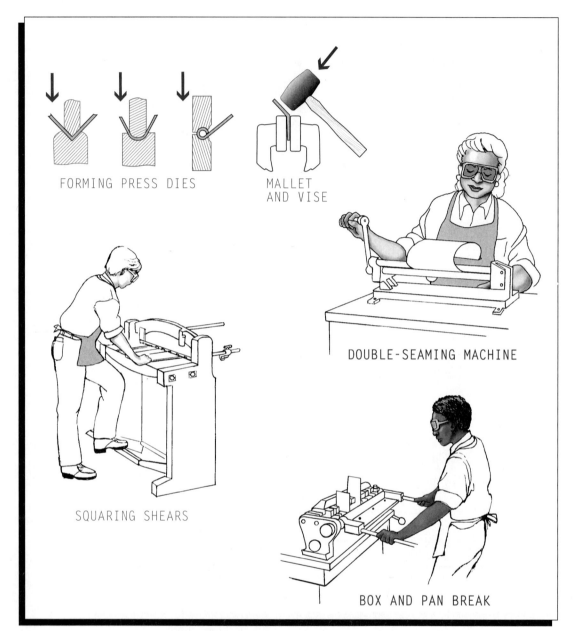

FORMING PRESS DIES

MALLET AND VISE

DOUBLE-SEAMING MACHINE

SQUARING SHEARS

BOX AND PAN BREAK

Fig. 14-5 Sheet metal forming equipment.

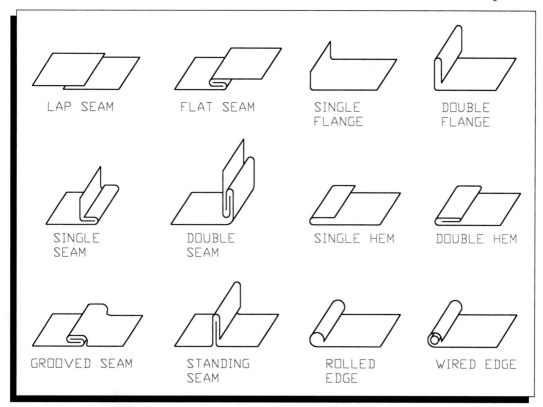

LAP SEAM

FLAT SEAM

SINGLE
FLANGE

DOUBLE
FLANGE

SINGLE
SEAM

DOUBLE
SEAM

SINGLE HEM

DOUBLE HEM

GROOVED SEAM

STANDING
SEAM

ROLLED
EDGE

WIRED EDGE

Fig. 14-6 Sheet metal seams and edge binding methods.

▶ Truncated Right Prisms

A **right prism** is a solid that has two parallel bases perpendicular to three or more sides. A **truncated prism** is one in which part of the prism is cut off at an angle. Fig. 14-7 shows an example of a right prism that has been truncated.

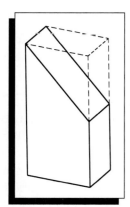

Fig. 14-7 A truncated right prism is one in which the top of the prism has been cut off at an angle that is not parallel to the base.

Use the following procedure to develop a truncated right prism using AutoCAD *(Fig. 14-8)*.

Note

Study each step in this and the next few procedures carefully. Later in this chapter, you will build on these techniques to create more complicated patterns. An understanding of the basics is essential.

1. Begin a new drawing called **PRISM**.
2. Create the front and top views as shown in Fig. 14-8A.
3. Draw the horizontal stretchout line. The stretchout line is the sum of the width of the sides. It lies at the base of the prism. Refer again to Fig. 14-8A.

Because the stretchout line begins with the back of the prism, the seam of the development will be between the back and the right side. Further, since the right side is 1.0 units high, 1.0 will be the height of the seam.

4. Draw a 1.0-unit vertical line at the beginning of the stretchout line. Draw a similar line at the other end of the stretchout line. These lines represent the seam where the development will be joined *(Fig. 14-8B)*.
5. Create the three vertical fold lines. Remember that the right side of the prism is 1.0 unit high, and the left side is 2.5 units high. Therefore, the fold line between the back and left sides will be 2.5 units high. The fold line between the left and front sides will also be 2.5 units high. The fold line between the front and right sides, however, will be 1.0 unit high. Refer again to Fig. 14-8B.
6. Connect the vertical lines to create the top of the side pieces *(Fig. 14-8C)*.
7. Add the bottom surface and the top surface of the truncated edge. Check the dimension on the top view of the prism to get the width of these surfaces: each should be 1.0 unit wide *(Fig. 14-8D)*.

Hot Tip

The easiest way to construct the top surface in AutoCAD is to offset the top of the side by the width of the top. In other words, offset the top of the side by 1.0 unit. Then use the LINE command and the ENDpoint object snap to join the two surfaces.

The important surfaces of a development are the ones that form the object. The placement, design, and fold line location of the cover surfaces—the top and bottom — may be optional, depending on the nature and application of the development. For example, the development for a cereal box has a split top and bottom with side flaps; a suburban mailbox has a bottom but no top; and some air conditioning developments have neither top nor bottom.

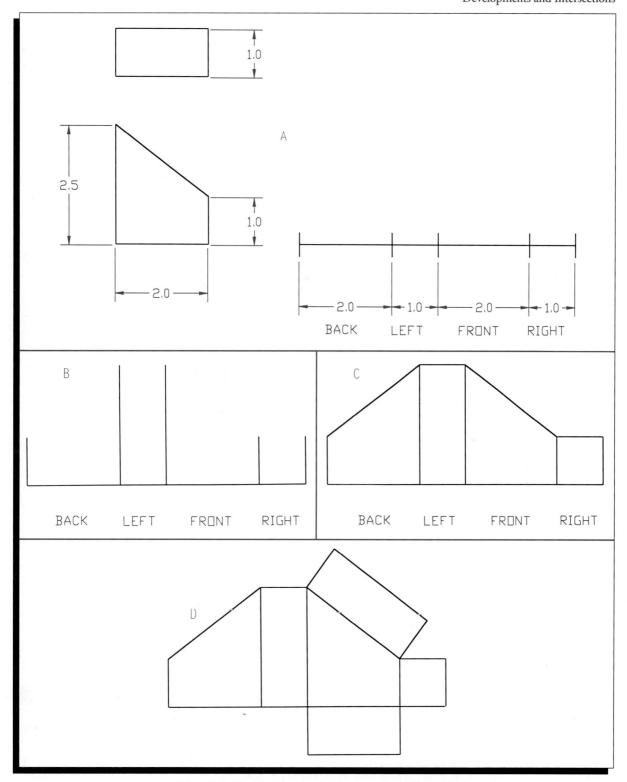

1.0

2.5

1.0

2.0

A

2.0 — 1.0 — 2.0 — 1.0

BACK   LEFT   FRONT   RIGHT

B

BACK   LEFT   FRONT   RIGHT

C

BACK   LEFT   FRONT   RIGHT

D

Fig. 14-8 Development of a truncated right prism.

▶ Oblique Prisms

An **oblique prism** is a prism in which neither the top nor the bottom (base) forms a 90-degree angle with the sides. It is a sloping prism *(Fig. 14-9)*. Since the base angle is not 90 degrees, the development does not unfold horizontally. Place the stretchout line through the center of the prism perpendicular to the oblique angle *(Fig. 14-10)*.

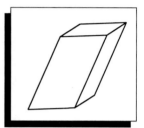

Fig. 14-9 An oblique prism is one in which the sides are not perpendicular to the bases.

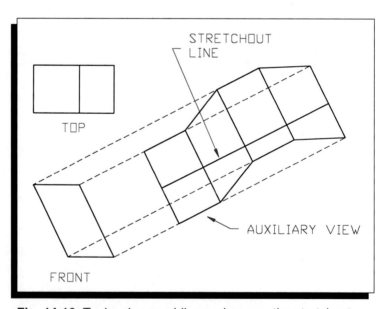

Fig. 14-10 To develop an oblique prism, run the stretchout line through the center of the development.

ACTIVITY

Use the following procedure to draw the development of an oblique prism. For simplicity, the covers will not be included *(Fig. 14-11)*.

1. Begin a new drawing called **OBLIQUE**.

2. Construct the front and top views as shown in Fig. 14-11A.

3. Use the **SNAP** command to rotate the grid to the oblique angle (**30°**).

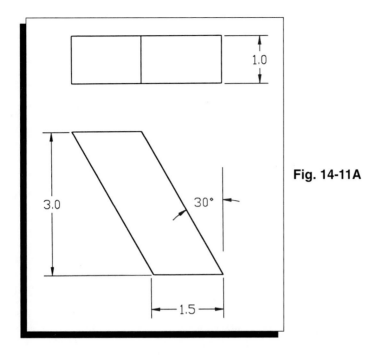

Fig. 14-11A

4. Draw a cutting plane (part of which will become the stretchout line) through the front view perpendicular to the oblique angle *(Fig. 14-11B)*.

5. Draw the auxiliary sectional view *(Fig. 14-11C)*.

The auxiliary sectional is necessary to establish the true width of the sides.

6. Extend the cutting plane to create the stretchout line.

7. Use a procedure similar to the one you used to make the development of the right prism in the previous "Your Turn" *(Fig. 14-11D)*.

The easiest way to create the pattern is to establish the top and bottom lines first. You probably extended a construction line from the corners of the front view to create your auxiliary view. Extend each of those construction lines and create additional construction lines as needed. Then all you have to do is figure out the angles at which to draw the lines.

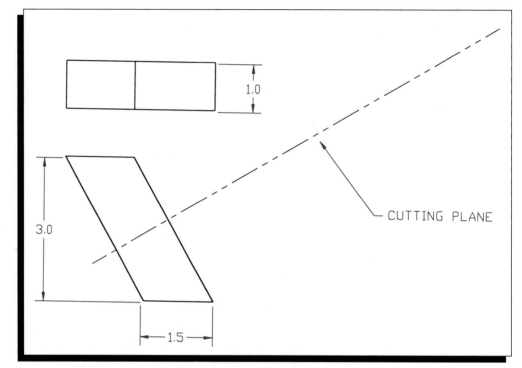

Fig. 14-11B

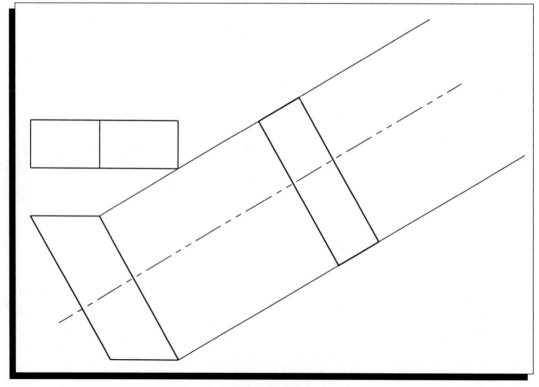

Fig. 14-11C

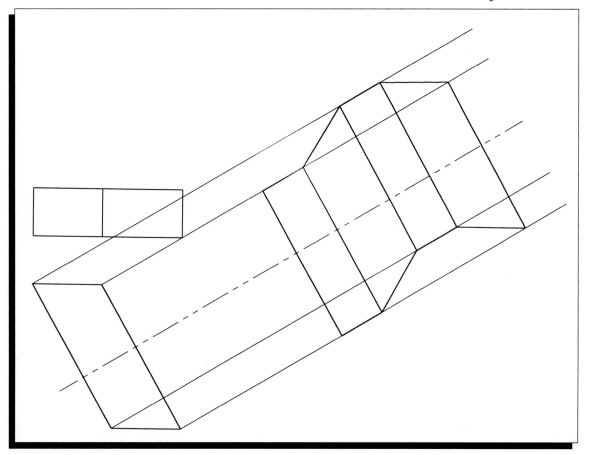

Fig. 14-11D

▶ Right Cylinders

A **right cylinder** is a solid with two bases perpendicular to one circular side *(Fig. 14-12)*. Two views are necessary to develop a right cylinder. The front view gives the height, and the top view gives the diameter. The length of the stretch-out line equals the circumference of the cylinder. You can find the circumference by multiplying the diameter by pi, which is approximately 3.1416. However, when you are working in AutoCAD, it is often faster and simpler to use the AREA command to find the circumference.

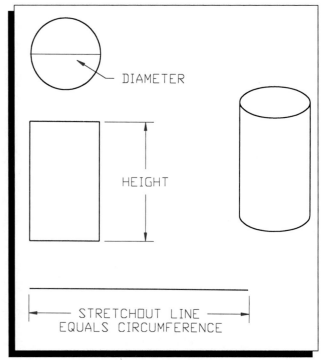

Fig. 14-12 A right cylinder.

ACTIVITY

Follow the steps below to make a development of a right cylinder *(Fig. 14-13)*.

1. Begin a new drawing named **CYLINDER**.
2. Draw the top and front views as shown in Fig. 14-13.
3. *Command:* **AREA** ⏎
4. *<First point>/Entity/Add/Subtract:* **E** ⏎
5. *Select circle or polyline:* Select the circle. AutoCAD calculates and displays the circumference and the area of the circle.
6. Draw the stretchout line using the circumference of the circle for its length.
7. Offset the stretchout line by 3 units to create the top of the cylinder.
8. Draw vertical lines to connect the stretchout line with the top of the cylinder at each end. These lines represent the seam edges at the ends of the development.

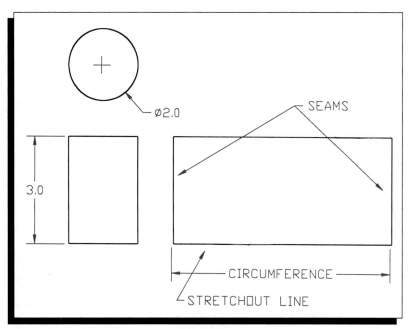

Fig. 14-13

▶ Truncated Cylinders

A **truncated cylinder** has one end cut at an angle other than 90 degrees *(Fig. 14-14)*. The cut end results in a curved line in the development. The curved line is an irregular curve; therefore, the development is only an approximation.

A cylinder may be considered a multi-sided prism *(Fig. 14-15)*. The sides are called elements. If there were an infi-

nite number of elements, the cylinder would be completely round. However, from a drafting standpoint, this is not practical. Drafters limit the number of elements to give a reasonable approximation of a cylindrical development. This approximation is used to establish the curve in the development of a truncated cylinder.

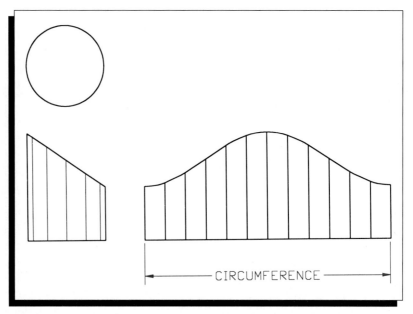

Fig. 14-14 The cut of a truncated cylinder will result in a curved surface in the development.

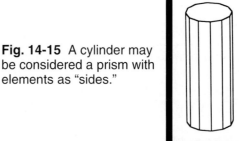

Fig. 14-15 A cylinder may be considered a prism with elements as "sides."

ACTIVITY

To make an approximate development of a truncated right cylinder, follow the steps below *(Fig. 14-16)*.

1. Begin a new drawing named **CYL2**.
2. Draw the front and top views of the cylinder *(Fig. 14-16A)*.
3. Change the **PDMODE** variable to 3, and change **PDSIZE** to **0.1**.
4. Use the **DIVIDE** command to divide the circle into twelve equal elements.

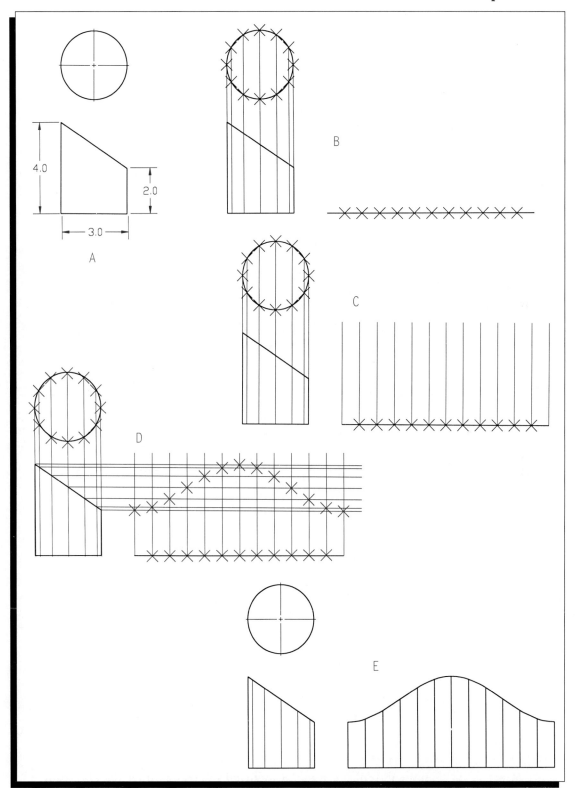

Fig. 14-16 The development of a truncated right cylinder.

When you work with similar developments in the future, remember that you can use any number of elements. Twelve is convenient for our application.

5. Project each point from the top view onto the front view. Notice that the left-most point corresponds to the left side of the front view, and the right-most point corresponds to the right side of the front view.

Begin at the left-most or right-most point and project lines from all the points on the top half of the circle first. Notice that the lines you create run directly through the points on the lower half of the circle, so you do not need to create additional projection lines through those points.

6. Draw the stretchout line and divide it into twelve parts representing the twelve elements from step 3 *(Fig. 14-16B)*.
7. Draw perpendicular (vertical) lines through each of the points on the stretchout line. Make sure each vertical line extends at least as high as the left side of the cylinder in the front view *(Fig. 14-16C)*.

The fastest way to do this is to draw a line through one end of the stretchout line. Then use the Through option of the OFFSET command to offset the line through each of the points on the line. You can do this quickly and easily by setting the running object snap to NODe.

8. Project the points on the sloping edge of the front view to the corresponding vertical lines on the stretchout *(Fig. 14-16D)*.

Remember that each projection line, except those from the left-most and right-most points, runs through two points in the top view — one on the top half of the circle and one on the lower half. Therefore, each of your projection lines, except for the line from the left-most point, corresponds to the position of a point on *two* of the vertical lines on the development. (The right-most point corresponds to both sides of the seam, so it must correspond to two points also.)

9. Use a polyline to connect each of the horizontal and vertical intersections on the stretchout. You should make this one continuous polyline.
10. Trim and erase the vertical lines as needed.
11. Set **PDMODE** back to **0**.
12. Smooth the polyline into a curve that fits the points using the **Fit** option of the **PEDIT** command *(Fig. 14-16E)*.

▶ Elbows

Elbows are short pieces used to change the direction of ducts and pipes *(Fig. 14-17)*. They are common in sheet metal duct work. The number of elbow sections in a bend is determined by the size of the duct, the angle of the bend, and the purpose of the duct. Develop elbows from the front view. This view will show the number of pieces and the true length of the elements of each piece.

Elbows have two primary radii. The heel is the larger, outer radius, and the throat is the smaller, inner radius *(Fig. 14-18)*. These determine the physical size of the bend.

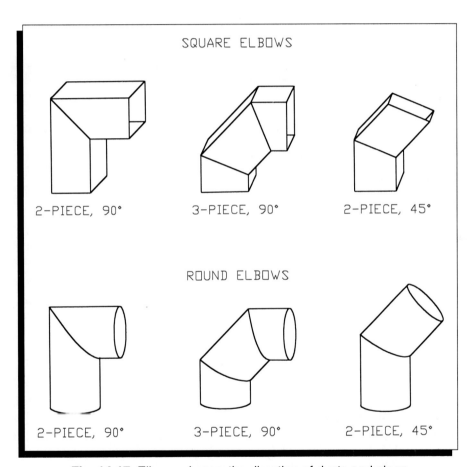

SQUARE ELBOWS

2-PIECE, 90°   3-PIECE, 90°   2-PIECE, 45°

ROUND ELBOWS

2-PIECE, 90°   3-PIECE, 90°   2-PIECE, 45°

Fig. 14-17 Elbows change the direction of ducts and pipes.

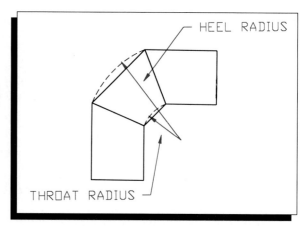

Fig. 14-18 The inner radius of an elbow is known as the throat radius. The outer radius is the heel radius.

ACTIVITY

Use the following procedure to develop a three-piece elbow pattern for a 90-degree bend. When you finish, your elbow and patterns will look like those in Fig. 14-19A.

1. Draw the throat radius. To begin, create a horizontal line from 3.5, 3.5 to 7,3.5 and a vertical line from 7,3.5 to 7,7. This will confine the boundaries of the 90-degree elbow.

2. Create the throat radius using the ARC command. For this example, use 7,4.75 for the starting point. Select the End option and select the point 5.75, 3.5. Then use the Radius option and select the intersection of the two boundary lines. This will ensure that your elbow will be regular in shape.

3. Offset the throat radius to create the heel radius. For this example, we want the diameter of the elbow to be 1.5, so use an offset distance of 1.5 units *(Fig. 14-19B)*.

Notice that the offset arc extends to the vertical and horizontal boundary lines.

4. Calculate the location of the joints using the reasoning that follows.

The middle piece of a three-piece elbow should be twice as big as each end piece. The angle to the joint is equal to the angle of the bend /(number of pieces × 2) − 2. In this case, 90/(3 × 2) − 2 = 22.5. One joint will be 22.5° from vertical, and the other joint will be 22.5° from horizontal. The midpoint of the middle piece will be 22.5° from both joints, or 45° from horizontal and from vertical *(Fig. 14-19C)*.

5. Extend a line from the intersection of your boundary lines 22.5° to the left of the vertical boundary line. Make the line 3.5 units long so it will intersect both radii. Since the vertical line is at 90°, you will extend the line @3.5<112.5. This marks the point on both arcs where the top joint of the elbow will occur.

6. Now do the same thing for the other joint. It should be 22.5 degrees above the horizontal boundary line, or 45° below the first joint, so you can calculate that the line should be at 112.5° + 45° = 157.5°.

7. Place a center line from the intersection of the boundary lines through the center of the middle section of the elbow (at 135°).

8. Draw tangents to the heel and throat radii at the horizontal, vertical, and 45° angle points. This is a front view of the elbow *(Fig. 14-19D)*.

Because the elbow extends exactly 90 degrees, the tangents at its ends are easy to find. The tangents at the intersection of the vertical boundary line and the arcs will be horizontal lines. The tangents at the intersection of the horizontal boundary line and the arcs will be vertical lines.

To find the tangent at the midpoint of the middle section, create a line from the intersection of the lower 22.5° line and the vertical tangent to the intersection of the upper 22.5° line and the horizontal tangent *(Fig. 14-19E)*. Remove the arcs used to create the throat and heel radii.

---

9. Draw a half view of the lower end of the duct.

Use the ARC command with the End option to create the half view.

---

10. Divide the arc into six equal parts. Remember to change PDMODE and PDSIZE to see the points AutoCAD creates.

11. Project each of the five points and draw elements on all three sections *(Fig. 14-19F)*.

Use the NODe object snap and be sure Ortho is on to construct the elements on the lower part of the elbow. For the middle part, use the DIST command to find how far apart the lines are. Offset the outer edge of the middle part of the elbow to create the element lines. Then use the INTersection object snap and Ortho to create the elements for the upper section of the elbow.

---

12. Draw one stretchout line equal to the circumference of the duct for each pattern.

The circumference of a circle with a 1.5-inch diameter is 4.71 inches. The fastest way to find the correct circumference is to create a circle with the diameter of the elbow and use AutoCAD's AREA command to find the circumference.

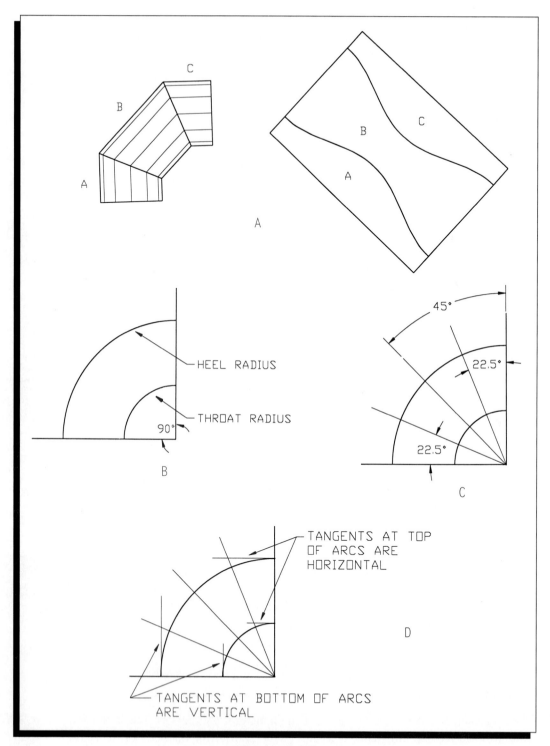

Fig. 14-19A-D

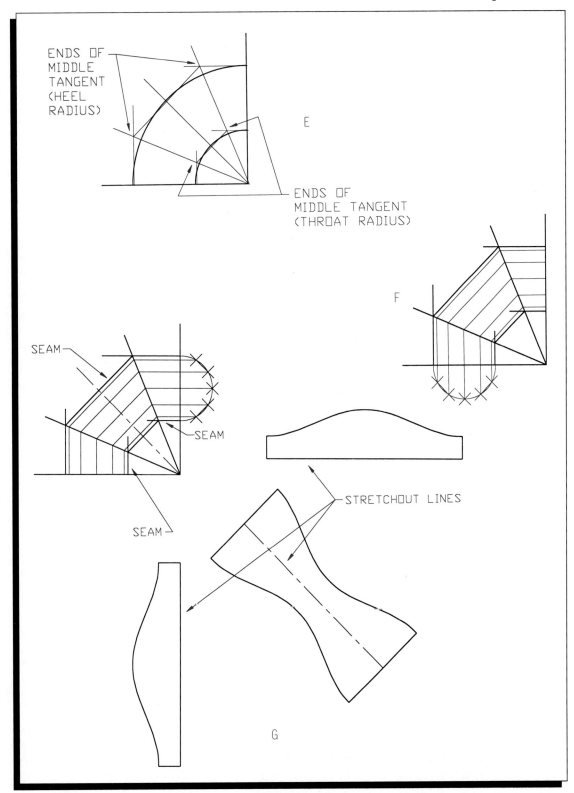

ENDS OF MIDDLE TANGENT (HEEL RADIUS)

E

ENDS OF MIDDLE TANGENT (THROAT RADIUS)

F

SEAM

SEAM

SEAM

STRETCHOUT LINES

G

Fig. 14-19E-G

431

13. Develop the patterns as truncated cylinders. To do this, extend projection lines from the elements on the elbow. Extend them down for the top part of the elbow, horizontally for the lower part, and at a 45° angle for the two ends of the middle section.

Rotate Snap to 45° to extend projection lines for the center section. Treat the center section as a double truncated cylinder with the stretchout line as a center line *(Fig. 14-19G)*.

14. Trim construction and boundary lines and erase projection lines to finish the drawing.

   Although all three patterns were developed separately, their curved surfaces are complementary. They can be nested and cut from one rectangular piece. The patterns were developed so that the seams on the end pieces are on the throat radius and the seam of the center piece is on the heel radius *(Fig. 14-19G)*.

Radial Line Developments

Radial line developments rotate from a single point and roll out into pie-shaped patterns. They include cones and pyramids. Fig. 14-20 shows examples of typical radial line developments.

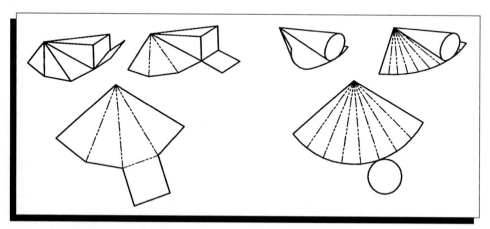

Fig. 14-20 Radial line developments for pyramids and cones.

▶ Right Rectangular Pyramid

A **right rectangular pyramid** is a pyramid built on a rectangular base with its vertical axis perpendicular to the base. The faces, or sides, of the pyramid intersect at a common point called a vertex *(Fig. 14-21)*.

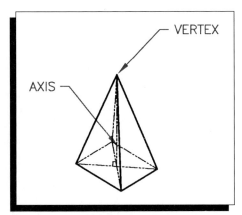

Fig. 14-21 A right rectangular pyramid.

ACTIVITY

Draw the development of a right triangular pyramid by following these steps. Refer to Fig. 14-22.

For this and the remaining "Your Turn" sections in this chapter, many of the steps you have already learned are not described in detail. For example, a step may say "Project the lines ..." without telling you precisely how to do so. If you need to refresh your memory, go back to earlier "Your Turn" sections.

For each of the "Your Turns" that follows, remember to:

- create a new drawing. Give the drawing a descriptive name.
- save your work often. These are fairly complicated developments, and you can lose a large amount of work if you do not save the drawings frequently.

1. Draw the top and front views. Use the dimensions shown in Fig. 14-22A, but do not include the dimensions on your drawing.
2. Find the true length of one edge by revolution *(Fig. 14-22B)*.
3. Swing a stretchout arc with a diameter equal to the true length of the edge.

The stretchout arc is the radial equivalent of a stretchout line in parallel line developments.

4. Draw a vertical diameter of the arc.
5. Starting at the intersection of the arc and the vertical diameter, step off chords equal to the four base edges *(Fig. 14-22C)*.
6. Connect the center of the arc with the ends of each chord.
7. Draw the bottom of the pyramid. You may connect the bottom to any of the base edges *(Fig. 14-21C)*.

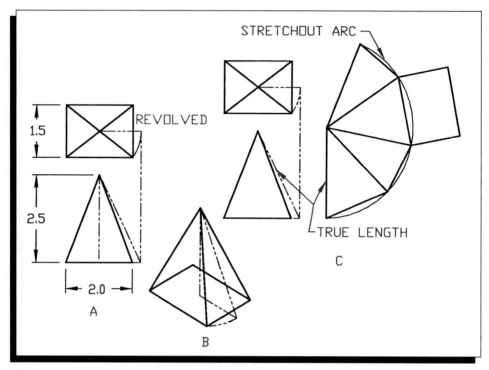

Fig. 14-22

▶ Truncated Right Pyramid

A **truncated right pyramid** is a right pyramid with the top cut off at an angle to its axis *(Fig. 14-23)*. The procedure for developing a truncated right pyramid is the same as that for developing a right rectangular pyramid except that you must first find the true lengths of the edges of the pyramid. Begin by developing the base of the development just as you did for a right pyramid. Then find the true lengths of the edges of the pyramid and transfer these lengths to the lines that connect the chords to the vertex in the development. Then draw a line to connect the points to show the top, or truncated, surface of the pyramid. Fig. 14-24 shows the steps in the development of a truncated right pyramid.

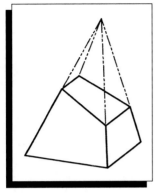

Fig. 14-23 A truncated right pyramid is one in which the top has been cut off at an angle to the base.

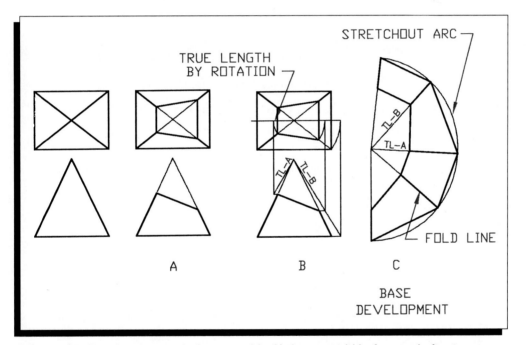

Fig. 14-24 The development of a pyramid: A) the pyramid before and after truncation; B) base development and rotation of truncated lines; C) the development.

▶ Right Cone

A cone is a cross between a pyramid and a cylinder; it is like a cylinder except it is pointed at one end. A **right cone** is one in which the axis is perpendicular to the base *(Fig. 14-25).*

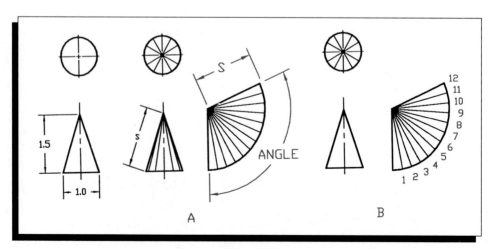

Fig. 14-25 In a right cone, the axis is perpendicular to the base.

Your Turn

ACTIVITY

Draw the development of a right cone by following these steps. Refer to Fig. 14-26.

1. Draw the front and top views using the dimensions shown in Fig. 14-26A.

2. Use the **AREA** command to find the circumference of the base in the top view.

3. Draw a stretchout arc with a radius equal to the true length slope (S) of the cone *(Fig. 14-26A)*.

4. Find the angle of the arc using this formula:

   angle = radius of base/(S) × 360

5. Divide the stretchout arc into 12 equal pieces.

6. Connect the 12 points on the arc to the center of the arc *(Fig. 14-26B)*.

Fig. 14-26

▶ Truncated Right Cone

A **truncated right cone** is a right cone that has been cut at an angle other than 90 degrees *(Fig. 14-27)*. The truncated right cone can be considered a pyramid with an infinite number of sides or elements, just as the truncated right cylinder can be considered a prism with an infinite number of elements *(Fig. 14-28)*. For ease of development, drafters restrict the number of elements.

The steps to develop a truncated cone are similar to those used to develop a truncated prism. However, to develop a truncated cone, you must find the true length of the elements instead of the true length of the edges. Fig. 14-29 shows the steps in the development of a truncated right cone.

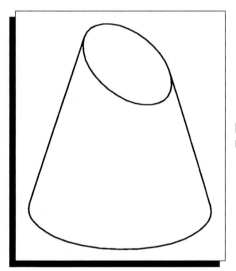

Fig. 14-27 A truncated right cone.

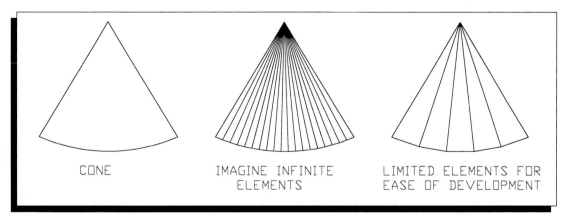

CONE   IMAGINE INFINITE   LIMITED ELEMENTS FOR
                   ELEMENTS   EASE OF DEVELOPMENT

Fig. 14-28 A cone can be considered a pyramid with an infinite number of sides. For convenience, drafters limit the number of sides, or elements, in a development.

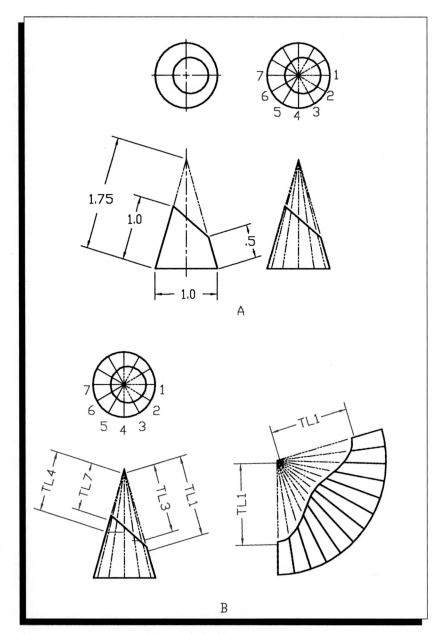

Fig. 14-29

Triangulation

Triangulation developments divide surfaces into a series of triangles. To create developments by triangulation, you must find the true length of each side of the triangles and then draw the triangles next to each other on a flat surface.

Triangulation developments are approximate. They are used for oblique cones and for transition pieces made up of curved and plane surfaces and different polygons. (You will learn more about transition pieces later in this section) *(Fig. 14-30)*.

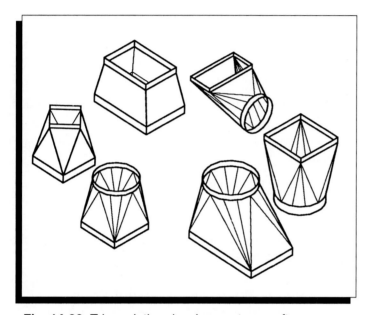

Fig. 14-30 Triangulation developments are often necessary for transition pieces. Transition pieces are connectors that allow you to join a piece of pipe or duct that has one shape or size to another piece that has a different shape or size.

▶ Oblique Cones

An **oblique cone** is a cone in which the vertical axis is not perpendicular to the base. Oblique cones can have a circular or elliptical base.

ACTIVITY

Cones with either circular or elliptical bases can be developed using the following procedure. This example assumes a cone with a circular base. Refer to Fig. 14-31.

1. Draw the front and top views *(Fig. 14-31A)*.
2. Divide the circular top view into 12 equal parts.
3. Transfer the points to the base of the front view.

It may help you to number the points as shown in Fig. 14-31 to keep track of the resulting triangles.

4. Draw the elements from the base to the vertex of the cone *(Fig. 14-31A)*.

This creates a series of triangles on the front view. Note that the elements on the left edge and the right edge are the only ones that show true length.

5. Find the true length of the five foreshortened elements by revolution *(Fig. 14-31B)*.

The following steps create the first triangle in the pattern. Refer to Fig. 14-31C as you perform these steps.

6. Draw a vertical center line and mark off the length of the longest element (element number 7). The top of this line will be the vertex of all the triangles transferred.
7. From the bottom of the element, strike an arc to the left equal to the chord of the points in the top view.
8. From the vertex of the pattern, strike an arc equal to the true length of the next longest element taken from the true length diagram.

These two arcs intersect at the end of the next element on the pattern.

9. Connect the point at the intersection of the two arcs to the vertex of the pattern.
10. Repeat steps 7 through 9 until all the triangles, or elements, are transferred to the pattern *(Fig. 14-31D)*.
11. Connect the ends of the elements with a polyline.
12. Use PEDIT and Fit to smooth the polyline into a curve *(Fig. 14-31E)*.

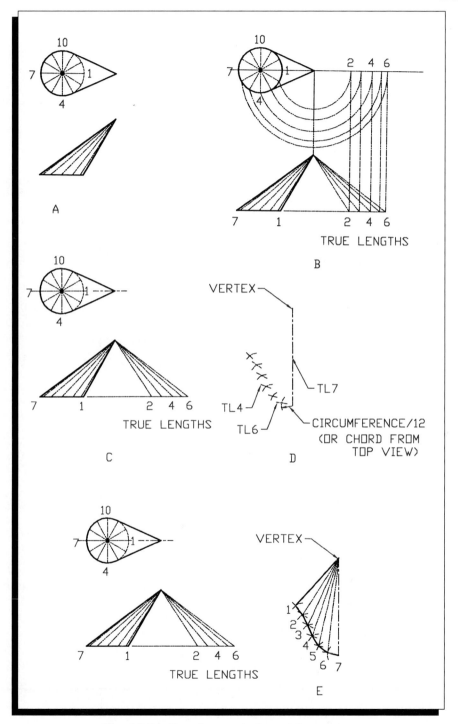

Fig. 14-31 Development of an oblique cone.

▶ Transition Pieces

Transition pieces are pieces that connect ducts and pipes of different shapes and sizes. They connect flat pieces, curved pieces, or a combination of flat and curved pieces. Refer again to Fig. 14-30.

The development of a flat-to-curved transition piece is similar to that of an oblique cone. The curved section resembles a cone, and the flat section resembles a pyramid. The following "Your Turn" will help you see the resemblance.

ACTIVITY

Follow this procedure to develop a symmetrical rectangular-to-round transition piece. Refer to Fig. 14-32.

1. Draw the top and front views as shown in Fig. 14-32A.
2. Divide the circular part of the top view into 12 equal parts.
3. Project the divisions to the front view, and draw the elements to their respective corners.
4. Revolve the elements to obtain their true lengths *(Fig. 14-32B)*.
5. Locate the seam in the center of the large triangle on the right side *(Fig. 14-32C)*.

The procedure works regardless of which side you use.

6. Draw the seam line to begin developing the pattern.
7. Draw the base of the triangle perpendicular to the seam line.
8. Draw the hypotenuse of the triangle. This is the first true length line *(Fig. 14-32C)*.
9. Lay out the three triangles, forming the corner.
10. Lay out the large triangle *(Fig. 14-32C)*.

The true length of the base comes from the top view. The sides are the same length. Since the transition piece is symmetrical, the true length lines of one corner are the same as the respective lines in each of the other corners.

11. Repeat steps 9 and 10 until the pattern is complete.
12. Draw a three-point arc for the circular opening at the top of the pattern *(Fig. 14-32D)*.

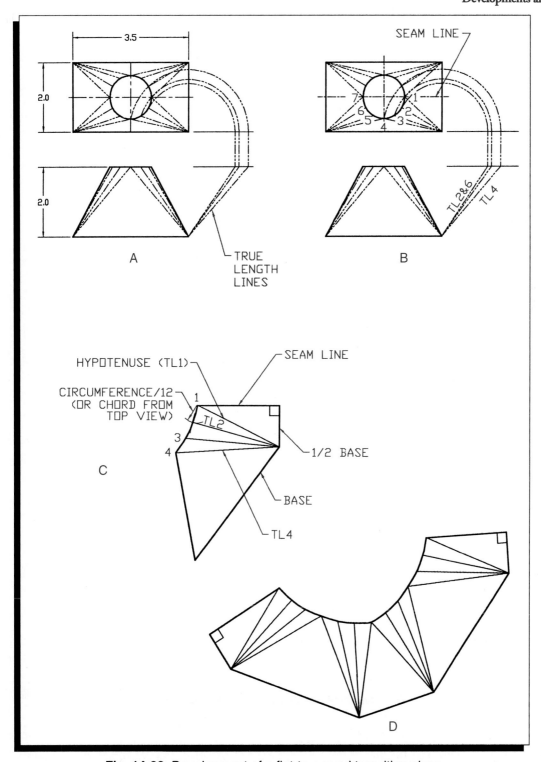

Fig. 14-32 Development of a flat-to-curved transition piece.

Intersections

Intersections are places where two or more surfaces meet. In other words, an intersection is formed by joining two or more surfaces *(Fig. 14-33)*.

The intersection of two plane surfaces is a straight line. The intersection of a plane surface and a curved surface or of two curved surfaces is a curved line. A **piercing point** is the intersection of a line with a straight or curved surface *(Fig. 14-34)*.

To create a pattern for an intersection, you must first draw the intersection. Then you can develop the pattern.

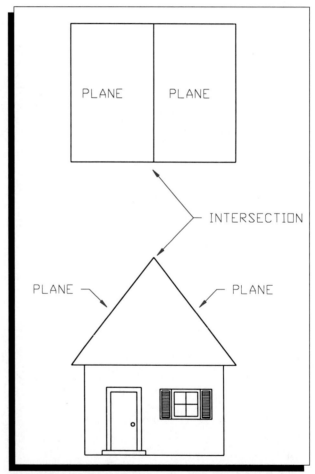

Fig. 14-33 Intersections are formed where two surfaces meet.

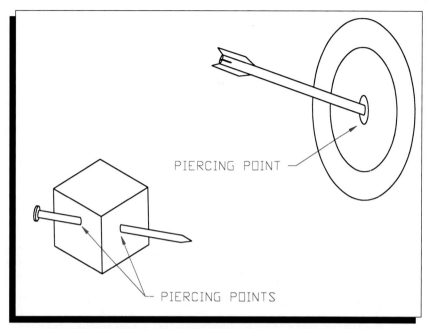

Fig. 14-34 A piercing point is formed when a line intersects a surface. Imagine the arrow and the nail to be lines. The point at which they intersect the planar surfaces are piercing points.

Prisms

To find the intersection of two prisms, locate the piercing points of one prism on the surfaces of the other prism *(Fig. 14-35)*. Draw straight lines to connect the points. These lines are the lines of intersection.

Fig. 14-36 shows the development of two prisms that intersect at right angles. Note that you can use the same procedure for prisms that intersect at oblique angles. For the intersection shown in Fig. 14-36, you need the top, front, and left-side views. Draw separate stretchout lines for the large and small prisms. Develop the prisms, being sure to use the true length of each line.

For the large prism, offset the height. Then draw the seam edges and the true width of the sides. (Take the true width from the top view.) Then locate the piercing points. You can locate them vertically by projecting the edge intersections from the front view. Locate them horizontally by measuring their true distances in the top view. Connect the piercing points with straight lines.

Follow the same procedure for the small prism. Use the appropriate views to find the true lengths and piercing points.

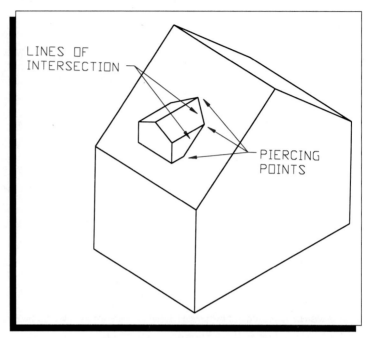

Fig. 14-35 Intersecting prisms.

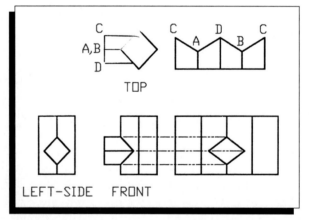

Fig. 14-36 The development of two intersecting prisms.

Cylinders

To find the points on a line of intersection, sometimes you need to use cutting planes similar to those used in sectional views. Cutting planes are especially useful with cylindrical objects.

The line of intersection of two cylinders can be found by using a series of cutting planes *(Fig. 14-37)*. The planes are equally spaced, and one plane usually passes through the center line of the cylinders.

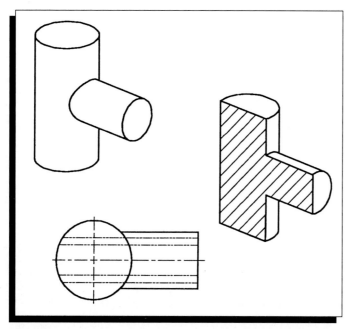

Fig. 14-37 A cutting plane for two intersecting cylinders.

ACTIVITY

The following procedure uses cutting planes to find the line of intersection of two cylinders. It also creates a development of the intersecting cylinders *(Fig. 14-38)*. This procedure can be used for cylinders at right angles or cylinders oblique to each other.

1. Draw the top and front views as shown in Fig. 14-38A.
2. Draw a half view of the small cylinder in the front view.
3. Pass five cutting planes through the top view. Remember that one of the cutting planes should be located through the center of the cylinder. Then pass two cutting planes through the front half view *(Fig. 14-38A)*.

The cutting planes will be used with the outer surfaces (diameter) of the small cylinder.

4. Project the intersecting points of the large cylinder and cutting planes in the top view vertically to the front view. Project the intersecting points of the half view arc and the cutting planes horizontally to the left *(Fig. 14-38B)*.

These lines cross at points on the line of intersection.

5. Connect the points with a polyline.
6. Smooth the polyline into a curve using the **Fit** option of the **PEDIT** command.
7. Draw the stretchout line for the large cylinder *(Fig. 14-38C)*.
8. Draw the vertical center line and lay out the cutting planes from the top view.

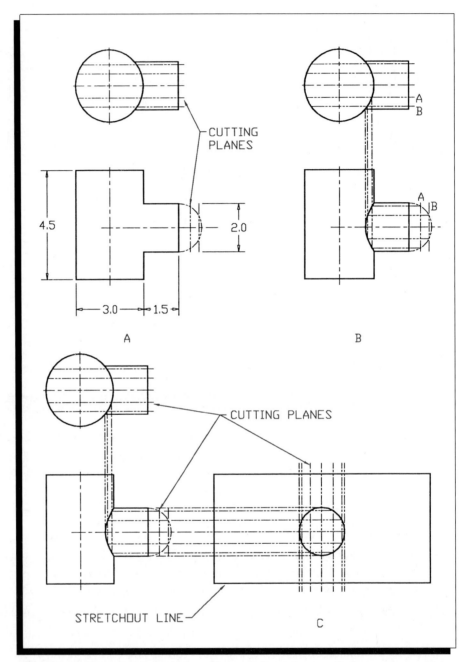

Fig. 14-38 The development and intersection of two cylinders using cutting planes.

9. Project the points from the line of intersection to their respective cutting planes on the development.

10. Connect the points with a polyline.

11. Use the **PEDIT** command to smooth the curve *(Fig. 14-38C)*.

Cone and Cylinder

The intersection of a cone and cylinder can be found by either the horizontal cutting plane method or the radial cutting plane method. The horizontal cutting plane method cuts the cone and cylinder across the axis of the cone to form a series of circles in the top view.

The cutting planes are parallel to the base of the cone *(Fig. 14-39)*. The radial cutting plane method cuts the cone vertically radiating from its vertex. The cutting planes are perpendicular to the base of the cone *(Fig. 14-40)*.

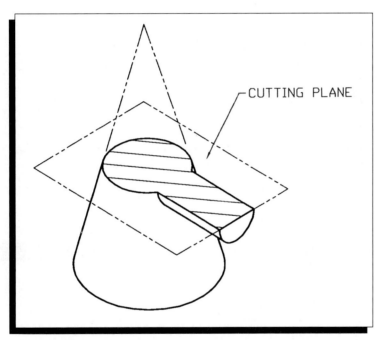

Fig. 14-39 A cone and cylinder with the cutting plane parallel to the base of the cone.

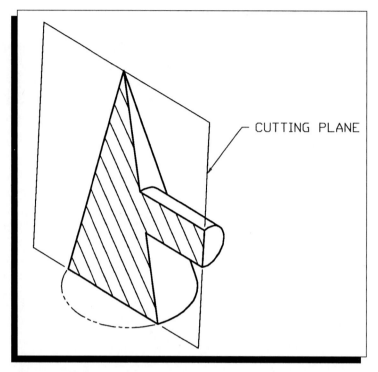

Fig. 14-40 A cone and cylinder with the cutting plane radiating from the vertex of the cone.

ACTIVITY

The following procedure shows how to find the intersection of a cylinder and a cone using first the horizontal cutting plane method *(Fig. 14-41)*.

1. Draw the top, front, and right-side views as shown in Fig. 14-41A.

You will need the same three views of this object later when you try the radial cutting plane method. To save time, save the drawing twice (with two different names) after you draw the three views but before you do any of the other steps. Call one drawing HORIZ and the other RADIAL. Use HORIZ to continue the steps in this procedure.

2. Draw seven horizontal cutting planes in the right-side view, including one on the center line and one each tangent to the top and bottom of the cylinder.

3. Project the cutting planes horizontally to the front view and then vertically to appear as circles in the top view *(Fig. 14-41A)*.

4. On the side view, measure the horizontal distance from the vertical center line to where each cutting plane crosses the cylinder. Lay out these distances on the cylinder in the top view *(Fig. 14-41B)*.

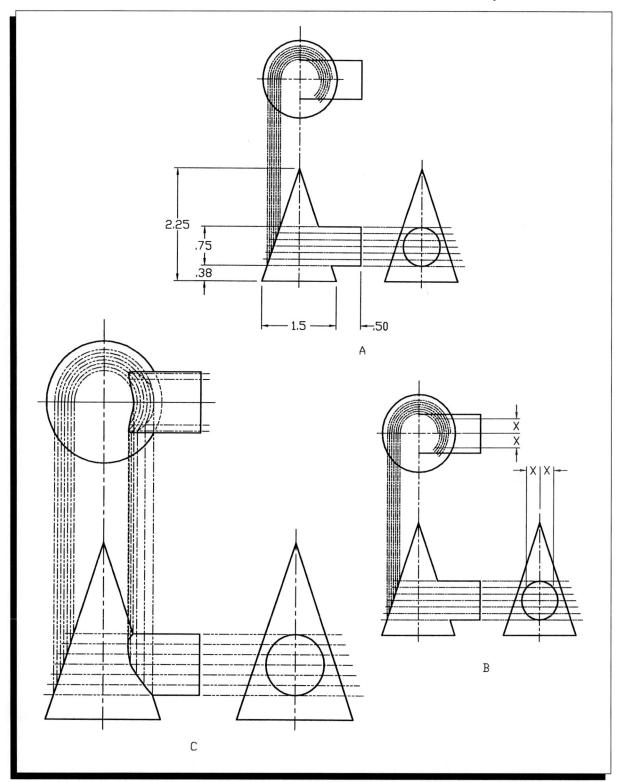

Fig. 14-41 The intersection of a cone and cylinder with horizontal cutting planes.

5. Project the points of intersection of the cylinder distances and the circular cutting planes in the top view to the front view.

6. Connect the intersections with the cutting planes in the front view using a polyline.

7. Use **PEDIT** to smooth the polyline into a curve *(Fig. 14-41C)*.
   Fig. 14-41C has been enlarged to provide more detail.

Now follow these steps to try the radial cutting plane method *(Fig. 14-42)*.

1. Open the RADIAL drawing you saved earlier.

2. Draw seven cutting planes from the vertex of the pyramid in the view containing the circle — the right side, in this case. One cutting plane should be tangent to each side of the circle, and one should pass through the center line. The others should be evenly spaced.

3. Locate the cutting planes in the top view.

4. Project the cutting plane points on the base of the cone in the top view to the base of the cone in the front view *(Fig. 14-42A)*.

5. Connect the points to the vertex of the cone.

6. Project the points where the cutting planes cross the circle in the side view to the front view.

7. Join the points where these projections cross the cutting planes in the front view with a polyline.

8. Use **PEDIT** to smooth the polyline into a curve *(Fig. 14-42B)*.

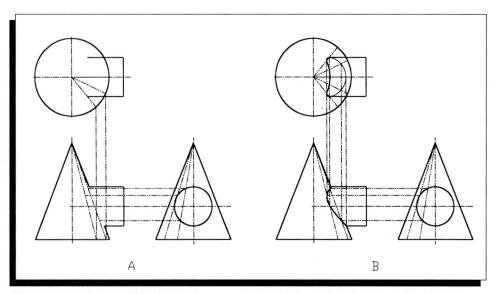

Fig. 14-42 The intersection of a cone and cylinder with cutting planes radiating from the vertex of the cone.

■ Chapter 14 Review

1. What is the difference between a pattern and a development?
2. If a right prism has been truncated, what has been done to it?
3. What is an oblique prism?
4. What is the purpose of sheet-metal elbows?
5. How can you connect ducts and pipes of different shapes?
6. What is triangulation?
7. How does the intersection of a plane surface and a curved surface appear?
8. What are the two methods of using cutting planes for determining the intersection of a cylinder and a cone?

■ Chapter 14 Problems

1. Create a development for each of the objects shown in Problem 1.
2. Find the intersections and create developments for the objects shown in Problem 2.

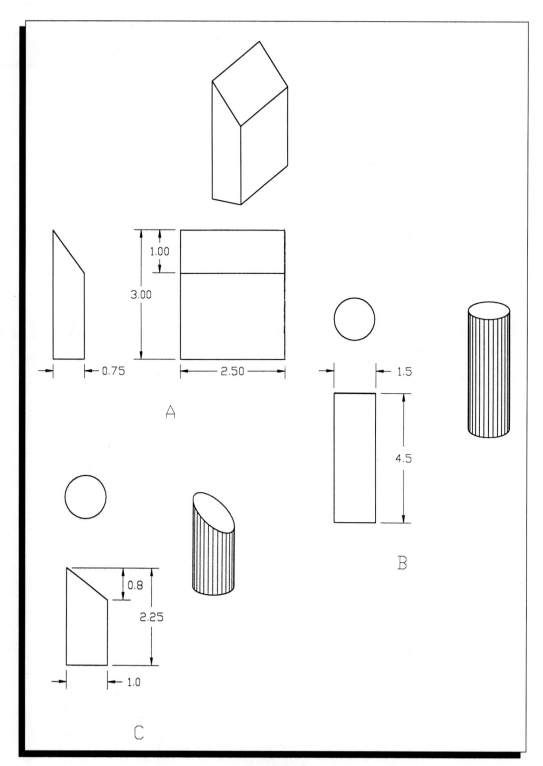

A

B

C

Problem 1

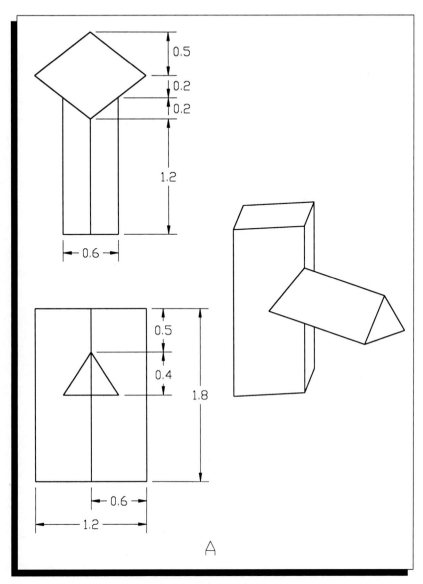

0.5

0.2

0.2

1.2

0.6

0.5

0.4

1.8

0.6

1.2

A

Problem 2A

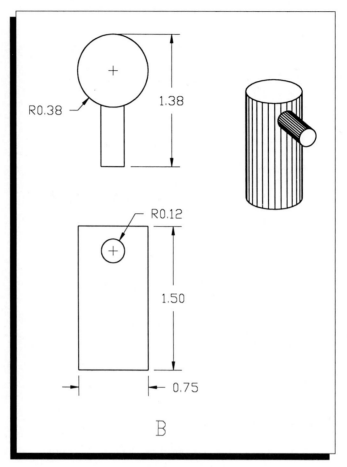

R0.38

1.38

R0.12

1.50

0.75

B

Problem 2B

Using CAD

Drafters Become Designers in the Sheet Metal Industry

"No matter what the final shape," says John Bridenbaugh, Plant Engineer at Defiance Metal Products, "everything we do in the sheet metal industry always starts out flat." Defiance produces many sheet metal parts for the automotive industry. "Generally we serve the heavy truck industry," says Bridenbaugh. "We make the non-glamorous parts: alternator brackets, chassis parts, and bumper brackets." Among Defiance's products are the roofs for the U. S. Army Hummer vehicle. The drawings for all these products also start out flat.

Drawings describe not only the complex sheet metal parts, but the production techniques used to make them. Machine tools have become faster and more accurate. Sheet metal technology has become more and more sophisticated. Now, computer software packages can take AutoCAD drawings and transfer them as instructions directly to the machine.

Jerry McMillan, Vice President of Operations at P. C. A. Sheet Metal, says the new software "has revolutionized the industry." The computer takes information that is constant about bending sheet metal and applies it to the development of a part. The AutoCAD drawing is critical to the entire process. McMillan says that parts are "nearly 100% rejection-free."

Bridenbaugh says: "We can take an AutoCAD model (a 3D drawing) and put it into a software program called Fabricam by Metalsoft and unfold it. Eliminate all the bends. Lay out the flat view. Productivity-wise, we're way ahead when we can get a 3D drawing because we don't have to mathematically calculate how much extra metal we have to leave for the bend radius and so forth."

Bridenbaugh believes his drafters are the key to producing error-free products. He says, "one thing we talk about with our drafters is that they're not drawing pictures any more, they are designing parts."

Bridenbaugh tells his drafters that all the hole diameters have to be labeled correctly. "If the drafter draws a 10 mm hole but labels it '6 mm,' the punch will make a 10 mm hole." Previously, an operator could look at a print and make the part correctly because the drafter's written instructions were clear. "But if you feed a print in electronically," says Bridenbaugh, "it goes through all the way untouched by human hands." The machine reads only the geometry, not the notes.

Because of computer processing, the original drafter has the ultimate control over what the sheet metal punch, programmable mill, or laser cutter does when it sees the program.

The Metalsoft program saves an immense amount of work. Don Newman, Executive Vice President of P. C. A. Sheet Metal, says that Metalsoft "builds all the integrity into the product." Bridenbaugh describes the process: "We put the AutoCAD drawing in and get machine code out the other end. Then we download it — no paper — right to the laser. So that is why it is very important for the drafter to build quality into the part with the original drawing."

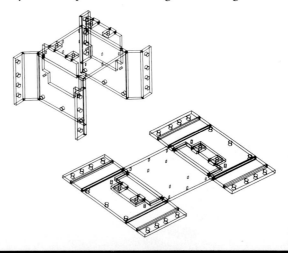

Chapter 15

▶

Key Terms

thread

helix

straight thread

tapered thread

external thread

internal thread

left-hand thread

right-hand thread

root

crest

pitch

lead

single thread

multiple thread

depth of thread

major diameter

minor diameter

pitch diameter

thread angle

thread form

thread classes

bolt

nut

screw

Threads and Fasteners

Objectives

When you have completed this chapter, you will be able to:

- understand thread terminology.
- draw and specify threads on drawings.
- describe various thread forms.
- draw and specify bolts and nuts on drawings.
- describe various types of screws and miscellaneous threaded fasteners.
- describe non-threaded fasteners and locking devices.

Fasteners hold parts together. They may be threaded or non-threaded *(Fig. 15-1)*. Threaded fasteners are bolts, screws, and studs. Non-threaded fasteners include keys, pins, and rivets. Some fasteners are permanent, and others are removable. Some fasteners, such as studs and rivets, can be removed, but they are destroyed in the process.

As a drafter, you may have to select a fastener for a specific application. You will draw fasteners and note the appropriate data on drawings. You will read and interpret thread and fastener specifications. Refer to Appendix A for thread data and fastener dimensions and specifications.

A major source of thread and fastener details and specifications is *Machinery's Handbook*, by Oberg, Jones, and Horton, published by the Industrial Press, Inc. In addition to threads and fasteners, *Machinery's Handbook* is a general reference manual for mechanical engineers, designers, drafters, toolmakers, and machinists.

Note

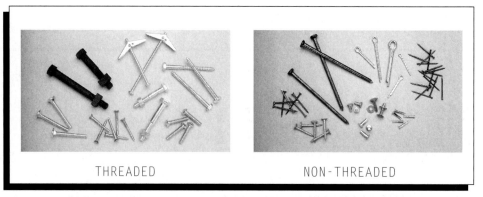

THREADED   NON-THREADED

Fig. 15-1 Examples of threaded and non-threaded fasteners.

Drawing thread and fastener details is very time-consuming. With AutoCAD, you can use library files to store ready-made symbols. You can call up the files and symbols whenever you need them. You may create your own symbols, or you may use library files created by third-party software companies. Ask your instructor about library files available at your school and for information on how to use them.

Threads

Threads are spiral grooves cut into the surface of cylinders or cones. The pattern of the grooves is a **helix**, which is the form made by the hypotenuse of a right triangle wrapped around a cylinder *(Fig. 15-2)*. Barber poles, springs, and screw threads are examples of helixes *(Fig. 15-3)*.

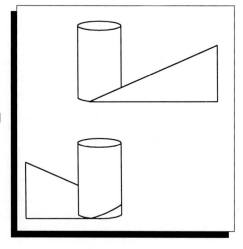

Fig. 15-2 The hypotenuse of a right triangle forms a helix as it is wrapped around a cylinder.

Fig. 15-3 Helixes are not uncommon among the items we use every day. When was the last time you saw or used one of these?

Terminology

Threads are common on fasteners. Because of their importance, you should understand the terms associated with them. Refer to Figs. 15-4 through 15-6 for illustrations.

- A **straight thread** is one formed on a cylinder. (The fastener shown in Fig. 15-4 has a straight thread.)
- A **tapered thread** is one formed on a cone.
- An **external thread** is on the outside surface of a cylinder or cone.
- An **internal thread** is on the inside surface of a cylinder or cone.
- A **left-hand thread** winds in the counterclockwise direction when viewed along its axis.
- A **right-hand thread** winds in the clockwise direction when viewed along its axis.
- The **root** is the deepest thread cut. It can be rounded or flat.
- The **crest** is the shallowest thread cut. It can be rounded or flat.
- The **pitch** is the crest-to-crest distance.
- The **lead** is the distance a thread moves in one revolution.
- A **single thread** has a lead equal to its pitch.
- A **multiple thread** has a lead equal to more than one pitch. Multiple threads permit rapid advancement of parts. For example, a double thread will advance twice as far as a single thread in one revolution.
- The **depth of thread** is the distance from crest to root.
- The **major diameter** is the largest diameter of a thread.
- The **minor diameter** is the smallest diameter of a thread.
- The **pitch diameter** is the average diameter between crest and root.
- The **thread angle** is the angle formed by the walls of the thread.

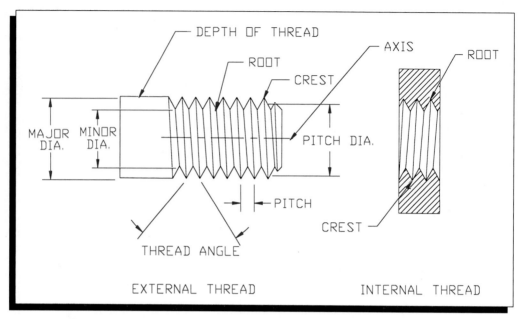

Fig. 15-4 Make sure you understand these terms before you continue.

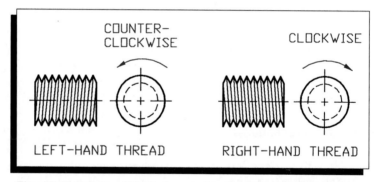

Fig. 15-5 Left-hand threads spiral to the left and right-hand threads spiral to the right when you look at them from the top.

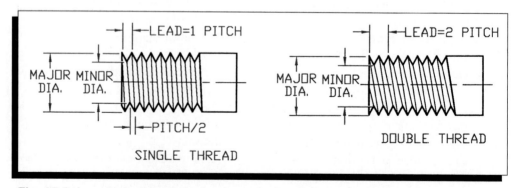

Fig. 15-6 In a single thread, the lead equals the pitch. In a double thread, the lead equals two times the pitch.

Thread Forms

Although all threads have a spiral form, there are a number of different thread shapes, or forms. The **thread form** is the profile of the thread in cross section *(Fig. 15-7)*. Each thread form is designed for a specific application.

Drafters generally specify thread data at the thread *(Fig. 15-8)*. Thread data includes the following information:
- nominal size (diameter) in inches
- threads per inch
- thread form
- thread class

For example, ¼ - 20 UNC 2A is an external thread with a nominal diameter of ¼ inch, 20 threads per inch, coarse thread form, and a class 2 fit. The paragraphs that follow explain the codes used in thread data.

Threads are assumed to be single and right-handed. If the screw is *not* single or if it is a left-hand screw, the left-hand (LH) and multiple thread designations are given after the thread class.

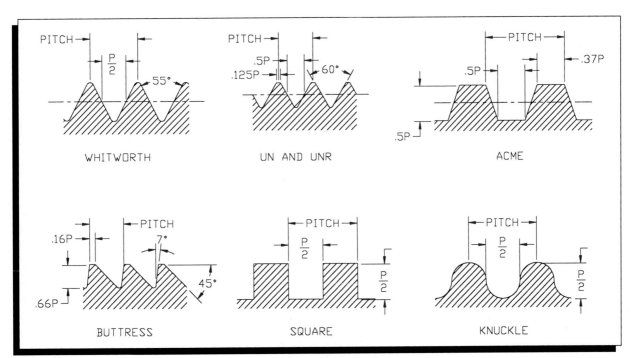

Fig. 15-7 Thread forms.

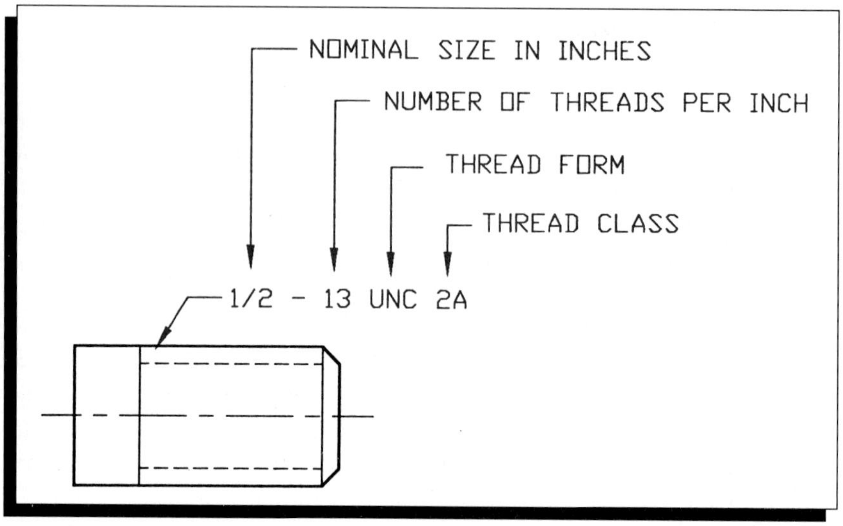

Fig. 15-8 Specify thread data (nominal size, threads per inch, thread form, and class) at the threads.

American National Standard for Unified Threads

The American National Standard Institute (ANSI) has established ANSI B1.1 as the American Standard for Unified Screw Threads. It is referred to as the Unified system because it has been agreed upon by the United Kingdom and Canada as well as the United States. Unified threads are the basic American standard for fastening type screw threads.

Fasteners with constant pitch threads are sized by the basic major diameter and the number of threads per inch. These two dimensions are combined to create the various thread series.

- Unified National Coarse (UNC) threads are the most common series. They are used for general purpose fasteners and allow rapid assembly and disassembly.
- Unified National Fine (UNF) threads have greater tensile strength than UNC threads, so they fasten more securely. (Tensile strength is a measure of resistance to lengthwise stress.)
- Unified National Extra Fine (UNEF) threads have the highest tensile strength. They provide the greatest security in high vibration situations.
- 8-thread series threads have 8 threads per inch regardless of the major diameter.
- 12-thread series threads have 12 threads per inch regardless of the major diameter.
- 15-thread series threads have 16 threads per inch regardless of the major diameter.

External threads and internal threads fit together. **Thread classes** are the allowances and tolerances that govern the fit between the external and internal threads. The allowance refers to how tightly a thread fits its mating part. The tolerance is the difference between the actual part size and the standard or exact size.

Class A threads are external and class B are internal. The number that precedes the A or B describes the tolerance and allowance. Class 2A and 2B threads are used for most general-purpose applications. Class 3A and 3B threads have closer tolerances and allowances, and class 1A and 1B threads have looser tolerances and allowances.

▶ Metric Threads

Metric threads used in the United States comply with American National Standard specification ANSI B1.13M. This standard is in basic agreement with the International Standards Organization (ISO) screw-thread specifications for standardization with other countries. The American National Standard M Profile corresponds with the ISO 68 profile.

The standard M profile for general-purpose mechanical applications is a coarse thread series. The simplified international designation is the thread form profile, M, followed by the nominal diameter in millimeters (mm). For example, an M14 is an M profile 14 mm in diameter.

A fine thread series can be specified for applications in which extra holding power is required. The simplified international designation for a fine thread requires the thread pitch in millimeters in addition to the thread form profile and nominal diameter. Thus, M14×1.5 refers to an M profile 14 mm in diameter with a pitch of 1.5 mm.

▶ Acme Threads

The Acme thread profile is used primarily for power transfer. A typical application of an Acme thread is in a vice *(Fig. 15-9)*. Acme threads have a broad flat crest and root with a sloping thread angle *(Fig. 15-7)*.

Acme threads comply with American National Standard ANSI B1.5. Drawing notes for Acme threads specify the nominal diameter, number of threads per inch, thread form (ACME), and the class of fit. For example, 1¾ - 4 Acme 2G describes an Acme thread that has a nominal diameter of 1¾ inches and 4 threads per inch.

Acme thread specifications do not differentiate between external and internal threads. Class 2G means a class 2 general-purpose fit. For less backlash and end play, specify class 3 or 4. There is no class 1.

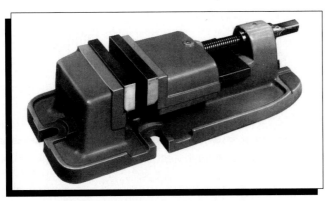

Fig. 15-9 A vice uses Acme threads.

▶ Other Thread Forms

Other thread forms include square, Whitworth, knuckle, and buttress *(Fig. 15-7)*. The square thread is difficult to machine and has few practical applications. The Whitworth thread is a British thread form which was never fully accepted in the United States.

Knuckle and buttress threads, while not common, have found specialized mechanical applications. Knuckle threads are used for items such as light bulbs and screw-type bottle tops. Buttress threads are used mostly on jacks.

Drawing Threads

Threads may be represented on drawings in detail showing their actual thread form *(Fig. 15-10)*. They may also be represented symbolically in schematic form or in simplified form *(Fig. 15-11)*. As with all objects, threads should be drawn as simply as possible without sac-

rificing important features. Draw detailed threads when details are necessary. Use the schematic form for less detail, and the simplified form for no visual details. A single drawing may contain all three forms.

▶ Simplified Method

The simplified method reduces the threads to horizontal lines and concentric circles. It can be used to represent all

thread forms, including the Unified, Acme, Whitworth, and square threads.

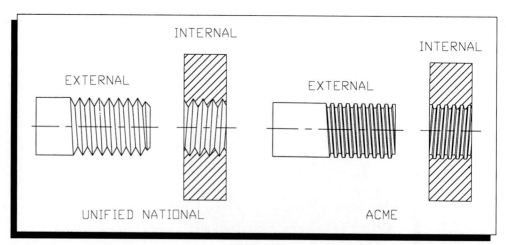

Fig. 15-10 Detailed drawings such as these include the actual thread form.

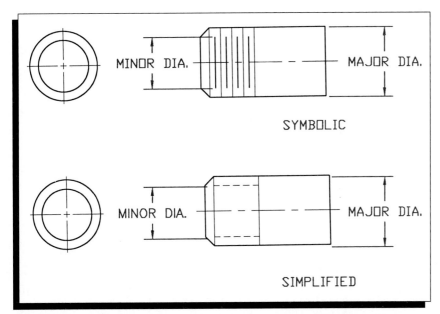

Fig. 15-11 In the symbolic method, straight lines are used to show the threads. In the simplified method, the minor diameter is shown using dashed (hidden) lines.

ACTIVITY

Follow the procedure below to draw the simplified front view of an external thread and the end view of an internal thread *(Fig. 15-12)*. In this procedure, both threads are drawn concurrently. It is common practice in AutoCAD to draw lines in groups by linetype so that layer changing will be minimized. Use the ZOOM command and object snap for accuracy. The threads are 1-8 UNC 2A and 2B. *Machinery's Handbook* gives an external minor diameter of 0.8512 and an internal minor diameter of 0.8647. However, for visual impact in this example, use a major diameter of 1 unit and a minor diameter of 0.75 unit for both external and internal threads.

1. Start AutoCAD and load your ASHEET prototype drawing. Begin a new drawing named **SIMPLIF**.
2. Make sure you have layers named **OBJECT**, **HIDDEN**, and **CENTER** containing the continuous, hidden, and center linetypes, respectively.
3. Make **CENTER** the current layer and draw horizontal and vertical center lines *(Fig. 15-12A)*. Make the horizontal line about 8 units long and the vertical line about 3 units long.
4. Change to the **OBJECT** layer.
5. Draw a **2**-unit square around the intersection of the horizontal and vertical center lines. This represents the material in which the internal threads will be created.

6. Draw a circle with a diameter of **0.75** units. Place the center of the circle at the intersection of the center lines. This is the minor diameter of the internal thread *(Fig. 15-12B)*.

7. Draw one line **0.5** units above the horizontal center line and one line **0.5** units below the horizontal center line to the left of the 2-unit square *(Fig. 15-12C)*. Begin the lines about 0.5 unit from the left edge of the square. Make each line about 4.5 units long. These lines represent the major diameter of the external thread.

8. Draw a vertical line on the right end of the external thread to connect the lines you drew in step 7.

9. Chamfer the upper and lower corners. Set the chamfer distance to **0.125**.

This is half the distance between the major and minor diameters.

10. Draw another vertical line to connect the major diameter chamfer points.

A chamfer on the end of the external thread will give it a positive entry into the internal thread *(Fig. 15-12C)*.

11. Change to the **HIDDEN** layer.

12. Draw a circle with a diameter of **1** unit. Place the center of the circle at the intersection of the horizontal and vertical center lines. This is the major diameter of the internal thread.

13. Draw two horizontal lines representing the minor diameter of the external thread *(Fig. 15-12D)*.

Hot Tip The bottom of the upper chamfer marks the position of the upper boundary of the minor diameter, and the upper end of the lower chamfer marks the position of the lower boundary of the minor diameter.

14. Break the left end of the external thread with three arcs and a hatch *(Fig. 15-12E)*.

15. Display the thread notes with a leader *(Fig. 15-12E)*.

Remember to save your drawing after you finish.

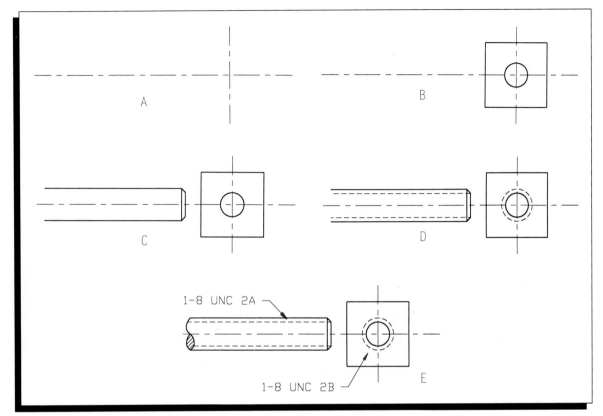

Fig. 15-12

▶ Drawing Detailed and Schematic Threads

The detailed and schematic methods of drawing threads simulate actual thread features. To draw them accurately, you must lay out the major diameter, minor diameter, and pitch *(Fig. 15-13)*. When you use the detailed method, angle the sides of the thread, crest to root to crest, 60 degrees. Draw the crest and root with a sharp V. In the schematic method, the full-length vertical line represents the crest and the short vertical line represents the root.

Threads present an excellent opportunity to create your own fasteners symbol library. Chamfer external threads 45 degrees, save them as blocks, and use them as a basis for your symbol library. Refer to Chapter 13, "Production Drawings," for a discussion of blocks and symbol libraries.

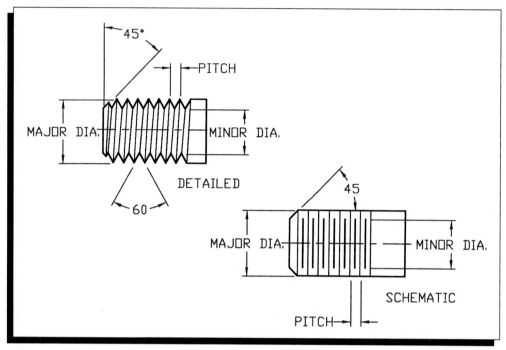

Fig. 15-13 Include this information when you draw detailed or schematic thread representations.

ACTIVITY

The following procedure draws the schematic representation of an external thread *(Fig. 15-14)*. As in the previous "Your Turn," use a modified 1-8 UNC 2A thread. Use a major diameter of 1, a minor diameter of 0.75, and a pitch of 0.125.

1. Begin a new drawing named **SCHEMAT**. Draw a horizontal center line and the major and minor diameters *(Fig. 15-14A)*.

Offset the center line by 0.375 (half of the minor diameter) above and below to create the two lines for the minor diameter. Use the CHANGE command to move the lines to the OBJECT layer, and trim the lines to match those for the major diameter.

2. Draw a vertical line to represent the end of the thread *(Fig. 15-14A)*.
3. Chamfer the end at a 45° angle from the minor diameter (set the chamfer distances to 0.125).
4. Draw a vertical line from the top of the upper chamfer to the bottom of the lower chamfer *(Fig. 15-14A)*.
5. Offset the first crest line from the line you drew in step 4. The distance from the chamfer to the crest equals the pitch (0.125) *(Fig. 15-14B)*.

6. Offset the first root line from the chamfer. The distance from the chamfer to the root equals one half the pitch (0.0625).

7. Trim the root to the minor diameter *(Fig. 15-14B)*.

8. Offset the root and crest lines for the length of the thread *(Fig. 15-14C)*.

Offset all the crest lines first. Each time AutoCAD prompts you to "Select object to offset," pick the leftmost crest line. At "Side to offset?" pick a point to the left. Then repeat the process for the root lines.

---

9. Erase the horizontal minor diameter lines and add the break at the end *(Fig. 15-14D)*.

10. Save the thread as a block for inclusion in a fasteners symbol library. Give the block a descriptive name, such as 1-8UNC2A.

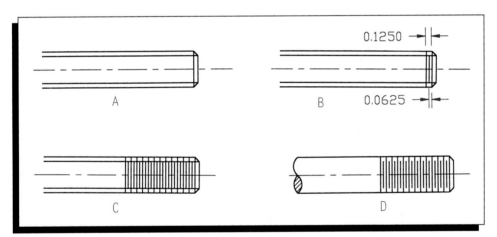

Fig. 15-14

Bolts and Nuts

A **bolt** is an externally threaded fastener which is normally tightened and released in its assembly by torquing a nut. A **nut** is a bolt retainer *(Fig. 15-15)*. Bolts and nuts are covered under ANSI specifications B18.2.1 and B18.2.2, respectively.

Both bolts and nuts are threaded. The specific thread depends on the bolt/nut designation and application. The bolt thread length is twice the diameter of plus ¼ inch for bolts under 6 inches and twice the diameter plus ½ inch for bolts over 6 inches. The thread length equals the bolt length only on short bolts.

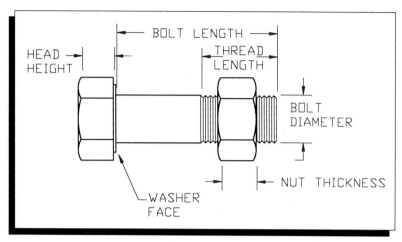

Fig. 15-15 Hex bolt with nut.

Bolt Forms

Three standard bolt forms are used: hexagonal (hex) head, square head, and round head. Hex bolts are made as hex, heavy hex, and heavy hex structural.

The terms used here for bolt strength are most commonly used in the mechanical trades. In other trades, other terminology may be used. For example, in the automotive industry, grades are used. Hex is the same as grade 5, heavy hex is the same as grade 8, and heavy hex structural is the same as grade 10. You may find that still other terms are used in other applications.

▶ Hex Head Bolts

Hex and heavy hex bolts have a flat head bearing surface. The bearing surface is the bottom part of the head, which comes directly in contact with the material being fastened. Heavy hex structural bolts have a head that contains a washer-like surface as an integral part of the head *(Fig. 15-16)*. All hex bolts may have Unified coarse, fine, or 8-thread series threads class 2A. The hex heads are chamfered at 30 degrees.

▶ Square Head Bolts

Square head bolts have flat machined head bearing surfaces *(Fig. 15-17)*. They may have Unified coarse, fine, or 8-thread series class 2A threads. Square heads are chamfered at 25 degrees but are drawn at 30 degrees like hex heads for standardization.

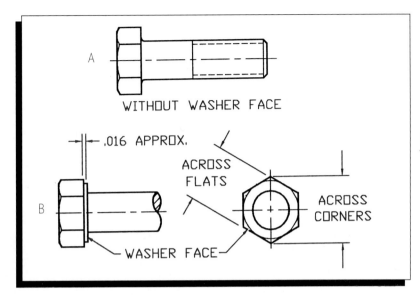

Fig. 15-16 Hex bolts A) without washer face; B) with washer face.

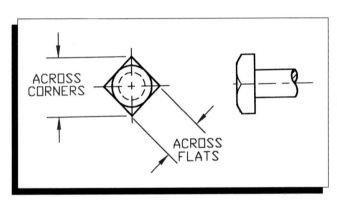

Fig. 15-17 Square-head bolt.

▶ Round Head Bolts

Round head bolts come in many configurations *(Fig. 15-18)*. Since they have no wrenching surface, they rely upon shoulder, or upper shaft, gripping devices such as a ribbed or fin neck. Round head bolts have Unified Standard Coarse class 2A threads only.

Bolt Designation

Designate bolts by giving:
- nominal size
- threads per inch
- length
- product name
- material
- protective finish, if required

For example, you could designate a bolt with this description: $\frac{3}{8}$ - 16 × 1$\frac{3}{4}$ square bolt, steel, zinc plated.

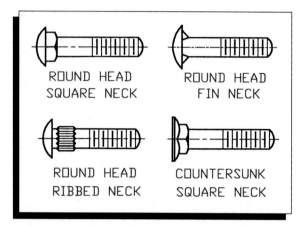

Fig. 15-18 Commonly used round-head bolts.

Nuts

Nuts are selected to match the bolts and the application. They are generally classified as flat, washer faced, plain, acorn, wing, and slotted *(Fig. 15-19)*. Each class has specific features and uses. A flat disk called a washer is placed under the nut to provide a bearing surface while the nut is being turned.

Designate nuts in the same manner as bolts, except of course you do not need to specify length. For example, you could designate a nut as ½ - 20 hex nut, steel, zinc plated.

Drawing Bolts and Nuts

Draw hex and square bolts and nuts dimensioned across the corners unless otherwise instructed. In general, bolt head and nut specifications are given across the flats circumscribed around a circle and across the corners inscribed within a circle. Bolt head dimensions can be approximated as relations to the bolt diameter *(Fig. 15-20)*.

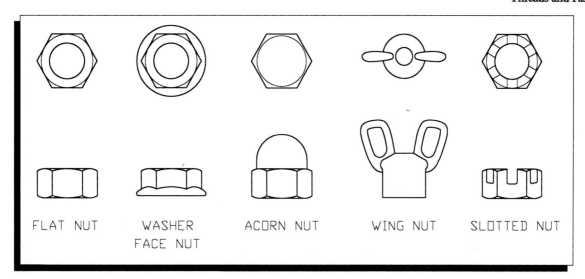

Fig. 15-19 Examples of nuts.

FLAT NUT

WASHER
FACE NUT

ACORN NUT

WING NUT

SLOTTED NUT

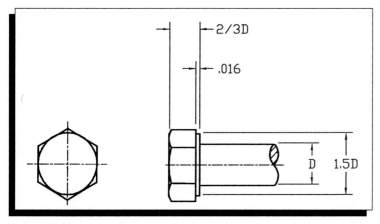

Fig. 15-20 Bolt head dimensions are often specified according to their size relative to the bolt diameter.

ACTIVITY

Your Turn

Follow the steps below to draw a hex bolt *(Fig. 15-21)*. The same general procedure can be used for a hex nut and a square bolt and nut. Use the ZOOM command and object snap as needed.

1. Start AutoCAD, load your ASHEET prototype drawing, and begin a new drawing named HEXBOLT.
2. Make the CENTER layer current, and draw horizontal and vertical center lines *(Fig. 15-21A)*.
3. Make the OBJECT layer current, and draw a circle with a radius of 0.75.
4. Draw a hexagon with the same center as the circle, circumscribed around a circle. Use a radius of 0.75. After the hexagon is in place, rotate it 90 degrees so that the edges are vertical *(Fig. 15-21A)*.
5. Outline the bolt head with two vertical and four horizontal lines *(Fig. 15-21B)*.

With Ortho on, project lines from each vertex of the hexagon to establish placement of the horizontal lines. Set the vertical lines 0.75 inch apart.

6. Trim *(Fig. 15-21B)*.

7. To establish the bolt head arcs, draw two construction lines 3 units long. Draw one line at an angle of 150 degrees from the intersection of the lower side of the bolt head and the top side of the bolt. Draw the other line at an angle of -150 degrees from the upper side of the bolt head and top side of the bolt. From the intersection of these two lines, create a circle tangent to the top of the bolt head *(Fig. 15-21C)*.

8. Trim the circle so that the arc shown in Fig. 15-21C remains.

9. Draw a vertical construction line through the intersection of the arc end points and the bolt head hex corners *(Fig. 15-21D)*.

10. Draw two circles using the 3-point method. Use the intersection of the construction line and the edge of the bolt as one point and the end of the arc you created in step 8 as the second point. Use the TANgent object snap to define the third point tangent to the top (bottom for the second circle) of the bolt head *(Fig. 15-21E)*.

11. Erase the construction lines and trim *(Fig. 15-21F)*.

12. To create the bolt shank, draw two horizontal lines: one 0.5 inch above the center line and the other 0.5 inch below the center line.

13. Add the two arcs and hatch for the break *(Fig. 15-21G)*.

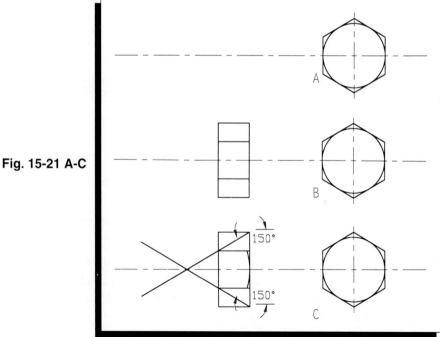

Fig. 15-21 A-C

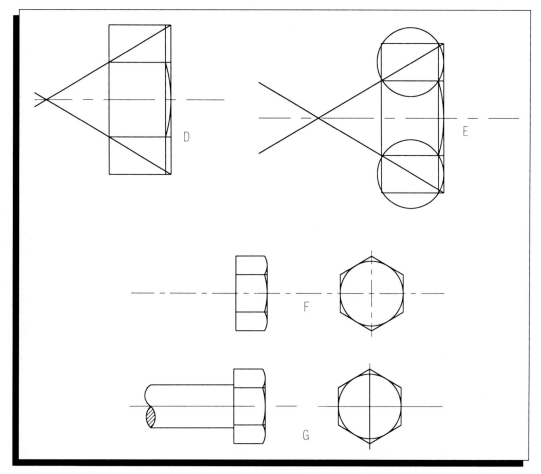

Fig. 15-21 D-G

Screws

A **screw** is an externally threaded fastener which is normally tightened and released in its assembly by torquing its head. Different types of screws have different characteristics. Set screws have a straight thread, and wood screws have a tapered thread. Machine screws mate to internal threads or nuts, but lag screws create their own internal threads.

Cap Screws

Cap screws fasten parts together by passing through a clearance hole in one part and seating into an internal thread in another part *(Fig. 15-22)*. The clearance of the hole is not usually shown on drawings. Cap screws have finished heads and are used when appearance and accuracy are important. They can have coarse, fine, or 8-thread series class 2A threads. Socket head cap screws have class 3A threads. Common heads for cap screws include round, flat, hex, socket, fillister, and 12-spline flange.

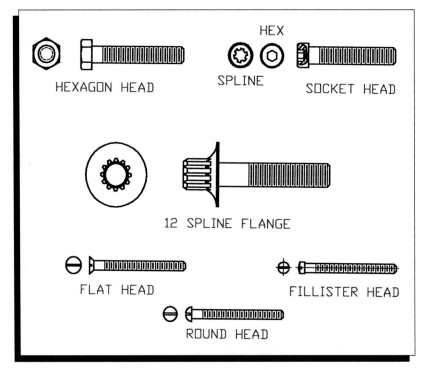

Fig. 15-22 Common cap screws.

Machine Screws

Machine screws resemble cap screws, but they are smaller *(Fig. 15-23)*. They are used frequently with nuts to fasten parts together, but they can also screw directly into tapped holes. Machine screws under 2 inches are threaded to their heads, and those over two inches have a minimum thread length of 1¾ inches. They may have coarse or fine class 2A threads. Typical head styles include flat, oval, round, fillister, truss, binding, and pan heads.

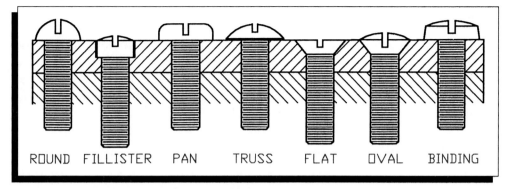

Fig. 15-23 Common machine screws.

Set Screws

Set screws lock one part against another. They screw into threaded holes and firmly "set" against the inner parts to hold the parts in place. With a set screw, the point is as important, and as varied, as the head *(Fig. 15-24)*. The style and combination of the points and heads depend on the screw's application. All set screws may have coarse or fine threads. Square-head set screws may also have 8-series threads. All set screws except socket heads have class 2A threads. Socket head set screws have class 3A threads.

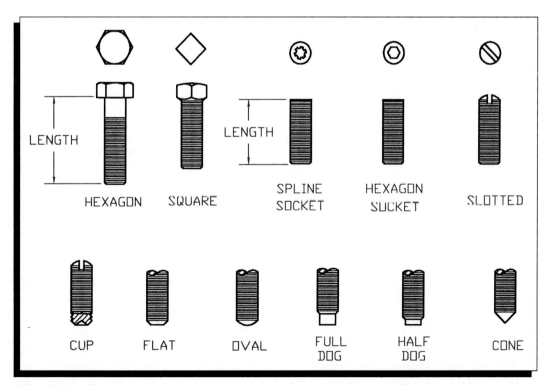

Fig. 15-24 Common set screws. Heads are shown in the top row. Points are shown in the bottom row.

Wood Screws

Wood screws create their own threads when you drive them into wood *(Fig. 15-25)*. Before you attempt to insert a screw into a hard wood such as oak or maple, you should drill a pilot hole. The wood screw can make its own hole in soft woods such as pine and spruce. Wood screws have flat, round, or oval heads with slotted or Phillips head configurations.

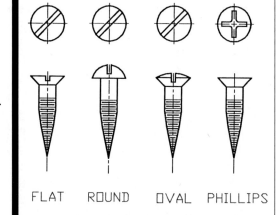

Fig. 15-25 Common wood screws.

ACTIVITY

1. Visit a hardware store and look at the selection of screws. Do you see any screws that are not discussed in the text? What are they? What type of threads do they have?

2. Talk to the person in the hardware store who buys the screws. How many varieties does the store buy? How does the buyer decide which screws to buy? What are the best-selling screws? The poorest-selling screws? Does the buyer have any idea why?

3. Write a summary of your findings for a class discussion.

Miscellaneous Threaded Fasteners

Other examples of threaded fasteners include studs, lag screws, turn buckles, hooks and eyes, stove bolts, and carriage bolts *(Fig. 15-26)*. Each of these threaded fasteners has a special job to do. For example, studs screw into metal or wood, leaving an exposed thread that can be used as an anchor. Lag screws hold metal to wood. They are also used for heavy wood construction. Turn buckles hold or increase tension on rods or cables. They can be used to take up slack on a bridge cable or to square up a sagging screen door.

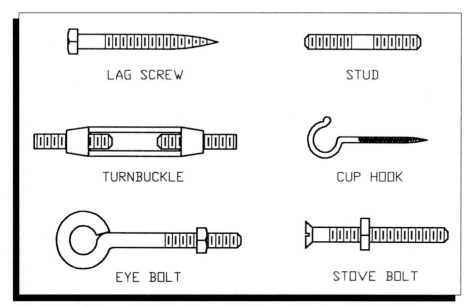

Fig. 15-26 Miscellaneous threaded fasteners have been designed to do special tasks.

1. Look for examples of miscellaneous threaded fasteners. Draw sketches of three of them using simplified threads. How would you specify them on a drawing?
2. Refer to the sketches you made in the first activity. Create AutoCAD drawings from the sketches. Block each fastener using a descriptive name to add to your fasteners symbol library.

Non-Threaded Fasteners

Not all fasteners use threads to hold parts together. Pins, keys, and rivets are examples of non-threaded fasteners.

Pins

Pins secure parts by passing through two or more holes *(Fig. 15-27)*. Pins are often held in place by a slight interference fit or by friction with the parts. They are usually designed for light work, and in some instances they are designed to shear if an excess load is placed on them. Common pins include taper, straight, cotter, and clevis pins.

- Taper pins fit into tapered holes. They are held in place by friction and are easily removed.

- Straight pins have a constant diameter. They fit into a hole with a slight interference.
- Cotter pins are made from a D-shaped piece of material bent to form a pin. After they are inserted into a hole, the ends are bent to prevent removal.
- Clevis pins are straight pins with a head on one end and a hole in the other end. They are held in place with cotter pins.

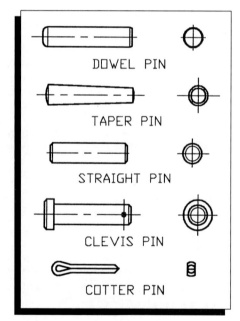

Fig. 15-27 Common types of pins.

Keys

Keys are metal pieces designed to prevent relative motion between parts *(Fig. 15-28)*. They fit into slots called keyseats which are machined into shafts. The upper part of the key fits into a keyway in the mating part. When the key is in place, the parts move together; if one part moves, the other part *must* move with it. Some keys are designed to shear if an excess load is placed upon them. Common types of keys include square, Woodruff, and Pratt & Whitney keys.

- Square keys generally have a width that is one quarter of the shaft diameter. The depth of the keyway is one half the key height.
- Woodruff keys are almost half circles. They fit into semicircular keyways machined into a shaft. The top of the key fits into a rectangular keyway.
- Pratt & Whitney keys are rectangular with rounded ends. They fit with two thirds of the key in the keyseat.

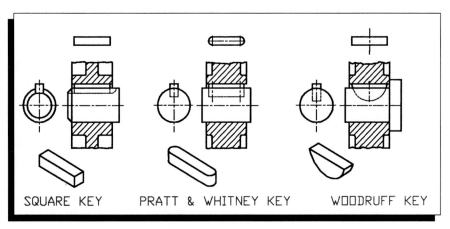

SQUARE KEY   PRATT & WHITNEY KEY   WOODRUFF KEY

Fig. 15-28 Common keys.

Rivets

Rivets are permanent fasteners designed to fasten sheet metal parts and steel plates. They are classified as ANSI large rivets (½ inch to 1¾ inch), and ANSI small rivets (¹⁄₁₆ inch to ⁷⁄₁₆ inch) *(Fig. 15-29)*. A rivet has a head and a shaft. The shaft is inserted into matching holes and is then "upset," or formed, to create another head.

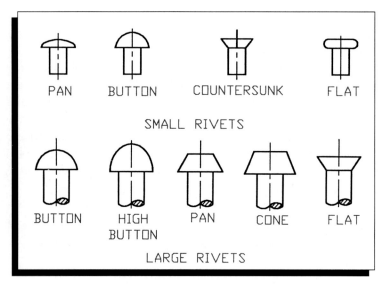

Fig. 15-29 Large and small rivets.

Locking Devices

Locking devices secure fasteners and minimize the accidental loosening of the fasteners by movement or vibration *(Fig. 15-30)*. Examples of locking devices include lock washers, slotted nuts with cotter pins, tab washers, jam nuts, castle nuts, and stop nuts.

- Lock washers may be split or have internal or external teeth *(Fig. 15-31)*. The split edges or teeth act to lock the nut against the base material.
- Cotter pins can be used with slotted nuts, castle nuts, and flat nuts. They fit through a hole in the bolt or screw and lock the nut in place.
- Tab washers are flat washers with tabs that can be bent to hold the nut in place.
- Jam nuts are flat nuts which are threaded on the shaft and jammed against the holding nut to lock it in place.
- Stop nuts have a plastic locking factor built into the top of the nut. Since they will lock anywhere on the threads, they can be used to hold parts together loosely under adverse conditions.

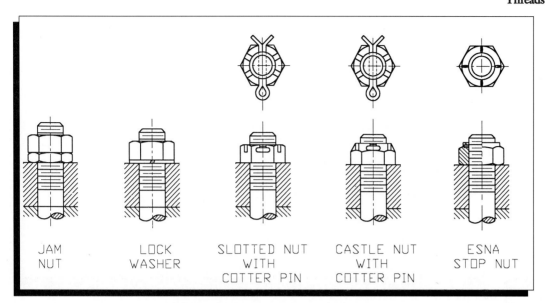

Fig. 15-30 Common locking devices.

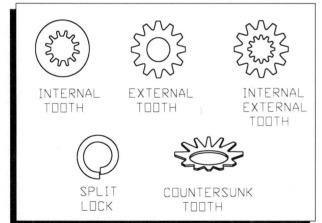

Fig. 15-31 Lock washers.

Look around your home or school for examples of non-threaded metal fasteners used with metal. How many different fasteners can you find? In how many different ways is the same fastener used? Would any of these fasteners work just as well if they were made of another material, such as plastic? Can you think of a better way to fasten parts together?

Chapter 15 Review

1. Name three thread forms.
2. What two dimensions are combined to create the various thread series?
3. What are thread classes?
4. A thread is designated as a ½ - 13 UNC 2A. What does this mean?
5. What is a bolt?
6. How does a screw differ from a bolt?
7. What is the purpose of a key?
8. Name four locking devices.
9. How does each of the locking devices you named in question 8 do its job?

Chapter 15 Problems

Use AutoCAD and the techniques you learned in this chapter to create drawings of these threads. Add the completed threads to the thread symbol library you began earlier in this chapter.

1. Create simplified views for the following threads. Use a front view for the external thread and a top view for the internal thread.
   a. ⅞-9 UNC 2A and 2B threads
   b. ½-13 UNC 2A and 2B threads
   c. ¾-16 UNF 1A and 1B threads
2. Create schematic views of the following threads.
   a. ¾-10 UNC 2A thread
   b. ¼-20 UNC 2A thread
   c. ¹⁵⁄₁₆-20 UNEF 3A thread
3. Create a ⅜ - 16 × 1¾ hex bolt.

Using CAD

Fast Fasteners

Displayed prominently on a wall in George Bockelmann's Virginia office is a technical drawing of a FAX machine he drew early in his career. If visitors look closely, and George makes sure they do, they can see that the drawing has been cut in half. He explains that "after I had drawn it several times and gotten it just right, I realized that the right half had to be moved a sixteenth of an inch upward. Thinking of all the hours it would take to re-draw it, I simply cut it in half and carefully taped it back together." Below that framed drawing, is a recent and perfect drawing of a similar item. Bockelmann's last-minute changes on that drawing are invisible — it was drawn in AutoCAD.

Bockelmann is an engineer who works along with the U.S. Navy to design models which are tested in the David Taylor model basin in Bethesda, Maryland. Bockelmann also is president of Electro-Pack Co., a mechanical engineering and product development company in Vienna, Virginia. Years of engineering and design experience give Bockelmann special skills with AutoCAD. "I did a lot of trial and error when I first started drawing, but I was learning the fundamentals," he said.

Bockelmann and Farshad Bon, a drafting and design supervisor for Spectra Physics, a company which makes lasers, are among many engineers and designers who use fasteners in their drawings. They draw these fundamental elements frequently and repetitively.

Both men discovered an AutoCAD add-on program called AimaFAST, an electronic toolkit that makes it easy to include fasteners in their drawings. Bon calls it "easy to use and very easy to learn." Bockelmann explains, "If you are drawing countersunk screw holes, rivet holes, or counterbore holes you simply pick the screw size. AimaFAST calculates countersink diame-

ters and depths instantaneously. Then you can place them repeatedly, wherever needed. Most amazingly, the program offers a wide range of hardware types and sizes. It takes away the tedious work of looking up each fastener in *Machinery's Handbook* or vendor catalogs."

Bon appreciates the ease with which he can examine his drawings. "You can grab any part that has fasteners: bolts, nuts, machine screws, rivets — any kind; and look at either the side view, top view, or 3-D view with one click," said Bon.

Once a drafter knows about all the fasteners and how to represent them, correcting a mistake is easy. "If you put in the wrong size," says Bockelmann, "it is no problem; just erase it and pop in a new one. Forget about how much time it takes to redraw it; it is already done for you, and the program gives you a complete description of the fastener."

Beginning drafters will appreciate Bockelmann and Bon's point of view after they learn what fasteners are and how they relate to a complete drawing. "Only then" says Bockelmann, "will they fully realize the benefits of an automatic fastener program like AimaFAST."

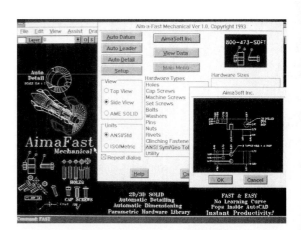

Chapter
16

▶ Key Terms

cam

follower

displacement
   diagram

dwell

gears

spur gear

pinion

involute

bevel gears

miter gears

rack

worm gears

worm

Cams and Gears

Objectives

When you have completed this chapter, you will be able to:

- describe the uniform, harmonic, and uniformly accelerated and decelerated motion of cams.
- draw the displacement diagrams for the cam motions.
- identify spur gears, bevel gears, worms and worm gears, and racks.
- use spur gear terms and formulas to complete data tables.
- draw spur gear teeth.

Cams and gears are mechanical parts that send motion from one place to another. Sometimes they change the direction of the motion, and sometimes they change its speed. Machine design relies heavily upon a knowledge of cams and gears. As a drafter, you will need to understand the function of cams and gears and know how to draw them.

Cams

A **cam** is a machine part that changes rotary motion (turning) into reciprocating motion (up-and-down or left-and-right) *(Fig. 16-1)*. The cam surface has an irregularly shaped outline. A **follower** rides on the cam surface and transmits the changes in the cam's surface into reciprocating motion. Common types of followers are shown in Fig. 16-2.

Cams are made as plate cams or as cylindrical cams *(Fig. 16-3)*. Grooved cams are a variation of plate cams; the follower rides in a groove instead of on the outer surface of the cam *(Fig. 16-4)*. Plate cams transmit motion perpendicular to the axis of the rotating shaft. The cylindrical cam follower rides in a slot, and the cam transmits motion parallel to the rotating shaft axis.

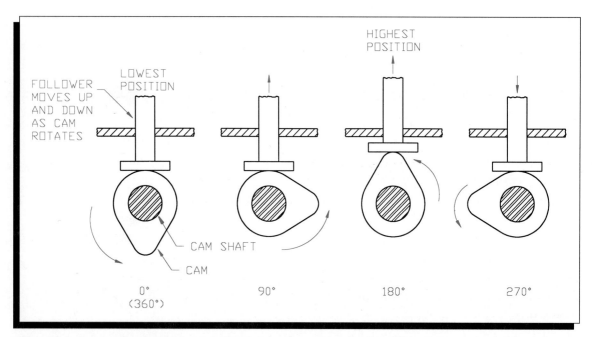

Fig. 16-1 A cam changes rotary motion into reciprocating motion. The rotary motion of the cam shaft is transferred to the up-and-down reciprocating motion of the follower.

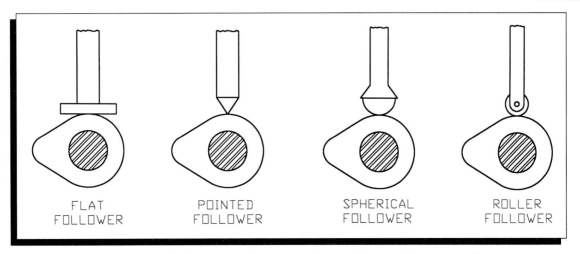

Fig. 16-2 A cam follower may have one of several types of contact surfaces.

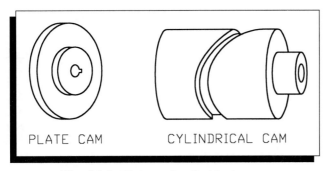

Fig. 16-3 Plate and cylindrical cams.

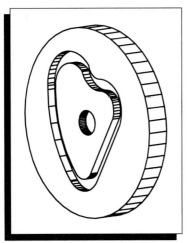

Fig. 16-4 Grooved cam.

Cam Motion

Cam motion is described relative to the movement of its follower. There are three types of motion: uniform motion, harmonic motion, and uniformly accelerated and decelerated motion. Displacement diagrams plot the three types of motion. A **displacement diagram** is a graph of the maximum rise of the cam follower on the Y axis against one complete rotation (360 degrees) of the cam on the X axis *(Fig. 16-5)*.

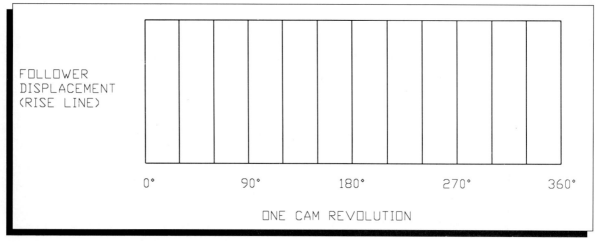

Fig. 16-5 An example of a displacement diagram form.

▶ Uniform Motion

Uniform motion is straight-line motion. The follower rises and falls at a constant speed. Uniform motion as a straight line is theoretical. In practice, the displacement diagram has a small arc at the beginning and at the end of the line to make a smooth directional change and avoid suddenly jarring the follower *(Fig. 16-6)*.

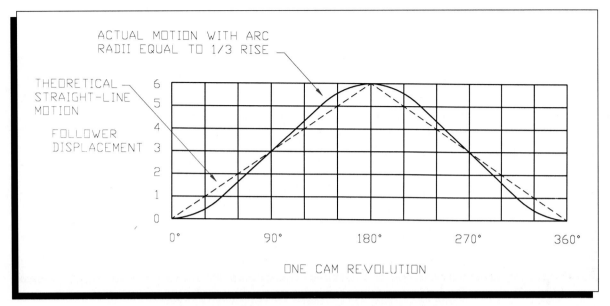

Fig. 16-6 Displacement diagram for uniform motion.

ACTIVITY

To draw a uniform motion displacement diagram, use the following procedure. For this exercise, the cam has a uniform rise of 2 inches through 180 degrees. Refer to Fig. 16-7 as you perform the steps. Fig. 16-7D shows the completed diagram.

1. Draw the length of the diagram. The length will represent 360 degrees — one complete rotation of the cam. For this diagram, let every 0.5 unit in AutoCAD represent 30 degrees. That will make the line length 6 units, which is a convenient length for working in AutoCAD.

Any length can be used for the displacement diagram. However, the length must represent one complete cam rotation, which is equal to the circumference of a circle drawn through the highest point of the cam.

2. Draw a 2-unit (2-inch) vertical rise line at the left end of the length line.

The rise line represents the maximum follower displacement. As you read in the introduction to this "Your Turn," the rise is 2 inches through 180 degrees. In other words, the maximum displacement is 2 inches and occurs at 180 degrees on the displacement diagram.

3. Offset the rise line at 0.5-unit intervals across the length of the horizontal line.

4. Divide the rise line into 6 equal parts and draw horizontal lines across the graph to create the diagram format. Label the diagram as shown in Fig. 16-7A.

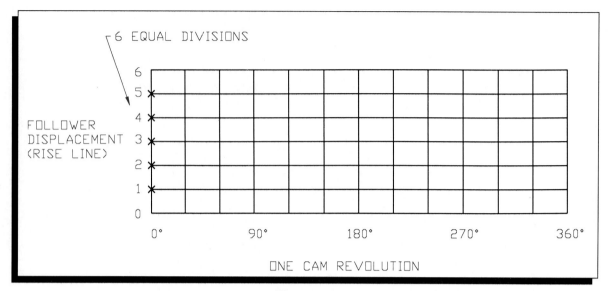

Fig. 16-7A

Remember to change PDMODE to see the division points created by the DIVIDE command.

5. Connect the lowest point (displacement line 0 at 0°) and the highest point (displacement line 6 at 180°) with a straight line.

This line represents the theoretical straight-line motion. Now follow these steps to create the line that represents the true motion. First, you will create arcs at each end with radii equal to ⅓ the length of the rise line you created in step 3.

6. Use DIST to find the length of the rise. Divide that number by 3 to get the radii of the two arcs.

7. *Command:* **ARC** ⏎

   Center/<Start point>: Use the INTersection object snap to snap to the bottom of the rise line you created in step 3.

   Center/End/<Second point>: **E** ⏎

   End point: Use INTersection to snap to the point one third of the way up the rise line. This point should fall at the vertical line representing 60 degrees and displacement line 2.

   Angle/Direction/Radius/<Center point>: **R** ⏎

   Radius: Enter the radius you determined in step 6.

8. Repeat step 7 to create another arc. Specify the starting point at the top of the rise line (displacement line 6 at 180 degrees), the end point ⅔ of the way up the rise line (displacement line 4 at 120 degrees), and the radius you determined in step 6.

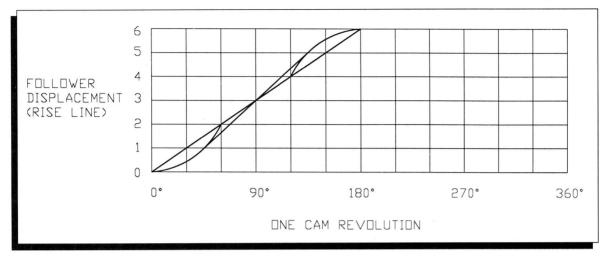

Fig. 16-7B

9. Draw a tangent to the arcs to represent the actual follower rise *(Fig. 16-7B)*.

10. Erase the theoretical line and trim the arcs to create a smooth curve *(Fig. 16-7C)*.

11. Complete the diagram by using the MIRROR command to create the second half of the cam's rotation *(Fig. 16-7D)*.

Use the 180-degree vertical line as the mirror line.

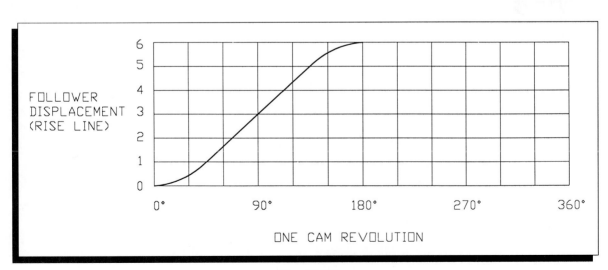

Fig. 16-7C

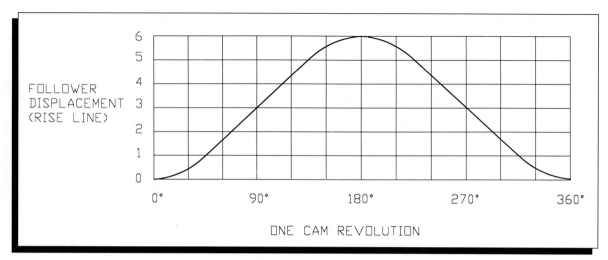

Fig. 16-7D

▶ Harmonic Motion

Harmonic motion is a non-linear motion that provides for a smooth start and stop. It is based upon trigonometry. Each point on the curve is the sine of its respective angle *(Fig. 16-8)*.

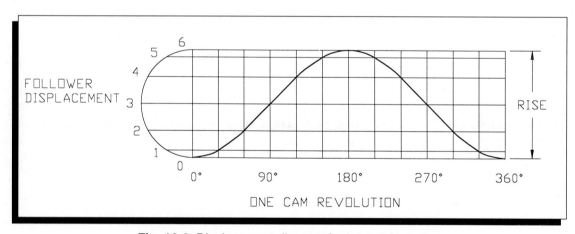

Fig. 16-8 Displacement diagram for harmonic motion.

ACTIVITY

To draw a harmonic motion displacement diagram, use the following procedure. For this exercise, the cam has a harmonic rise of 2 inches through 180 degrees.

1. Draw the length of the diagram and the vertical lines as you did in the previous "Your Turn."

2. Draw a 180-degree arc on the left end of the diagram. The radius of the arc equals one half of the rise, or 1 inch. The endpoints of the arc are at the top and bottom of the first vertical line.

3. Use the **DIVIDE** command to divide the semicircle into 6 equal parts. Each part will represent 30 degrees (180 divided by 6 = 30). Draw horizontal lines at each division point and label the diagram *(Fig. 16-9A)*.

4. Project each point on the arc to the displacement diagram. Mark the 30-degree intersections with the vertical lines. Note that each point on the arc will correspond to two points on the diagram *(Fig. 16-9B)*.

5. Connect the points with a polyline and fit the polyline into a smooth curve *(Fig. 16-9C)*.

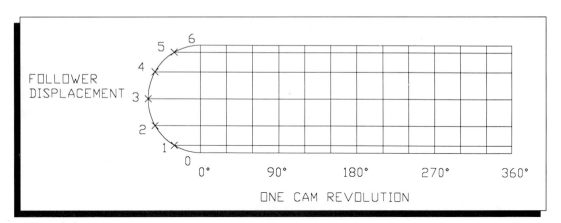

Fig. 16-9A

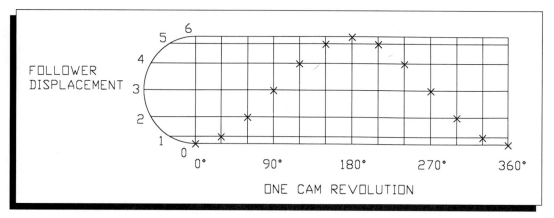

Fig. 16-9B

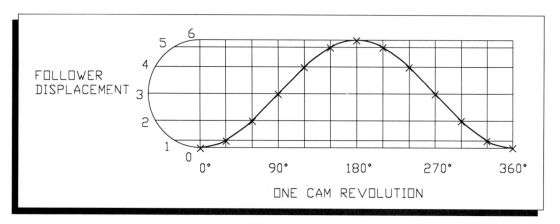

Fig. 16-9C

▶ Uniformly Accelerated and Decelerated Motion

A cam that has uniformly accelerated and decelerated motion has a steadily increasing and decreasing rise. The motion produced is very smooth. The action is similar to harmonic motion, but the displacement diagram curve is a parabola, not a sine curve *(Fig. 16-10)*. A parabola is formed when a cone is cut vertically at any place other than through its center *(Fig. 16-11)*.

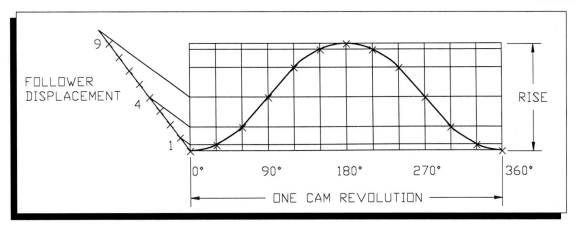

Fig. 16-10 Displacement diagram for uniformly accelerated and decelerated motion.

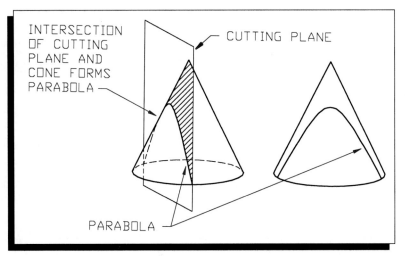

Fig. 16-11 A parabola results when a cone is sliced vertically at any plane except the one that goes through its vertex.

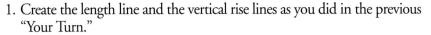

ACTIVITY

To draw a uniformly accelerated and decelerated motion displacement diagram, use the following procedure. This cam has a rise of 2 inches over 180 degrees.

1. Create the length line and the vertical rise lines as you did in the previous "Your Turn."

The rise increments are based upon the square of the number of length segments in half the rise. Since the length represents 360 degrees, the rise equals 180 degrees and half the rise equals 90 degrees or 3 segments.

$$3^2 = 9$$

499

2. Draw a line at a convenient angle and length from the lower left or right corner of the rectangle.

3. Mark the line into 9 equal parts *(Fig. 16-12A)*.

4. Connect the end of the line (point 9) to the midpoint of the vertical rise line.

Use the ENDpoint and MIDpoint object snaps to do step 4 accurately.

5. Draw a line parallel to the line in step 4 from point 4 of the angled line to intersect the vertical rise line.

The easiest way to do this is to use the OFFSET command to offset the line through point 9. Use the Through option with the NODe object snap to offset the line through point 4. Note that you will have to extend the line to the right to meet the rise line. You may also want to trim the line to the left of the angled line to reduce confusion.

6. Draw a line parallel to the lines in steps 4 and 5 from point 1 of the angled line to intersect the vertical rise line.

7. Draw horizontal lines from the three vertical rise line intersection points you established in steps 4 through 6. These are the rise increments *(Fig. 16-12B)*.

8. Complete the grid pattern by mirroring the lines at points 1 and 4 using the line at point 9 as the mirror line.

9. Mark the successive rise and length points for the complete revolution, connect them with a polyline, and fit the polyline into a smooth curve *(Fig. 16-12C)*.

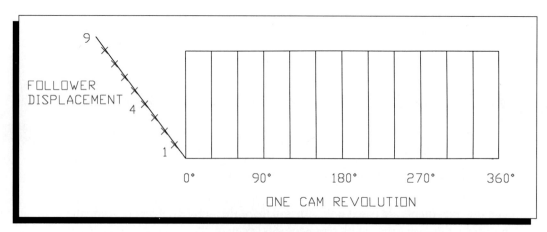

Fig. 16-12A

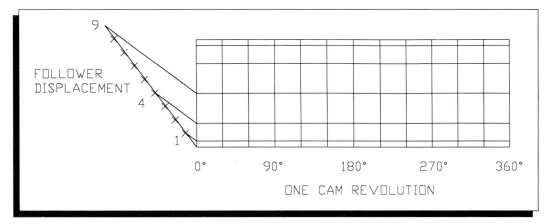

Fig. 16-12B

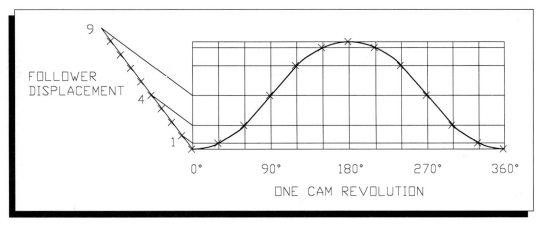

Fig. 16-12C

▶ Combined Motions

Simple cams are cut for one motion. Complex cams combine the motions. A complex cam may have a time period, called **dwell**, during which the follower does not rise or fall. The dwell frequently forms a transition between different types of motion *(Fig. 16-13)*.

Drawing a Cam Profile

A cam's profile is the shape of the surface on which the follower rides. To understand the concepts of drawing a cam profile, you can work with a uniform motion cam, as demonstrated in the following "Your Turn."

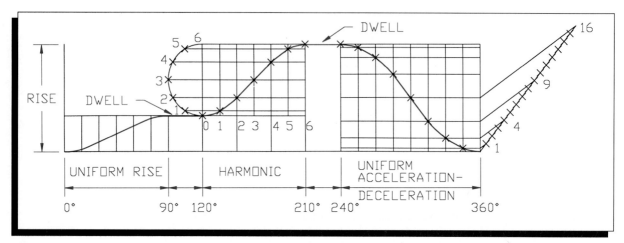

Fig. 16-13 This is an example of a displacement diagram for a complex cam. In the first 90 degrees, the cam uniformly rises through one third of its total rise. Then it dwells for 30 degrees. Beginning at 120 degrees, the cam moves through the rest of its rise using harmonic motion. After another dwell for 30 degrees, the cam fails using uniform acceleration-deceleration.

ACTIVITY

In this procedure, you will draw a cam that rotates in a counterclockwise direction and has a roller type follower *(Fig. 16-14)*. This cam has a follower rise of 1.8 inches.

1. Create a layer called CENTER and assign the CENTER linetype to it. Make CENTER the current layer.

2. Draw horizontal and vertical center lines and add a base circle with a radius of 2 units *(Fig. 16-14A)*.

The radius of the base circle is the distance from the center of the cam to the center of the follower roller when the two are closest together.

3. Offset six circles on the outside of the base circle. Each circle should be concentric with the base circle and offset at a distance of ⅙th of the follower rise. In this instance, the follower rise is 1.8 units. Therefore, each circle will be offset 0.3 units from the previous one (1.8 divided by 6 = 0.3) *(Fig. 16-14B)*.

4. Array the vertical center line seven times for 180 degrees to divide the base circle into 30-degree segments. Use a polar array and rotate the objects as they are copied. The base point should be the center of the circles *(Fig. 16-14C)*.

You may use smaller segments for greater accuracy.

5. Create circles 0.125 units in diameter to represent successive positions of the roller. Start at the 12:00 position and move clockwise. Place the center of each circle at the intersection of successive 30-degree radial lines and concentric circles. Fig. 16-14D shows the first seven circles in the right half. Use the COPY Multiple option with the INTersection object snap to do this quickly.

6. Complete the 0.125-unit diameter circles in the left half by reversing the concentric circles in step 5. Use the COPY Multiple option or the MIRROR command. (If you use MIRROR, select only the circles on the right side of the drawing to mirror, and use the vertical center line as the mirror line.)

7. Draw a polyline tangent to the point on each 0.5-unit circle that is closest to the center of the base circle. Fit the polyline into a smooth curve *(Fig. 16-14E)*.

This represents the actual profile of the cam.

8. Erase the construction circles so that your drawing looks like the one in Fig. 16-14F.

The curve traced by the center of the roller is the path the follower will take.

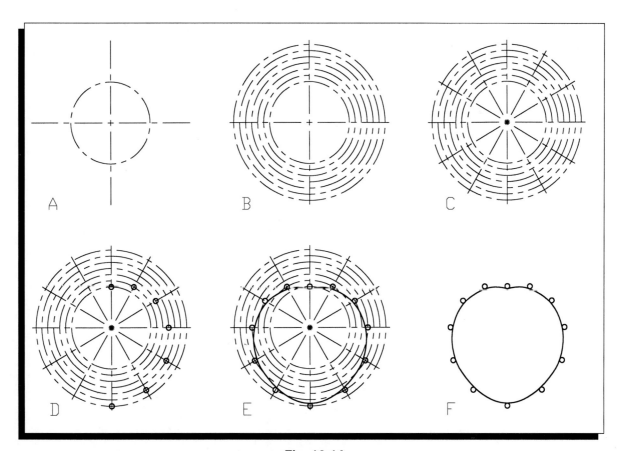

Fig. 16-14

Gears

Gears are mechanical devices that transmit motion and frequently change the speed and/or direction of the motion. Smooth gears operate by friction *(Fig. 16-15A)*. The surfaces of the gears are in firm contact. As one gear turns, friction turns the other. The major problem with smooth gears is slippage. To overcome slippage, teeth are added to gears *(Fig. 16-15B)*.

Gears transmit rotary motion. A shaft turns one gear, called the drive gear. The drive gear's teeth engage and turn another gear, called the driven gear. The driven gear rotates in the opposite direction from the drive gear. When two gears are not the same size, the larger gear is called a **spur gear**, and the smaller gear is a **pinion** *(Fig. 16-16)*.

The relationship between the speed of rotation of a spur gear and a pinion is directly proportional to their diameters. Since the teeth of mating gears are the same size, the diameters of the gears determine the number of teeth each will have. The number of teeth determines their relative speeds *(Fig. 16-17)*. For example, two gears with 12 teeth each will turn at the same speed. If a 24-tooth spur gear is the drive gear and the pinion has 12 teeth, the pinion will turn twice while the spur gear turns once. The gear ratio is 1:2 (drive gear to driven gear). If the 12-tooth pinion is the drive gear, the gear ratio is 2:1.

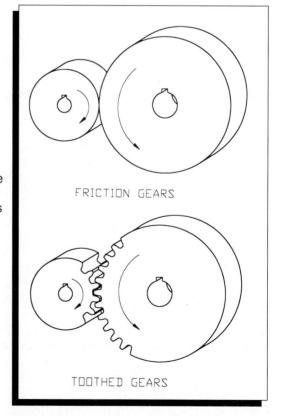

Fig. 16-15 Friction gears are prone to slippage. The addition of teeth to the working surface of the gears decreases slippage.

FRICTION GEARS

TOOTHED GEARS

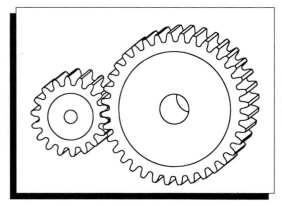

Fig. 16-16 The larger gear is a spur gear, and the smaller one is a pinion.

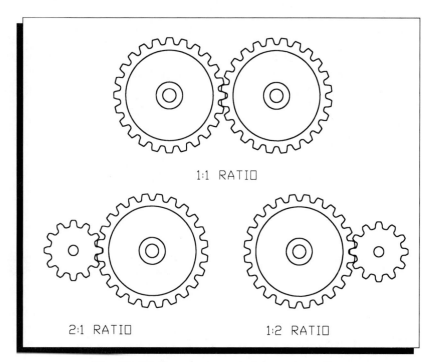

1:1 RATIO

2:1 RATIO

1:2 RATIO

Fig. 16-17 Spur gear and pinion gear ratios.

Gear Terms and Formulas

Drafters need to know the terms used to describe gears and the formulas for designing them. Refer to Fig. 16-18 as you read the following definitions and formulas. In the formulas, N equals the number of teeth. You will learn more about the terms as you read this chapter.

- *Addendum* (A) is the distance from the pitch diameter to the top of the teeth. $A = \frac{1}{DP}$.
- *Base circle diameter* is the diameter from which the tooth involute profile is developed.
- *Chordal thickness* (CT) is the thickness of the teeth at the pitch diameter. $CT = \frac{CP}{2}$.
- *Circular pitch* (CP) is the distance along the pitch circle from a point on one tooth to the same point on the next tooth. $CP = \frac{3.1416}{DP}$.
- *Dedendum* (D) is the depth of the teeth from the pitch diameter to the root. $D = \frac{1.157}{DP}$.

- *Diametral pitch* (DP) is the number of teeth per inch of pitch diameter. $DP = \frac{N}{PD}$.
- *Outside diameter* (OD) is the overall diameter of the gear. $OD = \frac{(N+2)}{DP}$.
- *Pitch diameter* (PD) is the diameter of an imaginary circle formed by the tangent of mating gears. This is the primary dimension of gear measurement. Most gear dimensions are taken from the pitch diameter. $PD = \frac{N}{DP}$.
- *Pressure angle* is the angle of contact or pressure between mating teeth. It is usually 14.5 or 20 degrees.
- *Root* is the area at the bottom of the teeth.
- *Root circle diameter* (RD) is the diameter of a circle drawn through the root. $RD = PD - 2D$.

ACTIVITY

A gear data table contains the following information:
- number of teeth
- diametral pitch
- pressure angle
- pitch diameter
- addendum
- chordal thickness
- outside diameter

Complete the gear data tables for the following sets of data.

1. A gear has 48 teeth, a diametral pitch of 8, and a 20-degree pressure angle.

2. A gear has 72 teeth, a pitch diameter of 6, and a pressure angle of 14.5 degrees.

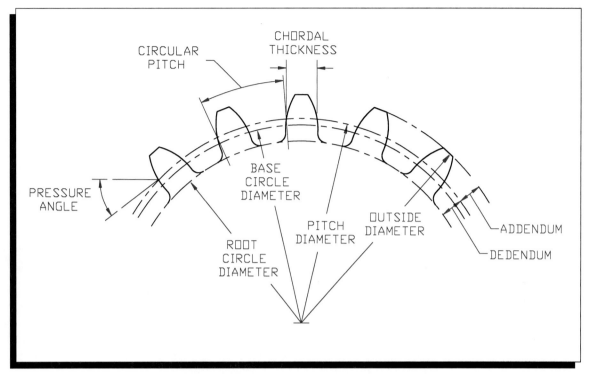

Fig. 16-18 Terms used to describe gear teeth.

Gear Teeth

The design of gear teeth is critical to overall gear operation. Gears should operate smoothly with a minimum of vibration. The most widely used tooth form is an involute. An **involute** is the path formed by a point on a line as it is unwound from the circumference of a circle *(Fig. 16-19)*. Imagine a taut string being unwound from a wheel. Drafters use an arc as an approximate involute.

Working drawings usually show gears in section, and the circular views often are omitted. Even when the circular views are included, they do not include the gear teeth *(Fig. 16-20)*. The gear tooth data is given in a gear data table.

Gear design drawings usually include gear teeth. In addition, working drawings may have one or two teeth drawn. As a drafter, you should be familiar with the procedure for drawing gear teeth.

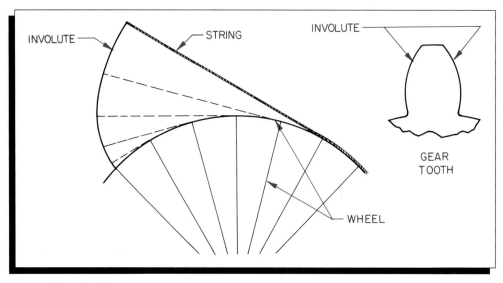

Fig. 16-19 An involute formed by a string being unwound from a wheel.

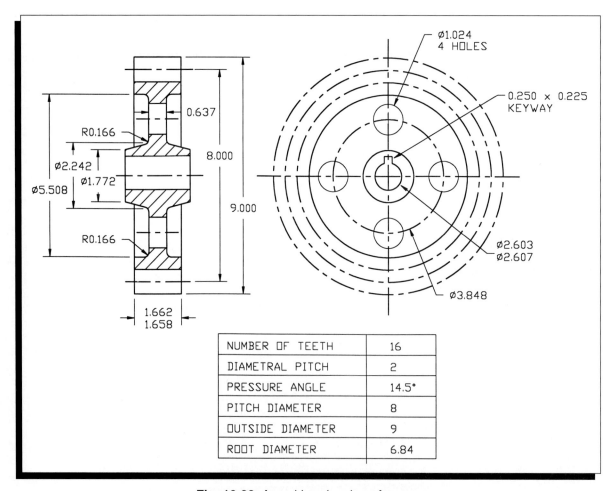

NUMBER OF TEETH	16
DIAMETRAL PITCH	2
PRESSURE ANGLE	14.5°
PITCH DIAMETER	8
OUTSIDE DIAMETER	9
ROOT DIAMETER	6.84

Fig. 16-20 A working drawing of a gear.

ACTIVITY

Follow the procedure below to draw gear teeth with the data table given.

Use the ZOOM command and object snap modes for accuracy. You may want to use two viewports to help you keep your perspective when you are zoomed in close.

Data Table	
Number of teeth	36
Diametral pitch	6
Pressure angle	14.5
Pitch diameter	6
Outside diameter	6.333
Root diameter	5.614

1. Draw the pitch diameter, outside diameter, root diameter, and horizontal and vertical center lines *(Fig. 16-21A)*. (Do not include the text.)

2. Draw a tangent to the pitch diameter where it intersects the vertical center line.

Since the tangent will be at the top quadrant of the circle, the fastest way to do this in AutoCAD is to use the QUAdrant object snap and the Ortho mode.

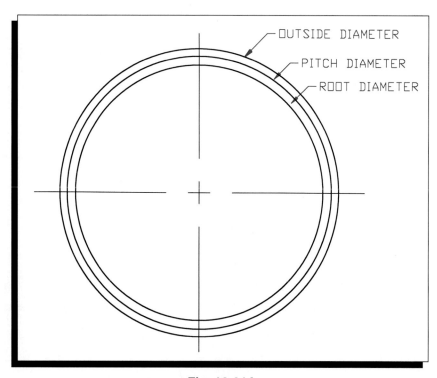

Fig. 16-21A

3. Draw a line from the intersection at a 14.5-degree angle below the tangent line.

4. Draw the base circle tangent to the line in step 4 *(Fig. 16-21B)*.

The first tooth is drawn on the vertical center line. Therefore the chordal thickness of the teeth must be calculated in degrees to the left and to the right of center. This is done using the formula 360 degrees/(4 × number of teeth).

5. Draw the chordal thickness of 5 degrees, 2.5 degrees left and right of the vertical center line. From the right side of the tooth, draw the 14.5-degree pressure angle to the left. From the left side, draw the 14.5-degree pressure angle to the right *(Fig. 16-21C)*.

Note

The formula given above calculates *half* of the chordal thickness:

360 degrees/(4 * number of teeth)
= 360/(4 * 36)
= 360/144
= 2.5

Therefore, the actual chordal thickness is 5 degrees.

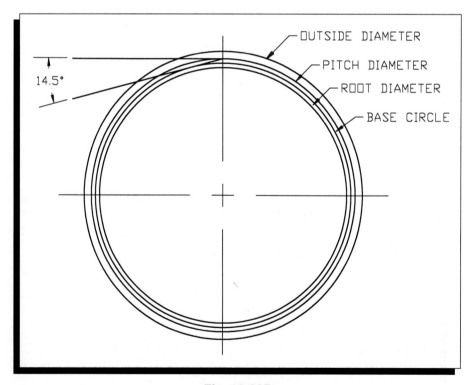

Fig. 16-21B

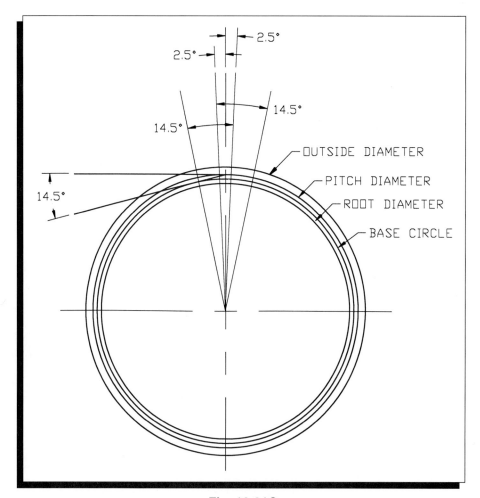

Fig. 16-21C

6. Draw circles with radii equal to ⅛ of the pitch diameter, in this case 0.75, from the intersection of the base circle and the side of the pressure angles *(Fig. 16-21D)*.

7. Trim both circles to the outside diameter and the root diameter *(Fig. 16-21E)*.

8. Erase, trim, and fillet to complete one tooth. The fillets can be any small radius *(Fig. 16-21F)*.

9. Array the tooth 36 times. Use a polar array and rotate the objects as they're copied. Trim as necessary *(Fig. 16-21G)*.

The tooth consists of five parts: two involutes, two fillets, and one top. Make sure you include all tooth parts in your array selection set.

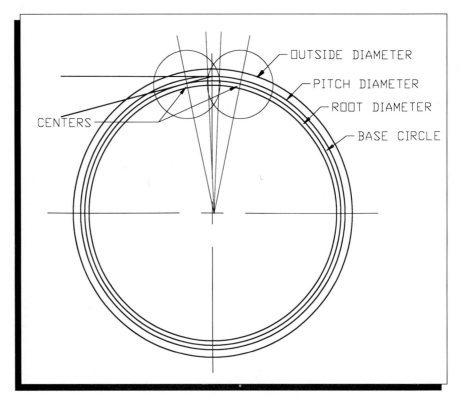

Fig. 16-21D

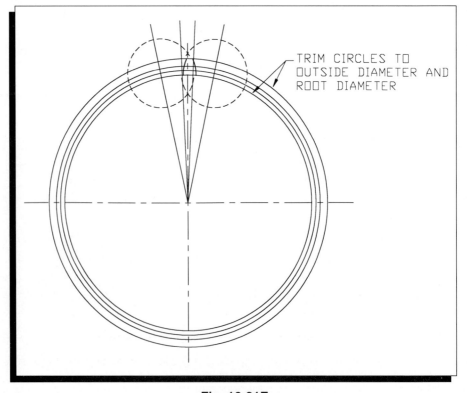

Fig. 16-21E

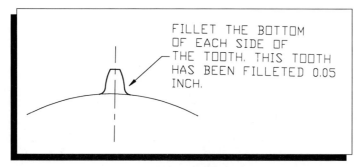

Fig. 16-21F

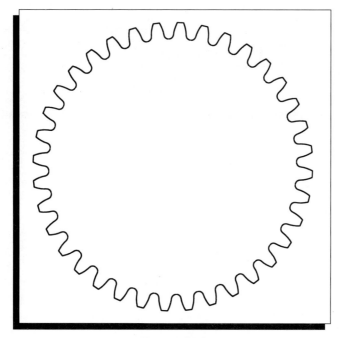

Fig. 16-21G

Bevel Gears

Bevel gears transmit motion between two intersecting shafts. The teeth are cut at an angle to the gear face *(Fig. 16-22)*. The shafts are usually at right angles.

These bevel gears are also called **miter gears**. Most of the spur terminology also applies to bevel gears.

Fig. 16-22 Bevel gears.

Rack Gears

A **rack** is a gear with a straight pitch rather than a circular pitch *(Fig. 16-23)*. The tooth profiles are straight lines. A rack usually operates with a pinion spur gear. The rack changes rotary motion to reciprocating motion. A typical application of rack and pinion gears is in the steering mechanism of automobiles. The same terminology and drawing techniques are used for racks as for spur gears.

Worm Gears

Worm gears transmit motion between shafts that are at right angles to each other but do not intersect *(Fig. 16-24)*. A **worm** is a form of a screw with threads the same shape as a rack tooth. A worm gear looks like a spur gear, but its teeth are fitted to the curvature of the worm threads. The worm gear uses the same terminology as the spur gear, but the worm is described in the same manner as a screw thread. Refer to Chapter 15 for more information about screw threads.

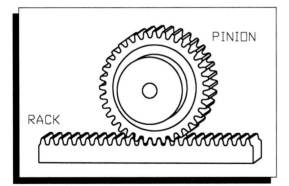

Fig. 16-23 Rack and pinion.

Fig. 16-24 A worm and worm gear.

■ Chapter 16 Review

Review

1. What is the function of a cam?
2. What is the function of a follower?
3. What are the three types of cam motion? Describe each.
4. How does a displacement diagram relate to cam motion?
5. If the drive gear is a 36-tooth spur gear and the pinion has 9 teeth, what is the gear ratio?
6. Which is smallest, the pitch diameter, the outside diameter, or the root diameter?
7. Describe an involute.
8. Which of the following are <u>not</u> gears?
   piston
   worm
   rack
   snake
   miter
   bevel

■ Chapter 16 Problems ▰▰▰▰▰▰

1. A gear data table contains the following information:
   number of teeth
   diametral pitch
   pressure angle
   pitch diameter
   addendum
   chordal thickness
   outside diameter

   Given the following data, complete the table for each of these gears.

   a. number of teeth = 16
      pressure angle = 14.5 degrees
      pitch diameter = 2
   b. number of teeth = 36
      pressure angle = 20 degrees
      diametral pitch = 6

2. Create a displacement diagram for a cam that has a rise of 3.5 inches over 180°. Assume uniform motion.

3. Create a displacement diagram for a cam that has a rise of 2.75 inches over 180°. Assume harmonic motion.

4. Create a displacement diagram for a cam that has a rise of 3 inches over 180°. Assume uniformly accelerated and decelerated motion.

5. Create a drawing of a gear (including gear teeth) using the data table below.

Number of teeth	16
Diametral pitch	2
Pressure angle	14.5°
Pitch diameter	8
Outside diameter	9
Root diameter	6.84

Using CAD

3D Studio Brings Designs to Life

Even though AutoCAD is powerful by itself, there are many ways to use it with other products that make it even more powerful. One such product is 3D Studio, which is used for rendering and animating 3D models created with AutoCAD. Renderings are drawings that look very real. They enable you to visualize what something looks like and how something will work. Animations show movement. For example, you might show how a moving part works or simulate taking a walk through a building. You have the ability to show your designs as they will actually appear after they are constructed or manufactured.

Michele Bousquet travels the world using her specialized 3D Studio skills. According to Bousquet, most 3D Studio work is divided into two types: architectural/engineering or broadcast/multimedia. The first concerns itself with a technical illustration of some kind, such as a walk-through of a house or an animation of how a ball bearing housing comes together. She finds that CAD renderings of engineering drawings are usually rather plain. With 3D Studio she is able to use dramatic lighting to make an impressive picture of an engineering model — rather like lighting a sculpture. The rendering shown here is one example.

She considers the second category to be more on the fun side. Animations created for broadcast are not created to any specifications. They are simply made to look good. Bousquet has done 3D Studio animation at a television station in Australia, the country where she currently works.

In Perth, Australia, Bousquet was once asked to make a presentation of a new engine design. An animation of a portion of the engine was required for a patent application. With a patent application, the designer usually submits paper sketches and descriptions to the patent office to give a detailed portrayal of the machine. In this case, the applicant wanted to be absolutely sure that his design would be described sufficiently so that there would be no ambiguity about what was patented or how the engine worked.

The design was submitted to Bousquet on paper sketches. It involved turning a cog with several smaller pieces that had to revolve with the cog and also move in and out of the center of the cog at a particular speed. Bousquet could not tell from looking at the sketches how the part actually worked. She started by creating the model as best she could in 3D Studio and made a few rough animations for the designer to see.

He was able to look at the animations and point out the exact spots that weren't right. After three or four such sessions, Bousquet could see exactly how the engine worked. The client was very pleased with the animation. She made three different animations, all from different views, then brought them into Animator Pro and added text and arrows to make a full presentation. The entire presentation fit on one 3½" high density disk so that the patent office could watch it on their own computer system.

Bousquet's dream for a long time was to develop a skill that she could use all around the world. She will continue fulfilling her dream with her next project — working with an architect in Belgium who wants to do more visualization and needs her help.

Chapter
17

Key Terms

working drawings   ferrous metal

plan   concrete

elevations   hydration

sections   traffic pattern

details   open floor plan

Architectural Drafting

▶ ▶ ▶

Objectives

When you have completed this chapter, you will be able to:

- understand construction materials and their applications.
- plan a house.
- draw simple plans, elevations, sections, and details.

Broadly defined, architecture includes the design of buildings, communities, and outside areas. Architectural drafters create drawings from the designs. As a drafter, you need to understand the planning, construction materials, and drawings necessary to design buildings.

Architectural Drawings

Architectural drafters begin their work with preliminary sketches *(Fig. 17-1)*. These are freehand sketches that allow drafters and designers to try various ideas of design and layout. When pre-liminary sketches come together with a pleasing and comfortable design and a workable floor plan, the drafter begins developing the working drawings.

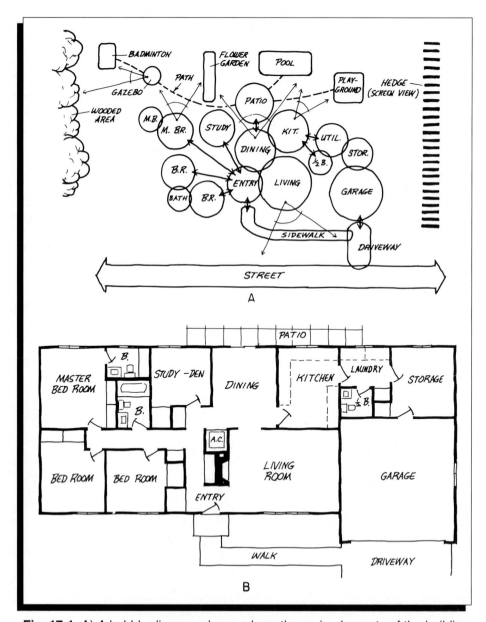

Fig. 17-1 A) A bubble diagram shows where the main elements of the building will go. B) A preliminary sketch of a floor plan.

Working Drawings

Working drawings are the drawings that guide construction. A complete set consists of many pages. To keep track of building parts, construction details are numbered and cross-referenced to specific pages. A typical cross-reference symbol is a ½-inch diameter circle containing the number of the detail in the top half and the page number in the bottom half *(Fig. 17-2)*. Other numbers on working drawings keep track of rooms, doors, and windows.

Working drawings contain three types of views: plans, elevations and sections, and details.

- A **plan** is a horizontal view that shows a building as if you were looking down from above it. A site plan is an aerial view of the location of a project *(Fig. 17-3)*. A floor plan is an arrangement of the rooms and areas on each floor *(Fig. 17-4)*. Floor plans are

the main working drawings. All the other drawings are cross-referenced to them.

- Elevations and sections are vertical views. **Elevations** describe the exterior of buildings and show locations of doors and windows *(Fig. 17-5)*. **Sections** are vertical cuts through complex building areas to show the relationships of spaces within the building *(Fig. 17-6, page 524)*. They perform the same job that any other type of section does, except in the case of architectural drafting, the object being sectioned is a building.

- **Details** are enlarged drawings of small areas of the plans. They show exactly how to construct difficult portions of buildings *(Fig. 17-7, page 525)*.

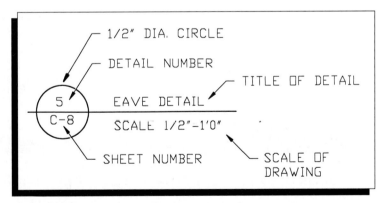

Fig. 17-2 A cross-reference symbol tells the detail number and the sheet or page on which the detail can be found.

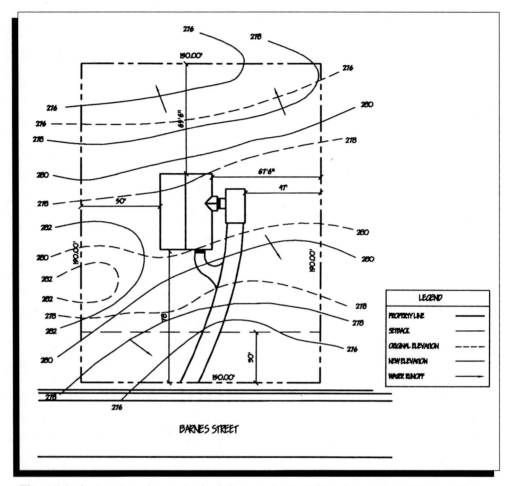

Fig. 17-3 A site plan shows the building on the building site, along with any easements or restrictions that may apply. Contour lines are also shown on some site plans.

ACTIVITY

Ask your instructor for a set of architectural working drawings. Then do the following exercises.

1. Identify each of the views described in this section.
2. Find some cross-reference symbols and identify the details on the floor plan or other views.

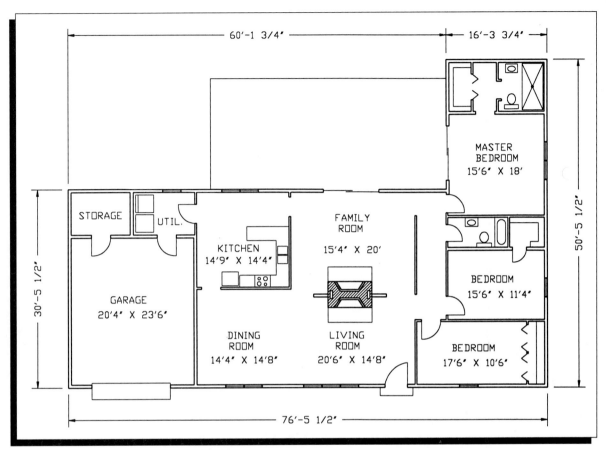

Fig. 17-4 A floor plan shows the layout of the rooms in a building.

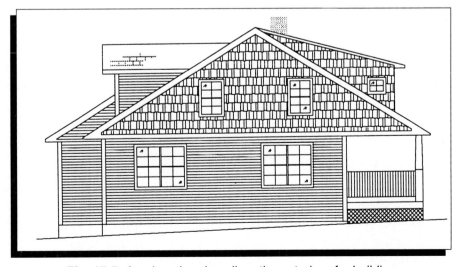

Fig. 17-5 An elevation describes the exterior of a building.

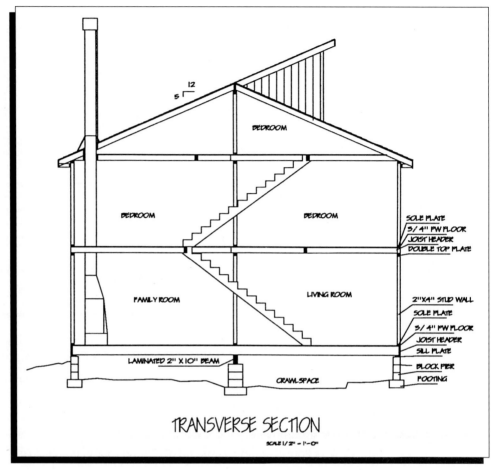

Fig. 17-6 A sectional view of a house is like any other sectional view — it allows you to see the inside of the building.

Architectural Models

Architectural models help both the architect and the client to visualize the final project. Models are three-dimensional renderings of the finished house. Both the interior and the exterior are made to scale and appearance *(Fig. 17-8)*.

Architectural models are often hand-made with cardboard, balsa wood, or other available products. However, this method of model making is a painstaking process that involves many tedious hours. Computer graphics have allowed drafters to use new ways of creating and viewing three-dimensional architectural models. Electronic models, like hand-made models, are based upon traditional architectural concepts of plans, sections, elevations, and details. Unlike hand-made models, however, computer models are easy to change. Several different versions of the model can be stored in computer memory so that a client can see the advantages and disadvantages of each. Electronic models are also extremely helpful in evaluating how projects will look and function when they are actually built.

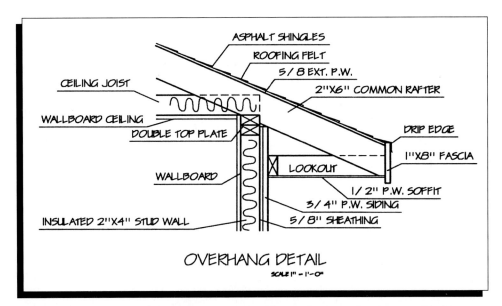

Fig. 17-7 A detail shows a closeup view of items that need to be specified in detail. This one shows details of a roof at the soffit. Note that the font used for this drawing is CITB (City Blueprint). You will find that this font is often used for architectural drawings.

Fig. 17-8 Models show both the interior and exterior to scale and appearance.

AutoCAD provides several methods of creating 3D models. The easiest way to create an architectural 3D model is to extrude a floor plan. You can create more detailed models by using extrusion, the 3DFACE command, and other basic AutoCAD features *(Fig. 17-9)*.

Models can be further enhanced through a variety of means. Several companies have developed specialized products for creating architectural models. One example is described in the story in this chapter featuring Ketiv Technologies and RenderStart Technologies. All of these products work to simulate the use of materials, lights, and cameras to show what a building will look like *(Fig. 17-10)*.

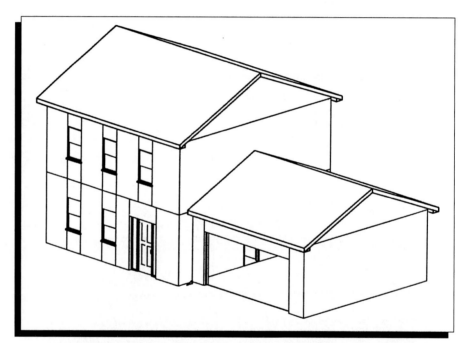

Fig. 17-9 A three-dimensional model created with AutoCAD.

Fig. 17-10 This image was produced in AutoCAD with AccuRender by Roy Hirshkowitz. The AccuRender image is untouched by any other graphics software.

Construction Materials

As an architectural drafter, you must understand construction materials and their applications. Major construction materials include wood, ferrous metal, and concrete *(Fig. 17-11)*. In addition, you should be familiar with the application of other materials such as masonry, glass, aluminum, copper, and bitumen.

Fig. 17-11 Three of the most common construction materials are wood, ferrous metal (usually steel), and concrete.

Wood

Wood is a cellular organic material containing fibers that are aligned in one direction. It is strongest when it is stressed perpendicular to the fibers, rather than parallel to them. Softwood (trees that retain their needlelike leaves all year) is used for framing and construction. Hardwood (trees that shed their broad leaves yearly) is used for finish materials such as floors and cabinets.

Moisture content is an important factor to consider when you are working with wood. Wood expands when moisture is added and contracts when moisture is removed. Lumber should have a moisture content compatible with the environment in which it is to be used. For example, a 9% moisture content is good for wood at 70 degrees with 50% humidity. Avoid wood that has been kiln-dried to 0% moisture content.

Wood is measured in board feet *(Fig. 17-12)*. One board foot can be 1 foot long, 1 foot wide, and 1 inch thick. It can also be 1 foot long, 6 inches wide, and 2 inches thick or 6 inches long, 6 inches wide, and 4 inches thick. Compare these three sets of measurements. In each case, length times width times thickness equals 144 cubic inches. One board foot is any combination of length, width, and thickness that equals 144 cubic inches.

Wood is easy to work with, but it has drawbacks. It rots, burns, and is subject to destruction by insects such as termites and carpenter ants. In most cases, these problems can be minimized, if not eliminated entirely, by treating the wood with preservatives, retardants, and pesticides.

Wood products have been developed for specialized applications. Plywood adds strength to large sheets and can provide a hardwood veneer to reduce finish costs *(Fig. 17-13)*. Particle board can reduce plywood costs for subflooring, sheathing, and roofing *(Fig. 17-14)*. Glue-laminated structural members add strength and reduce costs for beams and joists *(Fig. 17-15)*. Pressure-treated wood is being used for foundations, below-grade flooring, and extreme moisture conditions.

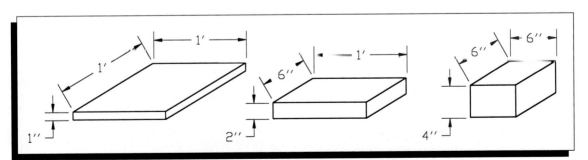

Fig. 17-12 One board foot equals 144 cubic inches.

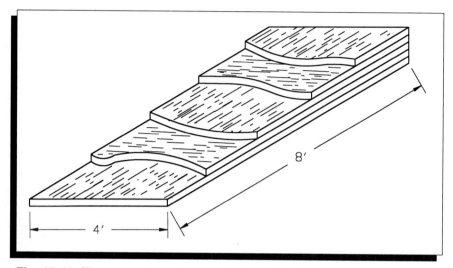

Fig. 17-13 The grain of each layer of plywood is at 90 degrees to the previous layer. This makes plywood a very strong substitute for wood in some situations.

Fig. 17-14 Particle board is often used for the subfloor of a house. It is made of small pieces of wood bonded together.

Fig. 17-15 Glue-laminated wooden structural I-beams or joists are actually stronger than their natural wood counterparts.

Ferrous Metal

A **ferrous metal** is iron or any alloy that has an iron base. Iron is used in construction in the form of cast iron, malleable iron, or wrought iron. Cast iron is made by pouring molten iron into sand castings. It is hard and brittle, and it is used extensively for fancy shapes that are not easily machined. Malleable iron is made softer than cast iron by the process of annealing, which is the controlled heating and cooling of iron. Iron hardware is made from malleable iron. Wrought iron is soft and is more than 97% pure iron. Decorative ironwork and sheet material are made from wrought iron.

Steel is an iron-based alloy with a low carbon content and fewer natural impurities than iron. Carbon content affects the strength and workability of steel. The higher the carbon content, the stronger the steel, but the lower its workability. The effective range of carbon in construction steel is 0.5 to 1.2 percent.

Steel designations are assigned by the American Society for Testing and Materials. A-36 is the most common type of steel. A-7 is slightly weaker. A-242 is stronger but more expensive.

Steel has many uses in construction *(Fig. 17-16)*. Joists, beams, decking, straps, plates, and boots are some applications. Steel is held together with bolts, rivets, and welds.

Two steel specialties are used for exposure to weather. Stainless steel has a high chromium content and provides an excellent surface that can be exposed untreated to the weather. Galvanized steel has a zinc coating that can be painted or surface-finished for improved appearance.

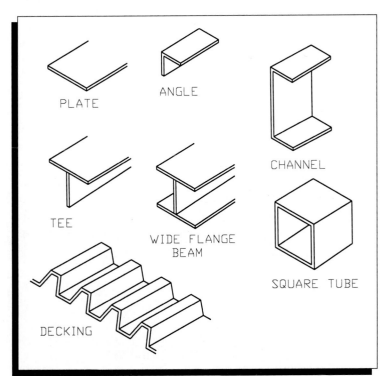

PLATE

ANGLE

CHANNEL

TEE

WIDE FLANGE BEAM

SQUARE TUBE

DECKING

Fig. 17-16 Examples of steel applications.

Concrete

Concrete is the final result of mixing cement, aggregate, and water. The standard for cement is called Portland cement and is established by the Portland Cement Association. The aggregate consists of sand and either gravel or stone, depending upon the intended application of the concrete. The water must be drinkable and free of high mineral concentrations.

An important rating of concrete is its compressed strength. A rating of 3,000 pounds per square inch (PSI) is considered standard. A 2,000 PSI rating is satisfactory for low traffic areas, whereas special structural members may have a rating as high as 7,000 PSI. The rating is controlled to a large extent by the water content of the concrete. Six gallons of water for a 94-pound sack (1 cubic foot) yields a compressed strength of about 3,000 PSI. Eight gallons of water will reduce the strength by about 1,000 PSI.

In addition to the quantity of water, two other factors are important to the compressed strength of concrete: temperature and the rate at which the mixed water is allowed to leave the mixture. The period of time it takes the concrete to reach its final strength is called its curing time. The optimum curing temperature of concrete is 74 degrees. Increasing the temperature to 105 degrees reduces the strength by 20 percent. Reducing the temperature to 40 degrees reduces the strength by 15 percent.

Cement is primarily lime. **Hydration** is the chemical reaction between water and lime while the concrete is curing. The longer the hydration time, the stronger the concrete. Hydration does not require air. If concrete is cured under water, hydration can take place for over a year with a resulting increase in strength.

Concrete works under compression, but it may fail when placed under tension *(Fig. 17-17)*. Steel works well under tension. When steel is set into concrete, the steel assumes the tension loads and reinforces the concrete. The two materials work well together because they have similar coefficients of thermal expansion. That is, they expand and contract at a similar rate when exposed to temperature changes.

The forms required to hold concrete while it is curing are a major expense in concrete construction. Concrete weighs about 150 pounds per cubic foot. Elaborate forms reinforced with steel are needed to hold the concrete during its curing process *(Fig. 17-18)*. Standard concrete sets up within hours, but it doesn't reach its full (100 percent) strength for 28 days. It reaches approximately 50 percent of its total strength after 3 days and 70 percent after 7 days. The final 30 percent takes 21 days. The forms are removed when the concrete is strong enough to support itself. This depends upon variables such as the mix, temperature, quantity, and application.

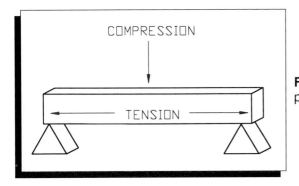

Fig. 17-17 Tension and compression are opposite forces.

Fig. 17-18 Forms have been built around the steel bars on the left side of this photograph. Concrete is poured into the forms to hold the shape of the concrete until it is solid enough to hold the shape by itself. The forms are then removed.

Miscellaneous Materials

Architects, builders, and drafters work with many other materials, including masonry, aluminum, copper, glass, gypsum, bitumen, sealants, and insulation. Different specialties may require in-depth knowledge of certain materials, but all architectural drafters should be familiar with the basic properties and applications of each.

1. Find two examples each of wood, ferrous metal, and concrete construction. Describe the examples with sketches or pictures.

2. Do you think the construction in your examples could be done realistically and safely with an alternate building material? For example, could you replace the wood with concrete or the ferrous metal with wood? Explain why you think so.

3. If you decided in question 2 that you could replace one material with another, how would the example be affected? If the example would not be affected, explain why not.

4. Start AutoCAD and enter **BHATCH** at the *Command:* prompt.

5. At the Boundary Hatch dialogue box, pick "Hatch Options ..." button.

6. At the Hatch Options dialogue box, make sure the "Stored Hatch Pattern" box is selected. Pick "Pattern..." button.

7. The Hatch Pattern dialogue box appears.

   A. Browse through the patterns.

   B. How many hatch patterns does AutoCAD have stored?

   C. Which patterns would you select for wood, ferrous metal, and concrete construction?

8. Cancel the dialogue boxes and exit AutoCAD.

House Planning

Architects usually design for a client. The architect's primary responsibility, therefore, is to satisfy the wants and needs of the client. To this end, the architect must determine:

- how much the client can afford.
- the style of the building.
- the intended use of the building.
- the site configuration.
- building codes and deed restrictions.

Although architects may design projects from skyscrapers to sheds and from mansions to condominiums, the balance of this chapter will concentrate on the design of houses. Some of the considerations involved in house planning include:

- the style of the house. Will it be a cape, saltbox, garrison, or split level? How many floors will it have? Will it have an attached garage, a detached garage, a carport, or none of these?
- the foundation of the house. Will it have a full foundation, a crawl space, or a slab? Will the construction be poured concrete or concrete block?
- the type of construction. Will it be wood or concrete?
- the type of roof. Will it be a gable, flat, hip, or gambrel roof?
- the type of windows and doors.

• the type of heating and cooling system. Will it have oil or gas, air or water? Will it use an alternate fuel such as electricity, wood, or coal? Will it have a backup system?

As a novice drafter, you may be confused by the terminology of architecture that applies to houses. Common parts of a house are identified by name in Fig. 17-19.

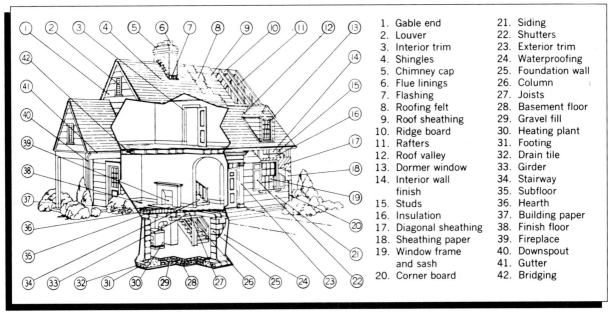

1. Gable end
2. Louver
3. Interior trim
4. Shingles
5. Chimney cap
6. Flue linings
7. Flashing
8. Roofing felt
9. Roof sheathing
10. Ridge board
11. Rafters
12. Roof valley
13. Dormer window
14. Interior wall finish
15. Studs
16. Insulation
17. Diagonal sheathing
18. Sheathing paper
19. Window frame and sash
20. Corner board
21. Siding
22. Shutters
23. Exterior trim
24. Waterproofing
25. Foundation wall
26. Column
27. Joists
28. Basement floor
29. Gravel fill
30. Heating plant
31. Footing
32. Drain tile
33. Girder
34. Stairway
35. Subfloor
36. Hearth
37. Building paper
38. Finish floor
39. Fireplace
40. Downspout
41. Gutter
42. Bridging

Fig. 17-19 Common terminology used in architecture.

Area Planning

Houses are designed around three areas: living, food preparation, and sleeping. The **traffic pattern** of a house consists of the main routes people take to get from one area to another *(Fig. 17-20)*. The dining room should be near the kitchen so food can be moved back and forth easily. The kitchen should have its own outside entrance so groceries can be brought in directly. Food preparation is messy; many people do not want it to conflict with living and sleeping, although others prefer to entertain in or from their kitchens. Living areas are noisy and sleeping areas are quiet. The traffic pattern should take all these needs — and the client's wishes regarding them — into consideration.

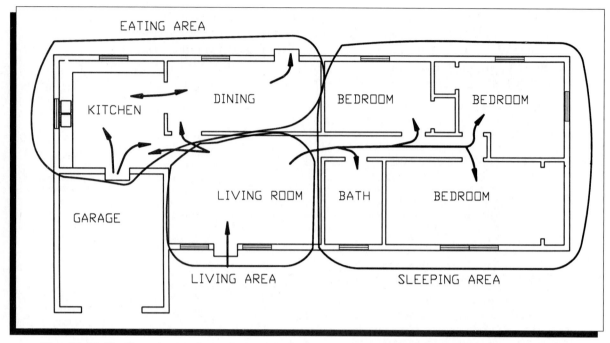

Fig. 17-20 Traffic patterns through the living, food preparation, and sleeping areas of a house.

▶ Living Areas

The main part of the living area is the living room *(Fig. 17-21)*. The living area may also include a family room, office, study, or den. Most waking hours are spent in these living areas. They should, therefore, have the best outside view. Large windows let the view in, but they reduce wall space and let heat out.

A living room should have a focal point such as a large window or a fireplace. It should be designed for furniture grouping to allow easy conversation. If possible, the front door should open into an entry hall rather than directly into the living room.

Older houses had both a "front parlor" and a living room. The front parlor had all the best furnishings and was used to entertain guests. The living room was for the family's day-to-day living. As houses became smaller, the front parlor disappeared — only to be replaced by the living room! And what took the place of the living room? The family room.

The family room, originally designed as a family recreation area, matured into a general family living space sometimes referred to as a "great room" *(Fig. 17-22)*. An **open floor plan**, a plan with a minimum of walls, often brings the kitchen-dining area into the family room so that the distinction between the living and food preparation areas becomes blurred.

Fig. 17-21 The main living area of a home should have a focal point, such as a fireplace, a large window, or a painting.

Fig. 17-22 A great room with an open floor plan is preferred by many people who like to entertain while they cook dinner. Other people like the openness of a great room.

▶ Food Preparation Area

The primary food preparation area is the kitchen *(Fig. 17-23)*. Eating areas include the formal dining room, dinette, "country kitchen" (sometimes called an eat-in kitchen), and a casual dining area taken from the family room or living room.

Kitchens are usually arranged in I-shaped, L-shaped, U-shaped, or corridor patterns. The range, refrigerator, and sink are the central pieces in the kitchen, and traffic among the three is usually arranged in a triangle *(Fig. 17-24)*. The walking space between the three appliances should be no more than 22 feet.

Kitchens need both overhead and under-counter cabinets as well as an ample pantry area for storage of appliances and utensils. There should be a minimum of 15 feet of free counter space. Both the refrigerator and the range need adjacent counter space.

Fig. 17-23 The kitchen is traditionally the site of food preparation, but today many people prefer to dine there, also.

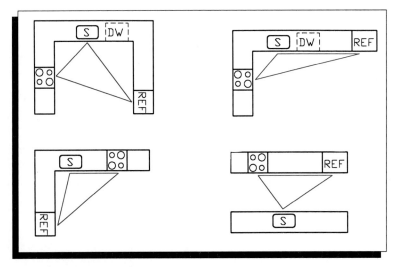

Fig. 17-24 The traffic between the range, refrigerator, and sink is arranged in a triangle.

▶ Sleeping Area

Bedrooms should be arranged so that they provide privacy and ventilation *(Fig. 17-25)*. The sleeping area should be isolated from both the food preparation area and the living area to minimize in-house noise. Bedrooms should be on the side of the house away from outside noises such as traffic and neighboring driveways. Windows are needed for ventilation, but locate them to ensure privacy and to maximize wall space for furniture.

Bedrooms need closets. Make sure the closets are large enough: 3 to 4 feet long for each person using the bedroom and at least 2 feet deep. Each closet should have a shelf above the closet rod.

▶ Bathrooms

Bathrooms belong to none of the three areas, but at the same time they belong to all of them *(Fig. 17-26)*. A four-bedroom house needs two full baths in the sleeping area. An additional half bath (one that has only a toilet and sink) should be located near the food preparation and living areas. When possible, locate bathrooms adjacent to other plumbing such as the kitchen plumbing or next to another bathroom. Design the bathroom around the fixtures, not the fixtures around the bathroom space. All fixtures should be easily accessible.

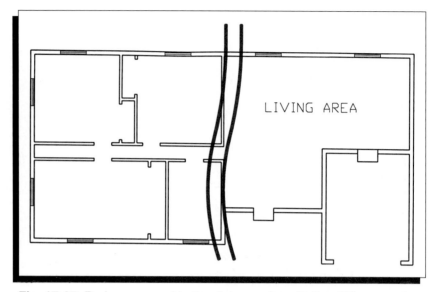

LIVING AREA

Fig. 17-25 Bedrooms need both privacy and ventilation. When possible, include windows in the bedrooms to provide natural ventilation.

Fig. 17-26 Bathrooms should be functional and convenient to the living and sleeping areas of a home.

▶ Stairs

Three common plans for stairs are straight, U-shaped, and L-shaped *(Fig. 17-27)*. Straight stairs may have a landing, but both the U-shaped and the L-shaped stairways must have landings to change directions. Triangular stairs may be used to make transitional turns when space is at a premium.

Stair design depends upon the space available and the location of the stairs within the house. The main stairs should be at least 2'8" wide and have a clear headroom of 6'8". Cellar stairs can be narrower and have less headroom. The sizes of the stair treads and risers determine how steep the stairs are *(Fig. 17-28)*. The minimum tread should be 9", and the maximum rise should be 8¼". To find the number of risers, divide the total stair height by the height of the riser.

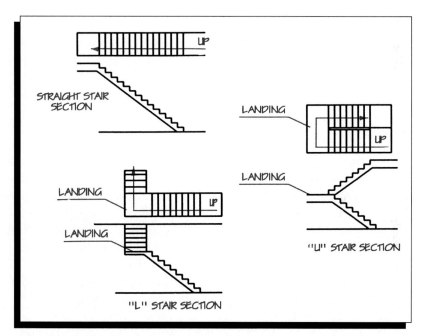

Fig. 17-27 The three most common stair layouts are straight stairs, U-shaped stairways, and L-shaped stairways.

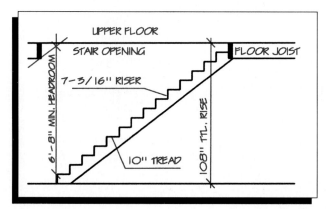

Fig. 17-28 A typical stairway.

1. Make a preliminary drawing of a floor plan for a one-story three-bedroom house with one and a half baths, a kitchen, and a living room.
2. Place the range, sink, and refrigerator in the kitchen and state the kitchen arrangement.
3. Indicate the traffic pattern for the house.
4. Place the fixtures in the bathrooms.
5. Assume that the house will have a basement, so it will need stairs. Find the number of stair risers if the total rise is 106" and the riser height is 7 ½".

Working Drawings

Working drawings begin after you develop a workable floor plan and a pleasing exterior with preliminary drawings. Before beginning your working drawings, you must understand the construction details of a house.

Symbols

Working drawings are drawn using standard architectural symbols. You may obtain these symbols from a library of architectural symbols developed by a third-party software developer, or you may draw the ones you need and save them in your own symbol library. Ask your instructor for the library of architectural symbols available in your school and how to use it. Standard symbols are used for windows and doors, plumbing and heating, electrical systems, and even landscaping *(Figs. 17-29 through 17-31)*. AutoCAD's HATCH and BHATCH commands provide material symbols for use on the drawings. Refer to Chapter 11, "Sectional Views," for more information about hatching.

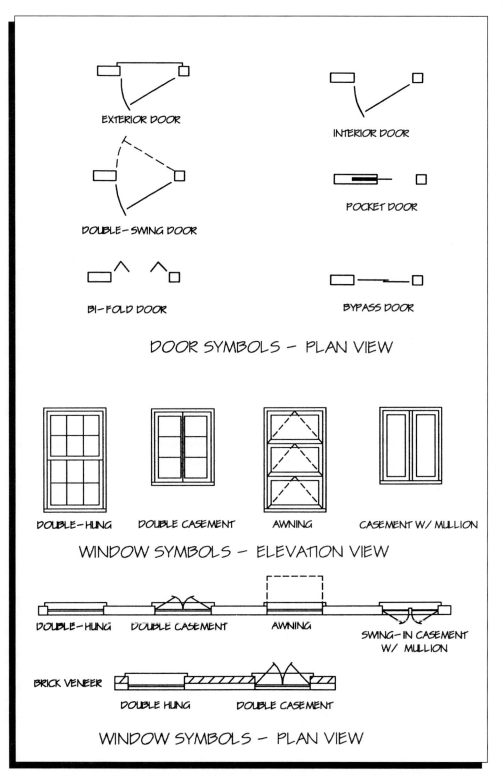

Fig. 17-29 Standard architectural symbols for windows and doors.

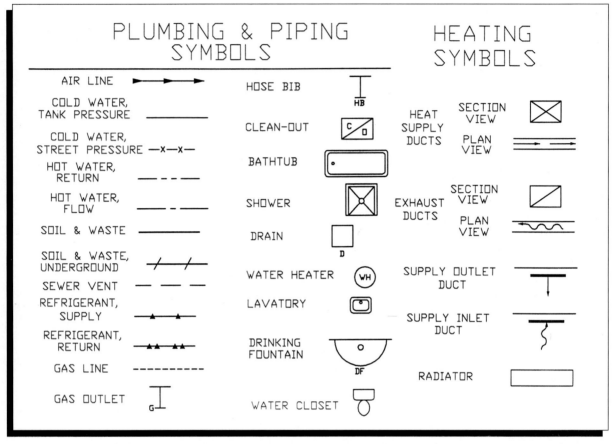

Fig. 17-30 Standard architectural symbols for plumbing and heating/air conditioning (HVAC).

Site Plans

The site plan, also called a plot plan, is a small scale layout of the house on the site *(Fig. 17-32)*. It shows the location of the house and other structures to be built relative to the property lines and existing trees and structures. The site plan also shows sidewalks, steps, and driveways outside the house.

Dimensions are an important part of the site plan. To locate a house cor-rectly, you must pay close attention to the minimum distance it can be built from the street as well as the required setback for the side and rear yards. All of these dimensions must be included on the site plan. The site plan may also include contour elevations as dimensions in feet above an existing local datum.

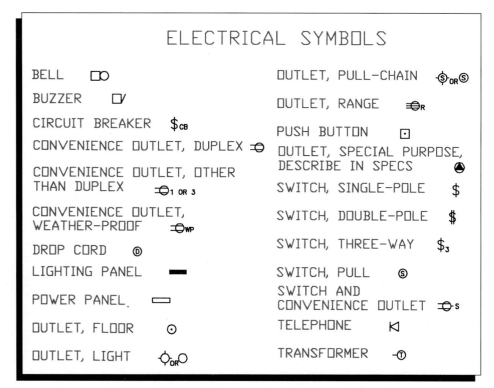

Fig. 17-31 Standard architectural symbols for electrical systems.

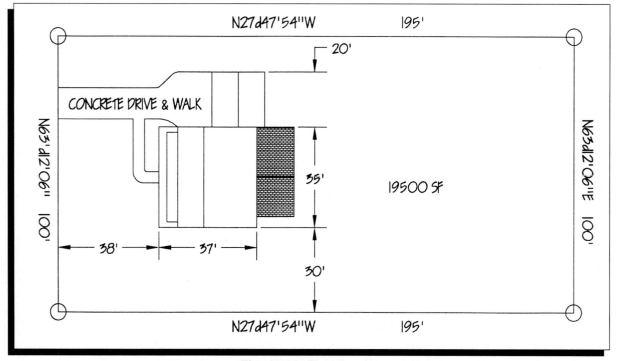

Fig. 17-32 The site plan.

Floor Plans

Floor plans form the core of architectural working drawings. You can think of them as horizontal sections through the house at the window level. They show the layout of each floor of the house *(Fig. 17-33)*. Walls appear as sections, and openings such as interior windows and doorways are blank spaces in the walls. Floor plans show:

- size and arrangement of rooms.
- location and type of windows and doors.
- wall thicknesses.
- kitchen cabinets and fixtures.
- bathroom cabinets and fixtures.
- stairs.
- plumbing and heating systems.
- electrical system.

Not all the information needed for one floor of a house is shown on one drawing. To avoid confusion, plumbing and electrical systems are often shown on separate floor plans. Window and door information may be shown in window and door schedules keyed by number to specific floor plans. Roof framing may require a separate plan.

The components of floor plans must be properly laid out and identified. In a well-executed floor plan, the rooms and appliances are identified. Dimensions are located correctly, and the type and location of all windows, doors, and other openings are identified.

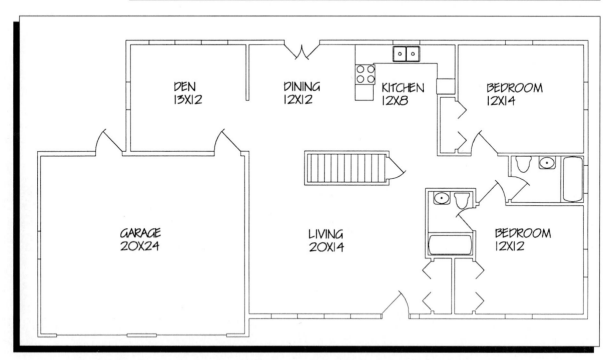

Fig. 17-33 The floor plan shows the layout of room, doors, and windows in a house.

Foundation Plan

The foundation plan is similar to a floor plan. It is the top view of the foundation before the house is built *(Fig. 17-34)*. Foundation plans show:

- size of footings and foundation walls.
- wall materials and openings.
- size and location of beams and columns.
- size and direction of floor joists.
- plumbing, heating, and electrical systems.
- stairs.

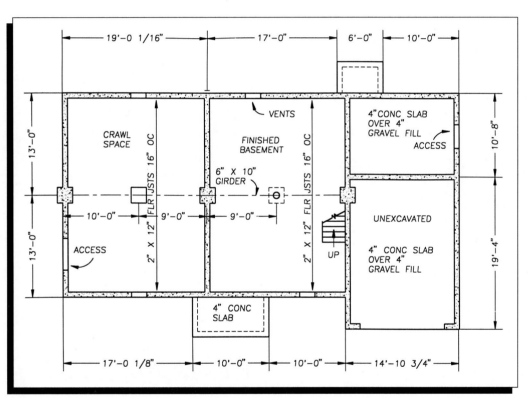

Fig. 17-34 A foundation plan shows the foundation of the house before the first floor is built.

Elevations and Sections

Elevations and sections show the vertical shape and dimensions of a house. An elevation is a view of the outside of the house *(Fig. 17-35)*. Each side of the house is usually shown and labeled by the direction it faces. All outside features are shown, including the roof line, type of siding, location of doors and windows, and grading of the soil. Very few dimensions are given. No horizontal dimensions should be placed on an elevation.

A section is a vertical slice through the entire house that helps clarify the features of a building under construction *(Fig. 17-36)*. The section shows all features on each floor, including foundation and roof, at that specific section. Sections are usually taken through a window.

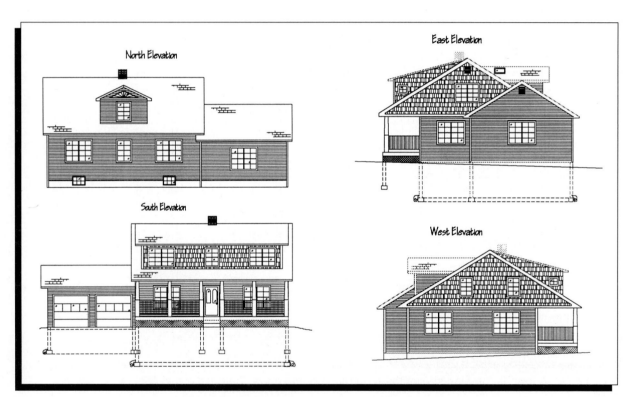

North Elevation

East Elevation

South Elevation

West Elevation

Fig. 17-35 Elevations are labeled by the direction they face.

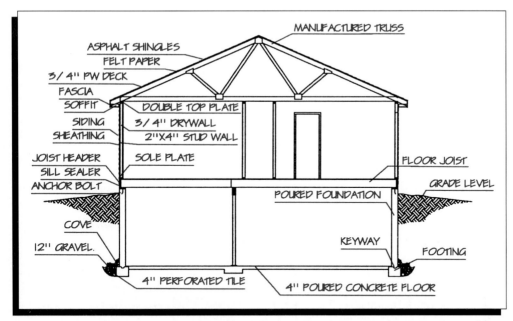

Fig. 17-36 A section of a one-story house shows features from foundation to roof.

Details

Details clarify the plans. Plan drawings are drawn to a scale that is too small to see small details. Whenever it is necessary to give localized information, a detail is drawn. Details are needed for:

- wall sections that show details from the footing to the eave *(Fig. 17-37)*.
- overhang detail *(Fig. 17-38)*.
- floor and ceiling framing. One of several types of framing may be used. In western framing, each floor is framed separately *(Fig. 17-39)*. Balloon framing has studs two stories high; the second floor hangs on a false girt inserted into the stud wall *(Fig. 17-40, page 552)*. Post and

beam framing has heavy posts and beams to carry a floor of planks *(Fig. 17-41, page 553)*. This allows less internal support for large, open areas.

- stairs *(Fig. 17-42)*.
- chimneys, fireplaces, and heating tracts *(Fig. 17-43, page 554)*.
- kitchen cabinets, bathroom cabinets, and other built-in installations *(Fig. 17-44, page 555)*.
- roof and roof framing *(Fig. 17-45, page 556)*. Common roof types include gambrel, shed, hip, Mansard, gable, and flat roofs *(Fig. 17-46)*.
- footing and foundation *(Fig. 17-47, page 557)*.

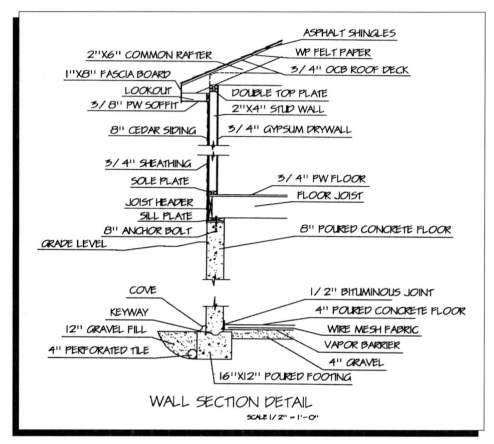

Fig. 17-37 Detail of a wall section.

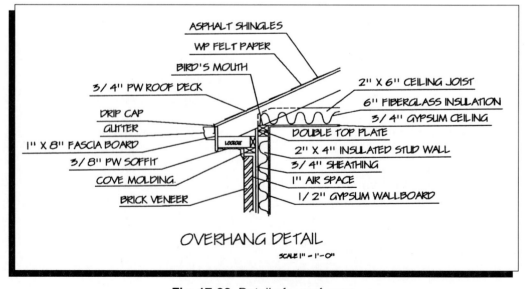

Fig. 17-38 Detail of a roof eave.

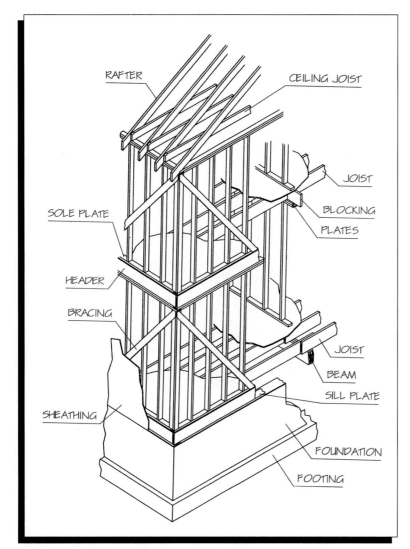

Fig. 17-39 Western framing, also called platform framing, frames each floor separately.

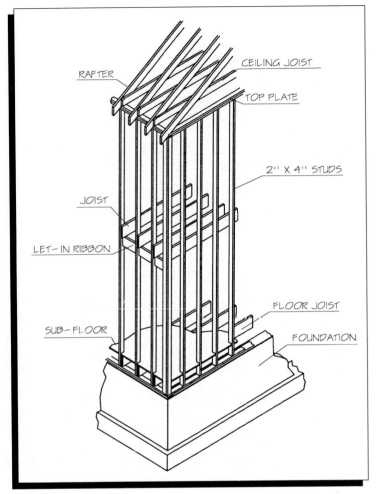

Fig. 17-40 Balloon framing uses studs that are two stories high.

ACTIVITY

Ask your instructor for a set of architectural working drawings. Then analyze the drawings by answering these questions.

1. Identify the type of drawing or drawings each page contains.
2. How many different floor plans can you find? Identify room layout, plumbing, electrical layout, and so on.
3. Do the doors and windows have a schedule or are they identified at their individual locations?
4. What type of roof does the house have?
5. Is the house built on a full basement, a crawl space, or a slab?

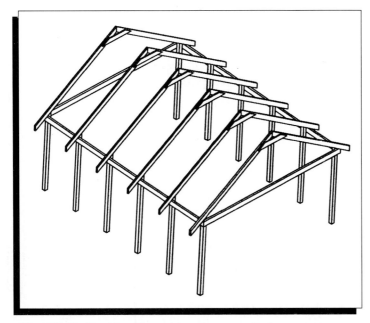

Fig. 17-41 Post and beam framing permits larger open areas.

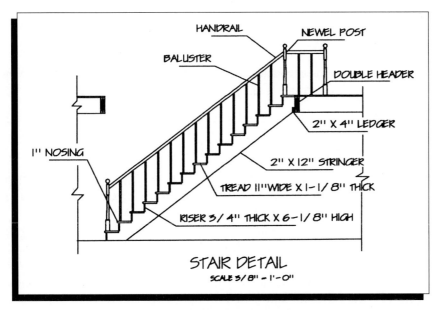

HANDRAIL

NEWEL POST

BALUSTER

DOUBLE HEADER

2" X 4" LEDGER

1" NOSING

2" X 12" STRINGER

TREAD 11" WIDE X 1-1/8" THICK

RISER 3/4" THICK X 6-1/8" HIGH

STAIR DETAIL
SCALE 3/8" = 1'-0"

Fig. 17-42 A detail of stairs.

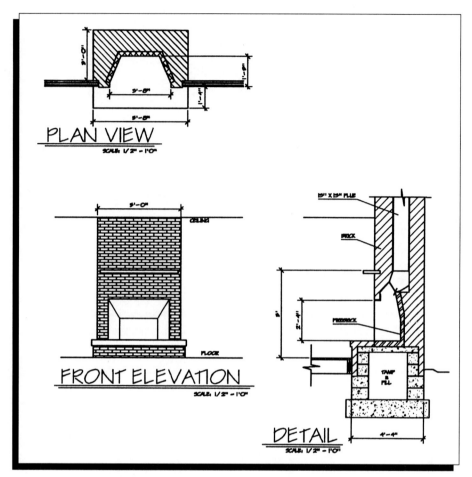

Fig. 17-43 Detail of a chimney with a fireplace.

Creating a Floor Plan with AutoCAD

Now that you are familiar with the basics, you can begin to work with AutoCAD to create floor plans. In the following "Your Turn," you will draw the floor plan of a three-bedroom ranch house. For simplicity, all the doors in this house are 28" wide, and all windows are 36" wide. You will draw a simplified window, a left-opening door, and a right-opening door and save each as a block. You can then insert the appropriate blocks into the drawing as many times as necessary.

Exterior walls, including the inside garage walls, are made of 2×6 lumber, which makes the actual wall framing thickness 5½". Interior walls are made of 2×4 lumber, so they are 3½" thick. Room dimensions, window locations, and door locations are approximate.

 Note

Wall coverings, including wall board, paneling, and siding, add to these dimensions. The following "Your Turn" uses wall framing thicknesses.

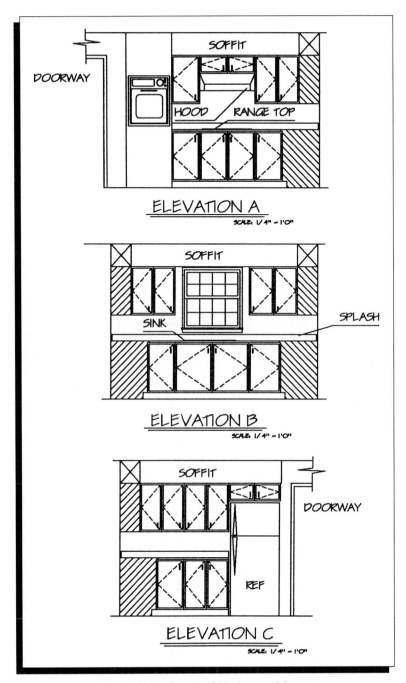

Fig. 17-44 Detail of kitchen cabinets.

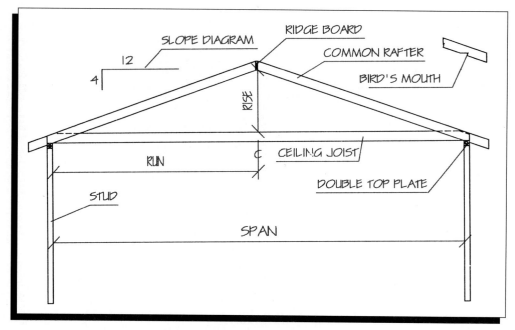

Fig. 17-45 Detail of roof framing.

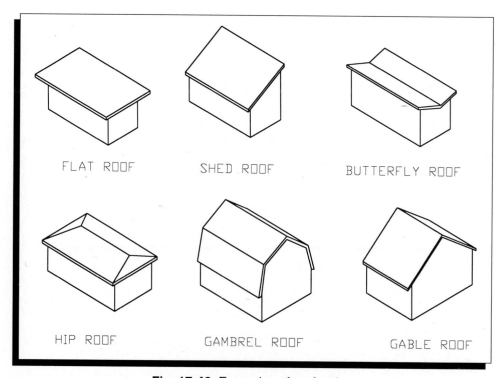

Fig. 17-46 Examples of roof styles.

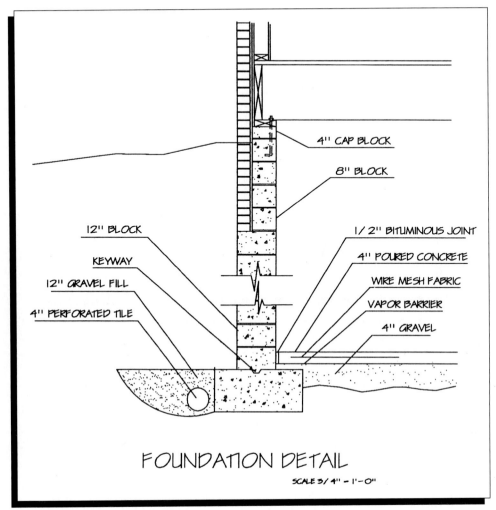

Fig. 17-47 Detail of footing and foundation.

ACTIVITY

Follow these steps to create a floor plan for a 3-bedroom house. Fig. 17-48A shows the result of this activity.

1. Create a new drawing called **FLPLAN** and set the following:

   Units: **Architectural at ¹⁄₁₆"**

   Limits: Lower left corner: **0,0**

           Upper right corner: **80',55'**

   Grid: **5'**

   Snap: **½"**

   Ortho: **ON**

   Zoom: **All**

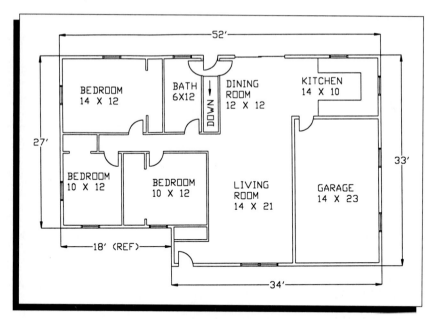

Fig. 17-48A

2. Make a new layer called **FLPLAN** and make it the current layer.

AutoCAD provides an AutoLISP routine that creates double lines. This routine is useful for creating the double lines for walls in floor plans. In Release 12, the double line (DLINE) routine is incorporated into the AutoCAD software. We will use this routine to create the walls of the house.

Refer to Fig. 17-48A as necessary while you follow these steps to create the polyline.

3. *Command:* **DL** ⏎

4. *DLINE Version 1.11, (c) 1990-1992 by Autodesk, Inc.*

Break/Caps/Dragline/Offset/Snap/Undo/Width/<start point>: **W** ⏎

Use the Width option to set the width of your double line to the thickness of the outside wall.

5. *New DLINE width <default>:* **5-½** ⏎

Be sure to include the hyphen in step 5. If you try to use a space, AutoCAD interprets your input as 5" and goes on to the next prompt. Note that you can use the decimal equivalent (5.5) if you prefer.

6. *Break/Caps/Dragline/Offset/Snap/Undo/Width/<start point>:* **65',45'**

Be sure to type the ' after each number to indicate feet. Without the ', AutoCAD will default to inches. Also, do not put a space after the comma between the numbers.

7. *Arc/Break/CAps/CLose/Dragline/Snap/Undo/Width/<next point>:* **@52'<180** ⏎
8. *Arc/Break/CAps/CLose/Dragline/Snap/Undo/Width/<next point>:* **@27'<270** ⏎
9. *Arc/Break/CAps/CLose/Dragline/Snap/Undo/Width/<next point>:* **@18'<0** ⏎
10. *Arc/Break/CAps/CLose/Dragline/Snap/Undo/Width/<next point>:* **@6<270** ⏎
11. *Arc/Break/CAps/CLose/Dragline/Snap/Undo/Width/<next point>:* **@34'<0** ⏎
12. *Arc/Break/CAps/CLose/Dragline/Snap/Undo/Width/<next point>:* **C** ⏎

Your drawing should now look like the one in Fig. 17-48B.

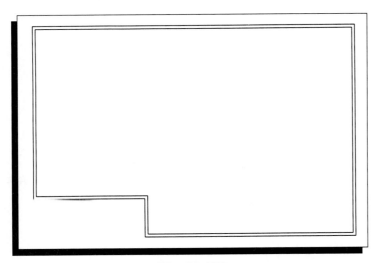

Fig. 17-48B

Since the garage probably will not be heated, the inside walls should be constructed as outside walls for the house. To draw the inside walls of the garage, you will use DLINE to offset the starting point and draw two double lines: one line 23 feet long and the other 14 feet long.

13. *Command:* **DL** ⏎

14. *Break/Caps/Dragline/Offset/Snap/Undo/Width/<start point>:* **OFFSET** ⏎

15. *Offset from:* **INT** ⏎

    of Pick the lower right-hand corner of the house (garage).

16. *Offset toward:* Pick any point inside the house.

17. *Enter the offset distance <default>:* **14'** ⏎

18. *Arc/Break/CAps/CLose/Dragline/Snap/Undo/Width/<next point>:* **@23'<90** ⏎

19. *Arc/Break/CAps/CLose/Dragline/Snap/Undo/Width/<next point>:* **@14'<0** ⏎

20. *Command:* **R** ⏎

Your drawing should now look like the one in Fig. 17-48C.

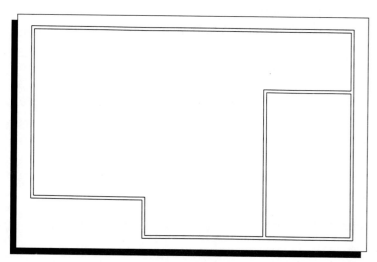

Fig. 17-48C

21. *Command:* **OFFSET** ⏎

22. *Offset distance or Through <Through>:* **12'** ⏎

23. *Select object to offset:* Pick the exterior wall segment on the 52-foot dimension.

24. *Side to offset?* Pick any point inside the house.

25. *Side to offset?* ⏎

26. *Command:* ⏎

Remember that pressing RETURN at the *Command:* prompt reenters the last command you used. In this case, it should recall the OFFSET command. If you used any other commands after step 21, such as REGEN or REDRAW, you will need to enter the OFFSET command instead of just pressing RETURN in step 26.

27. *Offset distance or Through <Through>:* **3½** ⏎
28. *Select object to offset:* Pick the new line that you just made.
29. *Side to offset?* Pick any point below the line.
30. Repeat steps 21 through 29 to offset the other wall of the central hall. This will be offset 12 feet up from the 18-foot dimension *(Fig. 17-48D)*.

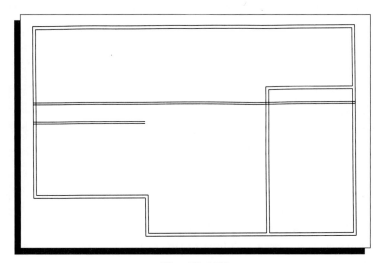

Fig. 17-48D

31. *Command:* **OFFSET** ⏎
32. *Offset distance or Through <Through>:* **28'** ⏎

This represents the width of the garage plus the living room.

33. *Select object to offset:* Pick the exterior wall segment on the 33-foot dimension.
34. *Side to offset?* Pick any point inside the house.
35. *Select object to offset:* ⏎

You may have noticed that you're getting a lot of practice using the OFFSET and TRIM commands. Since you should now be familiar with the prompts and how to use the command, the remaining steps in this procedure will tell you what to offset, trim, or extend without spelling out the procedure exactly.

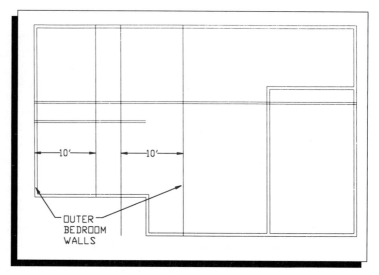

Fig. 17-48E

36. Offset the inner bedroom walls 10 feet from each of the outer bedroom walls *(Fig. 17-48E)*.

The space between the two walls will become a closet for the inside bedroom.

37. Zoom in on the bedroom area, and trim the walls and intersections *(Fig. 17-48F)*.

38. Offset each of the three vertical bedroom walls 3½ inches to give them thickness. Bring the inner walls into the closet and the living room wall into the bedroom. Trim the closet wall intersections *(Fig. 17-48G)*.

Your drawing should now look like the one in Fig. 17-48H.

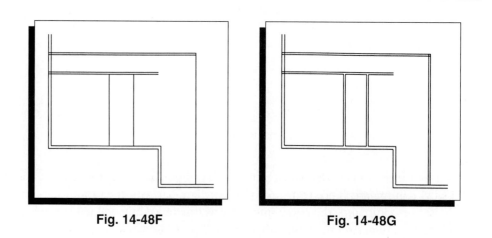

Fig. 14-48F   **Fig. 14-48G**

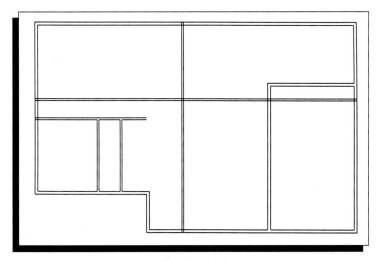

Fig. 17-48H

39. Complete the inside bedroom by extending the central hall and the front wall. Offset the front wall thickness 3½". Trim *(Fig. 17-48I)*.

40. Offset the outside kitchen wall 26' to create the bedroom side of the dining room wall *(Fig. 17-48J)*.

41. Offset the outside bedroom wall 14 feet to create the rear bedroom wall.

42. Offset one bathroom wall 17 feet and the other 23 feet from the outside bedroom wall.

43. Extend the central hall to the dining room wall and trim the walls you created in steps 40 through 42.

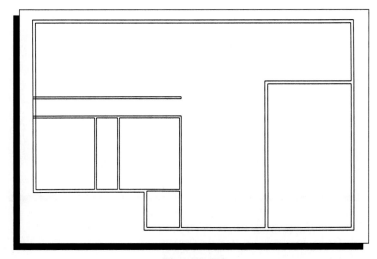

Fig. 17-48I

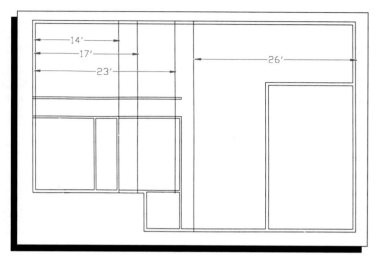

Fig. 17-48J

44. Offset the wall thicknesses 3½" and trim the wall intersections *(Fig. 17-48K)*.

45. Add the wall at the end of the central hall by offsetting the outside wall 6 feet. Then use an offset of 3½ inches to create the wall thickness and trim.

46. Add the front entry closet by offsetting the lower bedroom wall by 2 feet. Create the wall thickness and trim *(Fig. 17-48L)*.

47. Make a simplified window drawing 36" long and a left-hand and right-hand door each 28" *(Fig. 17-48M)*. Save each one as a block.

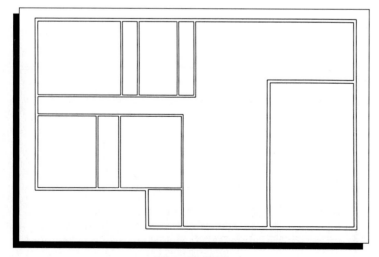

Fig. 17-48K

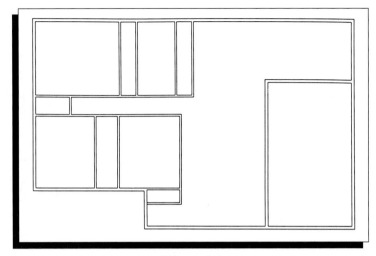

Fig. 17-48L

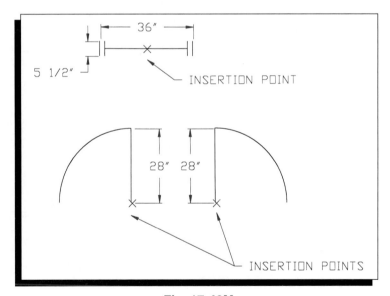

Fig. 17-48M

Use descriptive names when you name your blocks. For example, you might want to use WIN36 for the window and DR28L and DR28R. If you use these blocks later for other drawings, you will be able to tell what the block contains — and the size of the object — before you insert the block. Remember to use WBLOCK if you want to begin your own symbols library. The WBLOCK command allows you to insert the blocks into any drawing.

Hot Tip

48. Insert the windows approximately as shown *(Fig. 17-48N)*.

49. Insert the doors approximately as shown. Trim the walls from the doorways *(Fig. 17-48O)*.

50. Add the closet door openings approximately as shown.

51. Add the sliding door off the dining room and the kitchen counter *(Fig. 17-48P)*.

52. Create a new layer called **TEXT** and make it the current layer.

53. Add the text as shown in Fig. 17-48A.

54. Create a new layer called **DIM** and make it the current layer.

55. Dimension the floor plan as shown in Fig. 17-46A.

56. Plot your floor plan.

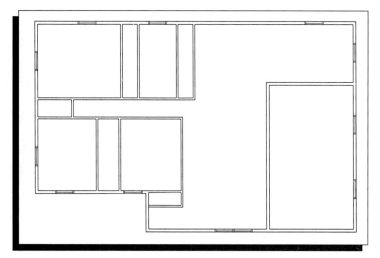

Fig. 17-48N

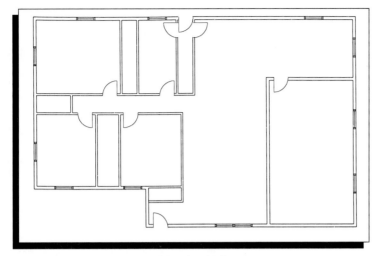

Fig. 17-48O

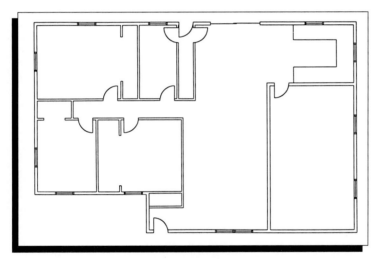

Fig. 17-48P

■ Chapter 17 Review ■

1. What three types of views do working drawings contain?
2. What purpose do architectural models serve?
3. Name three common construction materials and describe an application for each.
4. List five considerations involved with house planning.
5. When planning a house, why is the traffic pattern important?
6. What is the purpose of a site plan?
7. What is the difference between an elevation and a vertical section?
8. Why are detail drawings important?

■ Chapter 17 Problems ▬▬▬▬

1. Create a preliminary drawing of a small two-bedroom, two-bath house with a kitchen and a great room.

2. Plan the traffic patterns through the house you drew in problem 1. Rearrange the rooms if necessary.

3. After you have found a design that works, create a final floor plan for the house. Include windows and doors.

4. On the same scale, create and use blocks for basic furniture in the house. Also include blocks for bathroom and kitchen fixtures.

5. As a challenge, design and create the elevations for your house. Use Fig. 17-35 as a guide.

Using CAD

From Models to Virtual Buildings

As a building is designed, an architect often builds a model to get a realistic idea of what it will look like. The model is scaled to the same proportions that the final building will have and is changed often during the design process. It provides more information about the project because it's three-dimensional.

With CAD, designers can build similar 3D models with the computer and "take pictures" using a rendering package. Renderings are images created with AutoCAD that look very real. They are often described as photo-realistic because colors, materials, and shadows have been added that make the images look real.

AutoCAD has its own rendering capabilities. There are also application tools available for enhancing AutoCAD. One such tool is ARE 24 (Advanced Rendering Extension), produced by Ketiv Technologies and RenderStar Technologies. Application tools are comparable to a template used when drawing on a drafting board. The drawing could be completed without the template, but it would be much more difficult and time consuming.

To create a 3D computer model, designers first create designs with AutoCAD. They might also use an architectural application product such as ArchT, also produced by Ketiv. ArchT provides time-saving features for creating entities that are commonly required for architecture, such as wall, window, and door styles.

When the design is complete, the computer software organizes all the information about the building. ARE is then used to render different views. Views from the inside or outside of the building can be arranged like frames in a movie and placed on videotape. You can, for example, see what the building would look like if you walked through it.

According to William Holt, Applications Consultant at Ketiv, this information may one day be plugged into a virtual reality system. Virtual reality systems provide a setting where the computer controls what your senses are taking in. You look through special goggles that block out everything except the scene created by the computer. You can also interact with the environment through a data glove or other input device.

Holt paints this picture of how a house of the future will be designed and built: The family who will live in the house takes a walk through with the designer to see that it feels right, that all the windows are in the right places at the right height, and so on. The family realizes that the window overlooking the back yard is a little too high and the designer reaches out to move it down. They make all the final adjustments as they walk through. They choose items such as door knobs and linoleum as they go.

After the design is complete, the building begins. The framer tells the computer to show only the framing of the house. This person, wearing the headset as he or she works, sees exactly how long certain boards should be, how and where they fit, and so on. The painter then calls up only the information needed to paint. Colors will be labeled to show on which walls they are to be placed. All workers have access to each type of information they need to complete their work. Soon, the virtual building has become a reality!

Chapter 18

► Key Terms

isometric drawing   shading
modeling   rendering
wireframe   regions
surface modeling

►► Commands & Variables

SNAP   SHADEDIF
ISOPLANE   SOLMESH
3DFACE   SOLIDIFY
3DMESH   SOLSUB
REVSURF   SOLINT
EDGESURF   SOLEXT
TABSURF   SOLCYL
RULESURF   SOLBOX
SHADE   SOLUNION
SHADEDGE

Pictorial and 3D Drawings

Objectives

When you have completed this chapter, you will be able to:

- create isometric drawings.
- create surface models.
- shade models.
- create two-dimensional primitive regions.
- create simple solid models.

The real power of CAD begins where basic drafting leaves off. AutoCAD's three-dimensional (3D) and modeling commands and features allow you to show objects in realistic settings. You can turn, shade, and shape them to meet your needs. You can create models so realistic that they can be analyzed in ways that used to be possible only with the real thing. In addition, many of the activities described in earlier chapters can be done faster and more easily with the help of AutoCAD's 3D and modeling features.

Methods of analysis that used to be possible only on real (physical) models include finite element analysis and various stress tests. These methods can now be used on three-dimensional models built with AutoCAD. AutoCAD models can also be used to solve design interface problems that formerly required physical models.

Note

Isometrics

An **isometric drawing** is a two-dimensional (2D) pictorial drawing. In other words, it is a two-dimensional drawing that shows an object in three dimensions. The height axis of an isometric drawing is 90 degrees from horizontal, and both the width and depth axes are 30 degrees from horizontal *(Fig. 18-1)*. All straight lines are drawn to scale, and circles appear as ellipses.

AutoCAD provides an isometric grid and snap for use in creating isometric drawings. To activate the isometric style, use the Style option of the SNAP command and choose Isometric. Since the isometric grid is two-dimensional, an object created using it can be viewed only as a two-dimensional object. Unlike a 3D drawing, it cannot be rotated and viewed from any angle.

When the isometric style is activated, the ISOPLANE command allows you to switch the cursor among the top, right-side, and left-side planes of the isometric grid *(Fig. 18-2)*. You may also change the cursor setting at any time during the drawing session by pressing CTRL-E.

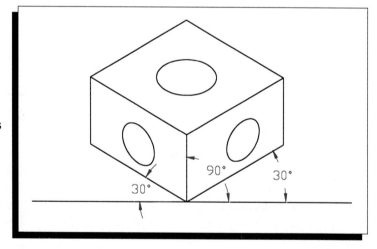

Fig. 18-1 In an isometric drawing, the height axis is vertical and the width and depth axes are 30 degrees above horizontal.

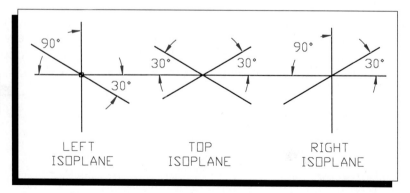

Fig. 18-2 AutoCAD's crosshairs can be set for the left, top, and right isometric planes (isoplanes). Notice that the top isoplane has no vertical component.

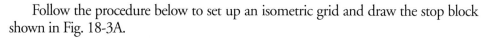

ACTIVITY

Follow the procedure below to set up an isometric grid and draw the stop block shown in Fig. 18-3A.

1. *Command:* **SNAP** ↵
2. *Snap spacing or ON/OFF/Aspect/Rotate/Style <default>:* **S** ↵
3. *Standard/Isometric <S>:* **I** ↵
4. *Vertical spacing <default>:* **0.5** ↵

By default, the crosshairs are set for the left-side isoplane (left mode).

5. Change to the top and right modes by pressing **CTRL-E**. Then return to the left mode.
6. Set the grid spacing to 1 unit.
7. Draw the block shown in Fig. 18-3B.
8. Modify the block as shown in Fig. 18-3C.

Make sure the crosshairs are set in the left isoplane mode. Add the ellipse as follows. Refer to Fig. 18-3D as you do these steps.

9. *Command:* **ELLIPSE** ↵
10. *<Axis endpoint 1>/Center/Isocircle:* **I** ↵
11. *Center of circle:* Select a point near the middle of the left side of the stop block.
12. Drag a radius approximately as shown in Fig. 18-3D.
13. Change the crosshairs to the top mode.
14. Repeat steps 9 through 12 to draw an ellipse in the approximate center of the top surface.

Your drawing should now look like the one in Fig. 18-3A. Notice the difference in the shapes of the two ellipses.

You may wish to create another ellipse to one side of the drawing with the crosshairs in the right mode. Doing so will allow you to compare the result of the ELLIPSE command in all three isoplanes.

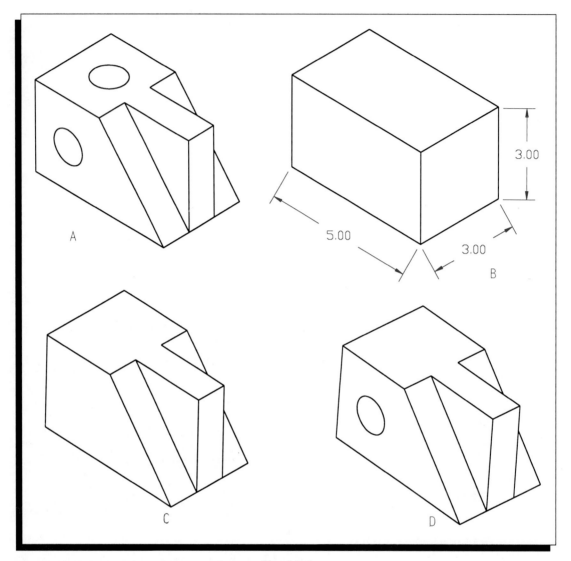

A

B

5.00

3.00

3.00

C

D

Fig. 18-3

Modeling

Modeling is the creation of a drawing in three dimensions. You can think of modeling as having a 3D object inside your monitor and viewing it through the screen. You can rotate the object so it can be seen from any angle. Models are usually viewed as pictorials. That is, they are turned so that all three dimensions can be seen on the screen.

Three types of modeling are available in AutoCAD: wireframe, surface, and solid. Wireframe and surface models can be made using the basic AutoCAD software. Solid modeling requires the use of AutoCAD's Advanced Modeling Extension, which is an option and may not be available on your computer.

Wireframe Modeling

Three-dimensional objects are generally drawn as wireframe models. **Wireframe** models are open models; you can see through a wireframe model as if it were made up of a wire frame. A bird cage is an example of what a wireframe model looks like *(Fig. 18-4)*. Fig. 18-5 shows a wireframe surface model created with AutoSurf. (AutoSurf works within AutoCAD.) The wireframe was exported as a DXF file and rendered in 3-D Studio, another product produced by Autodesk. You will learn more about rendering later in this chapter.

Chapter 8, "Views and Techniques of Drawing," discusses the creation of three-dimensional objects. The 3D drawing you made and viewed in Chapter 8 was actually a wireframe model. You can remove the hidden lines from a wireframe drawing by using the HIDE command. When the hidden lines are removed, the object takes on a familiar solid appearance.

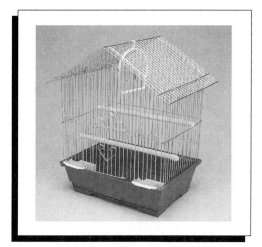

Fig. 18-4 A bird cage is a good example of a wireframe model.

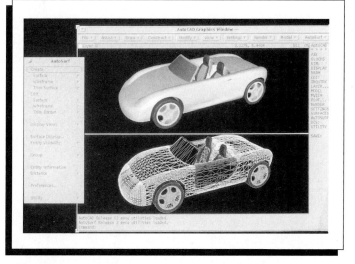

Fig. 18-5 Car design is one application of wireframe modeling. This model of a sports car was built in AutoCAD with AutoSurf. The model was then rendered to show what the finished car will look like.

3D Faces

You can draw closed 3D surfaces using the 3DFACE command and the X/Y/Z point filters. This method is useful when you want to specify a 3D point on the Z axis that you cannot pick with the mouse.

In AutoCAD's default (World) coordinate system, you can use the mouse to pick only those points that are flat to the screen. To specify 3D points, you can use 3DFACE, the point filters, or absolute coordinates. You can also change the viewpoint and the UCS (user coordinate system) so that a particular surface of a 3D object is flat to the screen. Then you can use the mouse as usual.

The 3DFACE command allows you to create a solid face using 3D coordinates. As you may remember, 3D coordinates include three values: the position on the X (width) axis, the position on the Y (height) axis, and the position on the Z (depth) axis *(Fig. 18-6)*.

The point filters allow you to specify one or more of the coordinates of an existing point automatically. You must then supply the remaining coordinates to complete the point description. For example, suppose you want to move an object 4 units back (away from you) on the Z axis. Enter the MOVE command and select the objects as usual. At the *Base point or displacement:* prompt, press .xy and RETURN. AutoCAD prompts you for the point from which you want to copy the X and Y values. Pick a point on the object you want to move. AutoCAD then prompts you for the missing coordinate, in this case Z. Enter the new value for the Z coordinate: -4. The object is moved 4 units away from you on the Z axis. The X and Y coordinates remain the same.

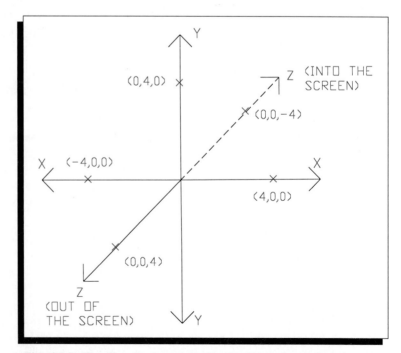

Fig. 18-6 The Z axis shows depth, the third dimension. In AutoCAD, positive Z coordinates come out of the screen (toward you) and negative coordinates go into the screen (away from you).

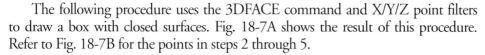

ACTIVITY

The following procedure uses the 3DFACE command and X/Y/Z point filters to draw a box with closed surfaces. Fig. 18-7A shows the result of this procedure. Refer to Fig. 18-7B for the points in steps 2 through 5.

1. *Command:* **3DFACE** ⏎
2. *First point:* Pick point P1.
3. *Second point:* Pick point P2.
4. *Third point:* Pick point P3.
5. *Fourth point:* Pick point P4.
6. *Third point:* ⏎

When you picked the fourth point, a line automatically connected the fourth point to the first point.

Steps 7 through 13 use the X/Y/Z point filters to copy the rectangular surface in the same XY position but 1.5 units from the original on the Z axis.

7. *Command:* **COPY** ⏎
8. *Select objects:* Pick the rectangular surface you just made.
9. *Select objects:* ⏎
10. *<Base point or displacement>/Multiple:* Pick the lower left corner (P1).
11. *Second point of displacement:* **.XY** ⏎
12. *of* Pick the same point as in step 10 (P1).
13. *of (need Z):* **1.5** ⏎
14. *Command:* **VPOINT** ⏎
15. *Rotate/<View point> <default>:* **-1,-1,1** ⏎
16. Use ZOOM All or ZOOM 1 to position the drawing on the screen.

Your drawing should look similar to the one in Fig. 18-7C. Steps 16 through 23 set a running INTersection object snap and draw the four sides using the 3DFACE command. Refer to Fig. 18-7D as you perform these steps.

17. *Command:* **OSNAP** ⏎
18. *Object snap modes:* **INT** ⏎
19. *Command:* **3DFACE** ⏎
20. *First point:* Pick P1.
21. *Second point:* Pick P2.
22. *Third point:* Pick P3.
23. *Fourth point:* Pick P4.
24. *Third point:* ⏎

You have now completed the top, the bottom, and one side. Repeat steps 18 through 23 three times, picking the four corners of each of the remaining three sides. Your result should look like the drawing in Fig. 18-7E.

Note Even though the object may look complete after you have created three of the four sides, the object will not be a solid block until you have created the fourth side.

25. Command: **HIDE** ⏎

Your drawing should now look like the one in Fig. 18-7A.

Note HIDE is a visual command. Even though hidden lines are not visible on your screen, they would appear if you plotted the drawing. To plot a drawing with the hidden lines removed, select "Hide Lines" from the Plot Configuration dialogue box before plotting.

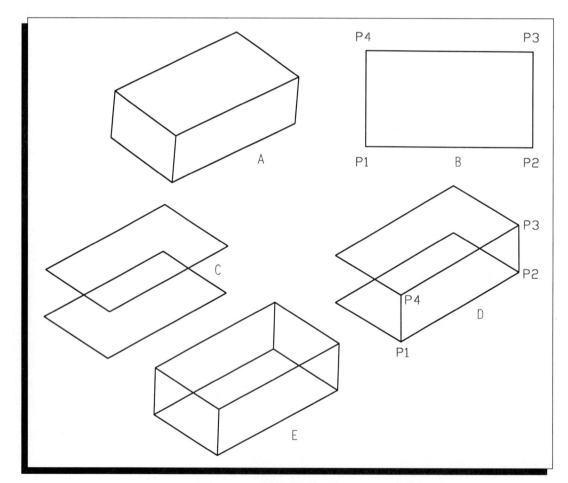

Fig. 18-7

Surface Modeling

Surface modeling creates 3D surfaces that are contoured to a shape you specify *(Fig. 18-8)*. Objects look more realistic than wireframe models. Meshes form the basis for many surface models. A polygon mesh can be created with the 3DMESH command. Other meshes can be generated and manipulated with commands such as REVSURF, EDGESURF, TABSURF, and RULESURF. The "Your Turn" activity in this section demonstrates the use of the REVSURF command. You may also wish to experiment with the other surface modeling commands listed above. For more information, refer to the AutoCAD Reference Manual.

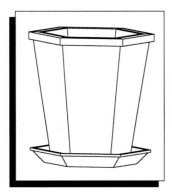

Fig. 18-8 A surface model such as this one of a flowerpot, exists in three dimensions in AutoCAD.

ACTIVITY

The REVSURF command can be used to create an object concentric around an axis of revolution. REVSURF stands for REVolution SURFace. The flowerpot in Fig. 18-8 was created using this command. Follow the procedure below to make a bowl with the REVSURF command. Fig. 18-9A shows the result of this activity.

1. Draw a vertical line for the axis of revolution *(Fig. 18-9B)*.
2. Use the ARC command to create the profile of the bowl.

The lower end of the large arc should touch the axis of revolution (vertical line). Use the CONTINue option to connect additional arcs to the first one to create the profile for the bowl.

3. *Command:* **REVSURF** ⏎
4. *Select path curve:* Pick the arc at the bottom of the curve.
5. *Select the axis of revolution:* Pick the vertical line.
6. *Start angle <0>:* ⏎
7. *Included angle (+=ccw, -=cw) <Full circle>:* ⏎

In step 7, "ccw" stands for counterclockwise and "cw" stands for clockwise. Arcs in AutoCAD are created in a counterclockwise direction by default. Therefore, if you enter a positive number, the angle is generated in a counterclockwise direction, and if you enter a negative number, the angle is generated in a clockwise direction. This applies only to revolutions that are not rotated a full 360 degrees.

8. Repeat steps 3 through 7 for each additional arc you used to create the profile of the bowl.

9. Erase the vertical line that served as the axis of revolution.

Your drawing should now look like the one in Fig. 18-9C.

10. Use the VPOINT command to rotate the bowl for viewing in different positions. To see the bowl clearly, remove the hidden lines *(Fig. 18-9A)*.

11. Save the drawing for use with the "Your Turn" in the next section.

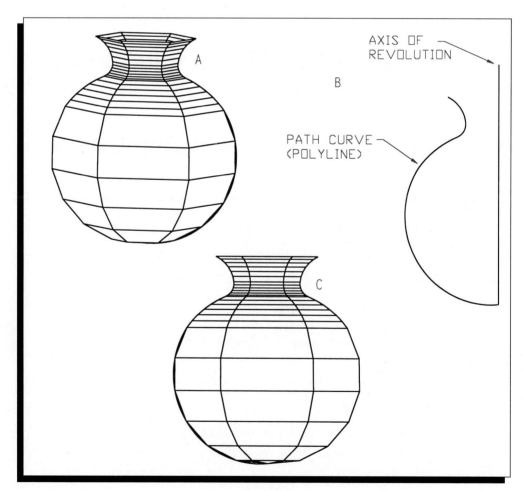

Fig. 18-9

Shading

Shading is the quick shadowing of three-dimensional objects to give them visual depth. When you enter SHADE at the command prompt, the 3D object on the screen is shaded using a quick method so that the shading is completed almost immediately.

The visual impact of shading is controlled by two system variables: SHADEDGE and SHADEDIF. SHADEDGE determines the kind of shading that will be produced. It has four possible settings: 0 through 3.

- 0 shaded faces, no edge high-lighting

- 1 shaded faces, edges high-lighted
- 2 simulated hidden-line rendering
- 3 no shading

When SHADEDGE is set to 0 or 1, SHADEDIF sets the percentage of diffused light reflected off the object and the percentage of ambient light. The value of SHADEDIF may be anywhere from 1 to 100; its default setting is 70. The higher settings increase the diffused light and add more contrast to the image.

ACTIVITY

In this activity, you will explore the effect of various settings of SHADEDGE and SHADEDIF on shading.

1. Bring up the 3D bowl which you saved in the previous "Your Turn."
2. *Command:* **SHADEDGE** ↲
3. *New value for SHADEDGE <default>:* **0** ↲
4. *Command:* **SHADE** ↲
5. *Command:* **SHADEDGE** ↲
6. *New value for SHADEDGE <default>:* **1** ↲
7. *Command:* **SHADE** ↲

Notice the difference in the way the bowl is shaded.

8. *Command:* **SHADEDIF** ↲
9. *New value for SHADEDIF <70>:* **100** ↲

Notice the effect of the increased diffused lighting on the shading of the bowl.

10. Vary the **SHADEDIF** value for each of the **SHADEDGE** settings, and notice the effect on the shading of the bowl.

Rendering

Rendering is the simulation of perspective, reflection, light, and shadow on three-dimensional objects. It is similar to shading, but rendering is a higher-quality process, and it takes longer to process *(Fig. 18-10)*. Although Render comes as an integral part of AutoCAD, you must load (initialize) it for each AutoCAD session. It does not load automatically because it is very memory-intensive. For the same reason, you

should unload it whenever you are not using it.

Loading Render is very easy. You simply enter the RENDER command. If Render has not yet been configured for your machine, a menu-driven configuration process occurs. After it is configured, Render is loaded automatically when you enter the RENDER command.

Fig. 18-10 This image of a teapot was produced in AutoCAD with AccuRender by Roy Hirshkowitz. The AccuRender image is untouched by any other graphics software.

ACTIVITY

Because rendering is a more advanced technique than shading, Render has many more options and characteristics that you can control. In this activity, you will explore some of the Render settings. After you perform the following steps, you may wish to experiment further with the Render options.

1. Bring up the bowl you created in a previous "Your Turn" activity.

2. *Command:* **RENDER** ↲

Notice that the bowl looks very much like it did when you shaded it with SHADEDGE set to 0, except that the edges are better defined.

3. Activate the **Render** pull-down menu and read through the various options.

4. Pick **Lights**... from the Render pull-down menu.

The Lights dialogue box appears. From this dialogue box, you can create new lights and control the ambient light that illuminates the bowl.

5. Create a new light by picking **New...**

The New Light Type subdialogue box appears.

6. Select **Point Light**.

The New Point Light subdialogue box appears. Notice the cursor in the Light Name box. You must assign a name to the new light before you can use it in the rendering.

7. Name the new light **TOP** and pick the **Modify** button to modify the position of the point light.

8. Notice the rubberband line that originates at the current light position. Move your pointing device (mouse) to the top of the bowl, pick a location there, and pick **OK**.

An icon appears with the word TOP inside. This is your indication of where the new light is positioned. Because you can have more than one light, the name appears in each light icon.

9. Pick **OK** again to close the lights dialogue box.

10. Render the bowl again and notice the difference made by the new point light.

11. Pick **Lights**... from the Render pull-down menu and pick **Delete** to delete the TOP light. Pick **OK** twice to return to the *Command:* prompt.

12. Pick **Finishes**... from the Render pull-down menu and then pick the **Modify**... box.

The Modify Finish dialogue box appears. From this dialogue box, you can control the color of the rendered objects and various characteristics of the lights used to illuminate the object.

13. Change the ambient light by moving the slider on the slide bar with your mouse.

You can see the effects on a rendered object without leaving the dialogue box. Notice the red sphere in the black box. Underneath it, there is a button labeled Preview Finish.

14. Pick the **Preview Finish** button and observe the change in the appearance of the sphere.

The Preview Finish feature may not work with some computer configurations.

15. Experiment with the remaining light characteristics and check the results using **Preview Finish**. When you find a combination you like, pick **OK** until you arrive at the *Command:* prompt.

16. Render the bowl. Notice the effect of the finish you chose.

17. Pick **Preferences**... from the Render pull-down menu.

The Rendering Preferences dialogue box appears. From this dialogue box, you can set general characteristics such as whether to apply finishes and smooth shading, and you can control the size of the light icons. You can also reconfigure Render from this dialogue box.

18. Pick **Smooth Shading** and pick **OK**.

19. Render the bowl again. What difference does the Smooth Shading option make?

The Smooth Shading option shades each entity separately. Because you used more than one entity (arc) to create the bowl, the appearance of the bowl may be a little odd. To prevent this from happening, you can use a polyline to create the original profile of the bowl. However, to make the bowl come out correctly, you would have to create both the inside and the outside of the bowl as a single polyline for the profile of the bowl. Otherwise, the bowl would appear to be solid.

20. Experiment further with the options available for rendering. For further information, refer to the AutoCAD Render Reference Manual supplied with Release 12.

Region Modeling

Regions are closed, two-dimensional areas that result when you combine two or more two-dimensional entities into a single entity. Region modeling is an advanced tool that may serve as an introduction to the AutoCAD Advanced Modeling Extension (AME). Region modeling is a standard part of AutoCAD Release 12; however, the commands are actually a subset of the AME package, which is optional.

ACTIVITY

The following procedure introduces the concept of regions. You will create a primitive, which is a basic region. Then you will subtract an area from the primitive, leaving a hole in it. The final region will be the primitive plus the hole. Fig. 18-11A shows the result of this procedure.

1. Draw a hexagon inscribed in a circle with a radius of 2.5 units *(Fig. 18-11B)*.

2. Command: **SOLIDIFY** ⏎

3. If your AutoCAD package does not include the Advanced Modeling Extension (AME), AutoCAD responds with "Initializing Region Modeler." If your package *does* include AME, AutoCAD gives the following message: "No modeler is loaded yet. Both AME and Region Modeler are available. Autoload Region/<AME>:." Press **R** to load the Region Modeler.

4. *Select objects:* Pick the polygon.

5. *Select objects:* ⏎

You have now created a solid 2D object, or primitive region. AutoCAD automatically crosshatches the area to show that this is a solid region.

6. *Command:* Draw a circle with a radius of 1 unit in the center of the polygon *(Fig. 18-11C)*.

Now you can use the Region Modeling command SOLSUB (SOLid SUBtraction) to subtract the circle's area from the primitive region.

7. *Command:* **SOLSUB** ⏎

8. *Source object...*
   Select objects: Pick the primitive (polygon).

9. *Select objects:* ⏎

AutoCAD responds with "1 region selected."

10. *Objects to subtract from them...*
    Select objects: Pick the circle.

AutoCAD responds with "1 found."

11. *Select objects:* ⏎

The crosshatching is removed from the circle because it is no longer part of the polygon. Instead, it represents a hole in the polygon. However, the polygon and the hole together make up a region; that is, they are considered a single entity *(Fig. 18-11A)*.

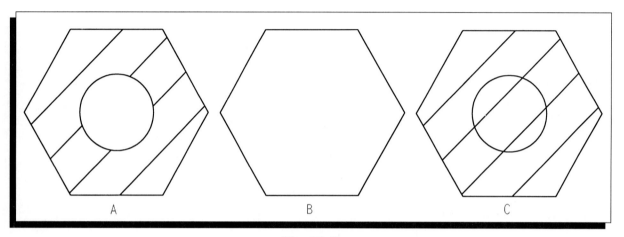

Fig. 18-11

Advanced Modeling Extension

Advanced Modeling Extension (AME) is an optional software package that runs under AutoCAD. With AME, you can produce sophisticated solid models using the Boolean operations of union, subtraction, and intersection. Boolean operations, named after the English mathematician George Boole, are based upon mathematical set theory *(Fig. 18-12)*.

- Union joins all the elements of one set with all the elements of another set. In AME, union is accomplished with the SOLUNION command.
- Subtraction removes the elements that one set has in common with another set. In AME, subtraction is accomplished with the SOLSUB command.

- Intersection creates a new set made up of the elements common to two or more other sets. In AME, intersection is accomplished with the SOLINT command.

In addition to the AME commands, you can use all of AutoCAD's standard 2D and 3D commands and features. As mentioned above, the commands used for region modeling are also used in AME. For details about the AME software program, refer to the AutoCAD AME Release 2.1 Reference Manual.

You can load AME from the Model pull-down menu by selecting Model, Utility, and Load Modeler. You can also load it from the keyboard by typing <u>(xload "ame")</u>. Type everything that is underlined exactly as shown (except for the underlining).

XLOAD is an AutoLISP command. AutoLISP is a computer programming language built into AutoCAD. With AutoLISP, you can write programs to customize AutoCAD to fit the way you work. AutoLISP is similar in design to the Lisp language used frequently in artificial intelligence.

Note

However, unless you have already loaded the Region Modeler, the simplest way to load AME is to enter the AME command you want to use. If the command is one that is also part of the region modeler, AutoCAD asks which one you want to load. However, if the command is not part of the region modeler, AutoCAD loads AME automatically and continues with the command you entered.

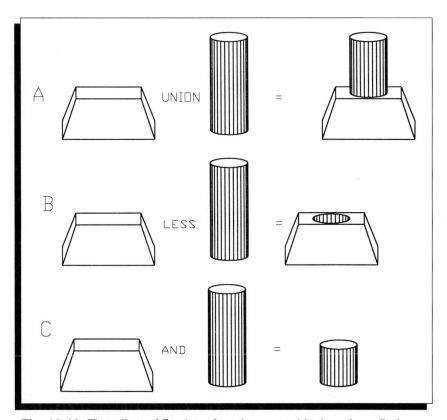

Fig. 18-12 The effect of Boolean functions on a block and a cylinder: A) Union results in a single entity that consists of a block with a cylinder on top. B) Subtraction results in a block with a hole the size of the cylinder. C) Intersection keeps only the parts of the original entities that intersect, or overlap, so a cylinder the height of the original block results.

ACTIVITY

Omit this "Your Turn" if AME is not installed on your computer. The following procedure uses AME commands to create the object shown in Fig. 18-13A.

1. *Command:* **PLINE** ⏎

Create a 6-unit by 4-unit rectangle *(Fig. 18-13B)*.

2. *Command:* **SOLEXT** ⏎

The SOLEXT command creates a solid extrusion, which turns a 2D polyline into a solid primitive.

3. *Select regions, polylines, and circles for extrusion…*

   Select objects: Pick the rectangle.

4. *Select objects:* ⏎

5. *Height of extrusion:* **1** ⏎

6. *Extrusion taper angle:* **5** ⏎

This sets the angle of the sides of the rectangular block at 5 degrees. Your drawing should now look like the one in Fig. 18-13C.

You will now create a solid cylinder inside the block. Before you do so, however, you will set the SOLWDENS variable, which controls the number of wires in the wire mesh that results when you create a solid primitive.

Note

It is not absolutely necessary to reset SOLWDENS. However, solids created at the default setting of 1 are harder to see clearly because each solid has only one pair of "wires" or lines. SOLWDENS is a system variable in AME. Its value can be anything from 1 to 12. The default setting of 1 generates very quickly, but higher settings create better-looking objects.

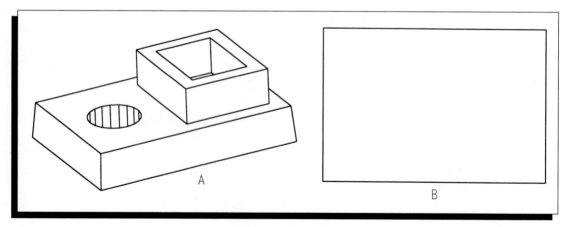

Fig. 18-13A and B

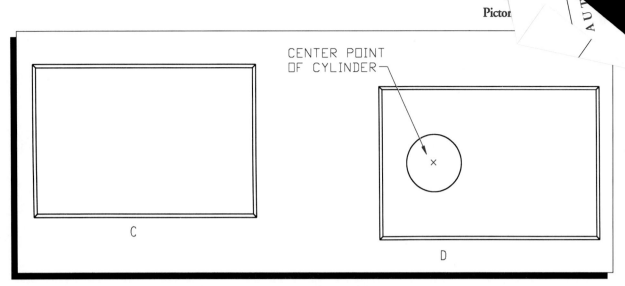

Fig. 18-13C and D

7. *Command:* **SOLWDENS** ⏎

8. *Wireframe mesh density (1 to 12) <default>:* **5** ⏎

9. *Command:* **SOLCYL** ⏎

SOLCYL creates a solid cylinder primitive.

10. *Baseplane/Elliptical/<Center point> <0,0,0>:* Pick a center point for the cylinder on the left side of the block *(Fig. 18-13D).*

11. *Diameter/<Radius>:* **0.75** ⏎

12. *Center of other end/<Height>:* **1** ⏎

Since the height of the cylinder is 1 and you extruded the block to a height of 1, the two are the same thickness, or height.

You may wish to use the VPOINT command to see what the objects you're building look like in three dimensions. Before you continue, however, make sure you return to the top view. From the VPOINT command, set the coordinates to 0,0,1. Then enter ZOOM All (or ZOOM 1).

You will now subtract the cylinder from the block, leaving a hole in the block.

13. *Command:* **SOLSUB** ⏎

14. *Source objects…*

    Select objects: Pick the block.

15. *1 found.*

    Select objects: ⏎

16. *1 solid selected.*

    Objects to subtract from them...

    Select objects: Pick the cylinder.

17. *1 found.*

    Select objects: ⏎

AutoCAD responds with "1 solid subtracted from 1 solid." You now have a hole in the block *(Fig. 18-13E).*

18. Use the VPOINT command to view the block from various points of view. Use HIDE to remove hidden lines. Notice that the block looks a bit odd.

19. *Command:* **SOLMESH** ⏎

20. *Select objects:* **ALL** ⏎

AutoCAD selects every entity currently shown on the screen when you enter "ALL" in response to the *Select objects:* prompt. You could also have used a window to select all the objects on the screen.

---

21. *Select objects:* ⏎

22. *Command:* **HIDE** ⏎

Your drawing should now look like the one in Fig. 18-13F. Return to the top view before continuing.

23. *Command:* **SOLBOX** ⏎

The SOLBOX command creates a solid box primitive.

24. *Baseplane/Center/<Corner of box> <0,0,0>:* Pick a point about ½ unit above the center of the lower edge of the block.

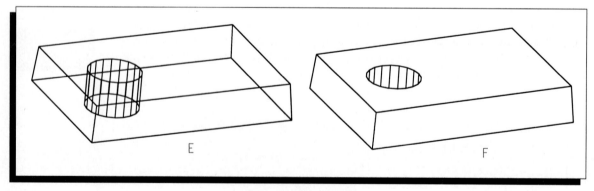

Fig. 18-13E and F

25. *Cube/Length/<Other corner>:* Drag the box to a point about ½ unit from the upper right corner and pick it.

26. *Height:* **2** ↵

Your drawing should now look like the one in Fig. 18-13G. Since all the heights are taken from the base of zero, a height of 2 will bring the box 1 unit above the top of the block.

27. Repeat steps 23 through 26 to create a smaller box 2 units high within the first box *(Fig. 18-13H)*.

28. Use the SOLSUB command to subtract the smaller box from the larger box.

The "source object" is the large box, and the "objects to subtract from them" is the small box.

---

29. Set the viewpoint so that you are looking down from the left front. Then enter **ZOOM 1**.

Use HIDE to remove hidden lines. Notice that the drawing looks a little strange from this point of view. It's time to join the boxes to the block.

30. *Command:* **SOLUNION** ↵

31. *Select objects:* **ALL** ↵

32. *Select objects:* ↵

33. *Command:* **SOLMESH** ↵

34. *Select objects:* **ALL** ↵

35. *Select objects:* ↵

36. *Command:* **HIDE** ↵

37. Remove the hidden lines.

Now the drawing should look much better *(Fig. 18-13A)*.

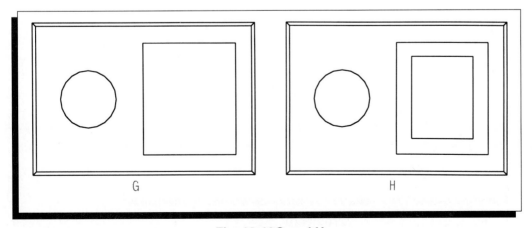

G   H

Fig. 18-13G and H

■ Chapter 18 Review ▬▬▬▬▬

1. In an isometric view, what are the angles of the height, width, and length axes from the horizontal?
2. What is the difference between an isometric drawing done in AutoCAD and a wireframe model?
3. What are the three types of modeling available in AutoCAD?
4. What command would you use to put a closed surface on a box?
5. What is the difference between shading and rendering?
6. What is a primitive?
7. Name the three Boolean operations and describe the function of each.
8. What is AME? What is its function?

■ Chapter 18 Problems ▬▬▬▬▬

1. Use AutoCAD's isometric grid to create the 2D isometric drawings shown in Problem 1.
2. Create the 3D surface models shown in Problem 2. Use the REVSURF command. Shade the models.
3. For each of the models you created in problem 2, use Render to create a high-quality shading.
4. Use the Region Modeler to create the region shown in Problem 4.
5. (Optional) If AME is installed on your computer, use AME to create the object shown in Problem 5.

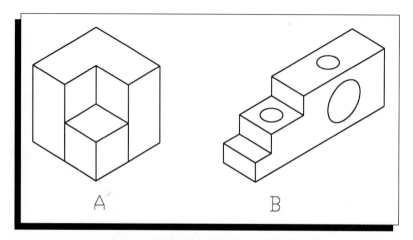

Problem 1

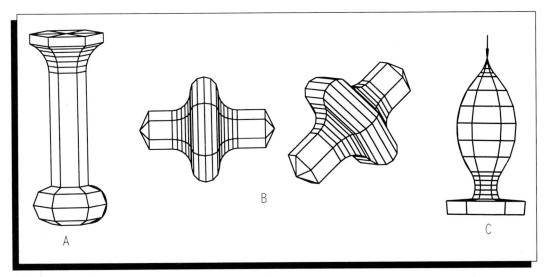

Problem 2

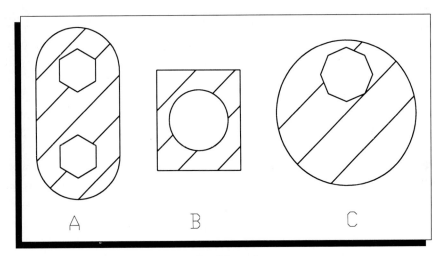

Problem 4

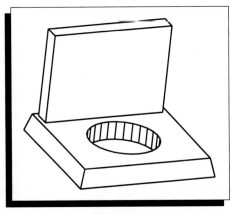

Problem 5

Using CAD

Engineer Finds AutoCAD Essential for Product Design

Manufacturing companies rely on designers and engineers to develop long lasting, attractive products that work well. GVO, Inc. of Palo Alto, California, is a 30-year old consulting firm that provides both industrial design and mechanical engineering to the manufacturing industry. Their specialty is consumer product design and packaging. They work with a variety of products, from shavers, irons, and garden sprayers to laptop computers.

Mechanical engineers such as Michelle Pillers, a Project Manager for GVO, find AutoCAD to be essential to the design process. Pillers' specialties are precision mechanization, automatic machinery, and consumer product design.

Pillers has been working with AutoCAD since Release 1.4. She says, "I feel lucky that I've been able to grow up with the program. As a result, I've been able to digest little by little as it has evolved." Her favorite feature is AutoLISP — the cornerstone of AutoCAD's open architecture. AutoLISP is a powerful programming language that makes it very easy to automate AutoCAD for many routine tasks: title block insertion, slide capture for animations, command shortcuts, and so on.

Pillers' second favorite feature is AME (Advanced Modeling Extension), which is an optional software package that works within AutoCAD. According to Pillers, AME delivers powerful solid models at a reasonable cost. Solid models are 3D drawings that represent objects with volume. They hold information such as physical and material properties of the object.

Even though Pillers uses AME daily for product design, she finds that at times 2D drawing is still necessary. For example, many parts require draft — a taper of 1 to 2 degrees applied to a surface to ease part removal from a mold. To help create 2D drawings from 3D AME models, she has devised a procedure for stripping away everything associated with the model and leaving only its 2D counterpart. The resulting drawing can then be updated for things like draft and threaded hole representation.

Pillers' models and final drawings help make production of consumer items possible. Look around your home or classroom and consider the role that AutoCAD may have played in developing the many products you see and use every day.

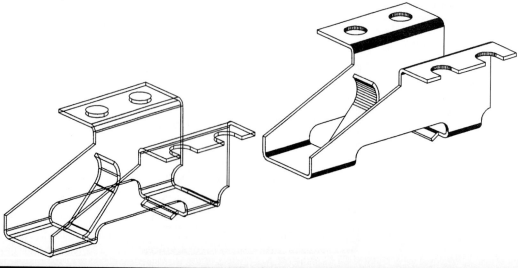

Pictorial Glossary

absolute coordinates coordinates based on actual (absolute) distance from the origin in the Cartesian coordinate system. See *Cartesian coordinate system.*

absolute zero another term for the origin in the Cartesian coordinate system; the point (0,0,0). See *Cartesian coordinate system.*

Acme thread a thread profile that has been adapted for use primarily for power transfer.

actual dimensions the measured dimensions of an object.

actual size the measured size of a part or object.

acute angle any angle that measures less than 90°.

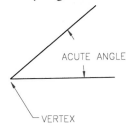

addendum on a gear, the distance from the pitch diameter to the top of the teeth.

aligned method of placing dimensions a placement method in which dimensions are placed in line with the dimension lines. See also *unidirectional method of placing dimensions.*

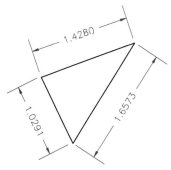

allowance in limit dimensioning, the maximum interference or minimum clearance between mating parts.

alphabet of lines the standard library of line symbols used in drafting and their accepted uses.

AME (Advanced Modeling Extension) an optional software package made by Autodesk; AME extends the modeling capabilities of AutoCAD.

angle a circular measure of two lines that have one point in common. The common point is called the *vertex,* and the two lines are called *rays or vectors.*

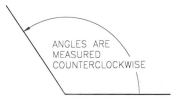

ANSI the American National Standards Institute, an organization that specifies standards for many areas of mechanical work, including drafting. ANSI Y14 is a collection of documents that contain the organization's drafting standards.

aperture the pick box that appears at the center of the crosshairs when you activate an object snap mode.

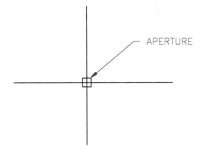

apparent dimensions the distance from one point to another as it looks from your point of view.

arc   a part of a circle. To draw an arc in AutoCAD, use the ARC command. The arc shown here was created by using AutoCAD's default method (starting point, second point, endpoint) and picking the points shown.

array   multiple copies of an object arranged in a pattern you specify. A **rectangular array** is one in which the copies are arranged in rows and columns. A **polar array** is one in which the copies are rotated around a center point that you specify.

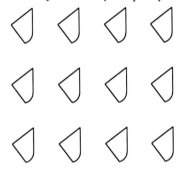

RECTANGULAR ARRAY

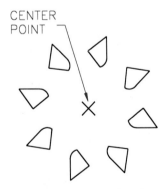

CENTER POINT

POLAR ARRAY
(TO FILL 360°,
ROTATED AS COPIED,
ARRAYED AROUND
CENTER POINT SHOWN)

ASCII   the acronym for the American Standard Code for International Interchange; an international standard for text characteristics that allows electronic text to be read by various software programs.

assembly drawing   a drawing that shows how parts should be put together to create a finished product.

attribute   text information that you can associate with an AutoCAD block; you can later extract the attributes in a drawing to create a document such as a bill of materials or an inventory. See also *bill of materials.*

auxiliary section   a section that is viewed at an angle.

auxiliary view   a view other than the six basic views (front, back, top, bottom, right side, left side); auxiliary views are used to describe an inclined surface of an object completely when the surface cannot be described using one of the basic views.

axis   (*plural: axes*) an imaginary line that shows an object's orientation in space. There are three axes in the Cartesian coordinate system. The X axis is horizontal, the Y axis is vertical, and the Z axis is at right angles to the other two. See *Cartesian coordinate system.*

basic hole system   a system of fit in which the hole is the controlling part; basic hole systems are very common because they can be machined using standard tools.

basic shaft system   a system in which the shaft is the controlling part; seldom used because special tools are required to machine them.

basic size   in limit dimensioning, the size to which allowances and tolerances are applied.

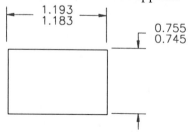

BASIC SIZE = 1.188 X 0.750

bevel gears   (also called *miter gears*) gears in which the teeth are cut at an angle to the gear face so that the gears can transmit motion between two intersecting shafts.

bilateral tolerance   in limit dimensioning, a tolerance that is allowed on both sides of the design size. In this case, a tolerance of 0.005 is allowed on both sides of the design size of 0.375.

0.375± 0.005

bill of materials   a document that lists all the materials needed to build a product, including the quantity of each material.

bisect   to divide something into two equal parts.

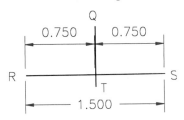

LINE QT BISECTS
LINE RS. NOTE
THAT LINE RS
DOES NOT BISECT
LINE QT.

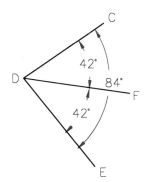

LINE DF BISECTS
ANGLE CDE

blips   small pick points that are left when you use the pointing device to pick a location on the screen. Blips are not permanent, and they do not print. To clear the screen of blips, use the REDRAW or REGEN command. The small, irregular marks below are examples of blips.

block   a group of entities stored under a single name in AutoCAD. After you have stored the entities as a block, you can insert the block as many times as you need to in a drawing.

bolt   an externally threaded fastener that is designed to be tightened and released by turning a nut.

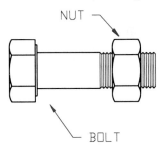

NUT

BOLT

boundary definition error   an error that occurs in AutoCAD when you try to hatch an area that is not completely enclosed.

broken-out section   an irregularly shaped section in which a small portion of the object is removed so that the viewer can see inside.

cabinet oblique   a type of oblique drawing in which the depth is drawn at exactly half its actual value.

cam   a machine part that changes rotary motion into reciprocal motion.

Cartesian coordinate system   A system of three axes used for drafting: X, Y, and Z. The X axis is horizontal and is used for showing width. The Y axis is vertical and is used for showing height. The Z axis runs from a point directly behind the origin to a point directly in front of the origin. It is at right angles to the X axis and the Y axis.

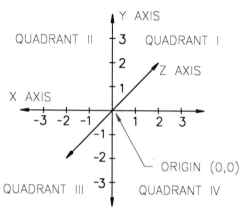

CARTESIAN COORDINATE SYSTEM

cartographer   a drafter who specializes in drawing maps.

cavalier oblique   a type of oblique drawing in which the depth is drawn at full size.

center line   a line used to define the center of an object or feature. Center lines are drawn as line segments: long-short-long-short-long-short. AutoCAD provides a standard center line. To load this linetype into your drawing, enter the LINETYPE command and specify CENTER.

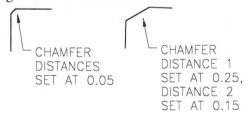

CENTER LINE

chamfer   a small, angular edge on a corner of an object. In AutoCAD, you can create a chamfer using the CHAMFER command.

CHAMFER DISTANCES SET AT 0.05

CHAMFER DISTANCE 1 SET AT 0.25, DISTANCE 2 SET AT 0.15

chord   any line segment that joins two points on a circle.

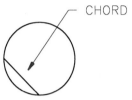

CHORD

circle   a closed, curved line that contains 360 degrees measured in a counterclockwise direction from the 3 o'clock (east) position.

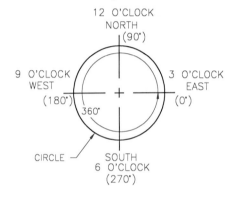

circular pitch   on a gear, the distance along the pitch circle from a point on one tooth to the same point on the next tooth.

circumference   the total distance around a circle.

circumscribed   a polygon that completely encompasses a circle whose radius is equal to the distance across the flats of the polygon.

clearance fit a fit in which a clearance will always occur between a shaft and hole because the maximum diameter of the shaft is smaller than the minimum diameter of the hole.

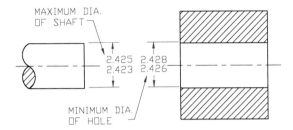

clockwise moving in the same direction that the hands of a clock rotate.

CLOCKWISE

CNC (Computer Numerical Control) a machine that accepts input directly from a CAD computer system to operate various machinery.

coincidental lines lines that fall one on top of the other so that there appears to be only one line present.

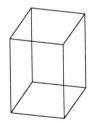
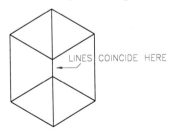

THIS PRISM IS TRANSPARENT, SO YOU CAN SEE ALL OF ITS SIDES.

WHEN THE PRISM IS ROTATED TO CERTAIN POSITIONS, SUCH AS THIS ONE, TWO OR MORE OF THE LINES COINCIDE.

command line the text line near the bottom of the AutoCAD screen at which you can enter commands and choose options.

communication the process of relating information to others, both verbally and nonverbally.

compass a drawing instrument used to draw circles accurately. Several kinds of compasses are available to draw circles of different sizes.

complementary angles two angles that, when placed so that they share one common side (vector), measure 90°.

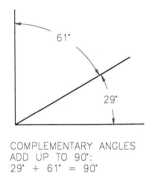

COMPLEMENTARY ANGLES ADD UP TO 90°: 29° + 61° = 90°

computer-aided drafting (CAD) the use of a computer to create drafted documents.

concentric having the same center point.

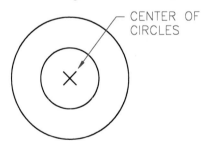

CENTER OF CIRCLES

concrete a mixture of cement, aggregate, and water that is used in many applications for structural members of a building or structure.

construction lines temporary lines made as a reference for the purpose of constructing other lines and features accurately.

continuous line an unbroken linetype used to show object and part outlines. See *visible line*.

control sequence a special sequence of characters that tells AutoCAD to insert symbols that cannot be entered directly from the keyboard, such as a diameter symbol (∅), and that controls features such as underscoring and overscoring.

counterbore   an enlargement of a hole to a specific depth; usually used to position the heads of screws and bolts level with or below the surface of an object.

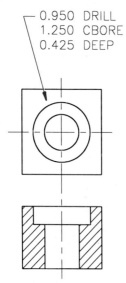

0.950 DRILL
1.250 CBORE
0.425 DEEP

countersink   a conical enlargement at the entrance of a hole; usually used to allow the head of a screw to lie flush with the surface of an object.

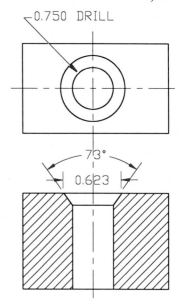

0.750 DRILL

73°

0.623

counterclockwise   moving in a direction opposite from the direction that the hands of a clock rotate.

COUNTERCLOCKWISE

crosshatching   a pattern of lines that show a solid surface that has been cut by a cutting plane in a sectional view.

crossing window   a selection box created on the screen from right to left. A crossing window is different from a regular window in that every entity that is even partially included in a crossing window becomes selected. The crossing window (dotted box) below selects the whole figure, even though two of the lines are not entirely within the window.

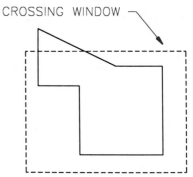

CROSSING WINDOW

cursor a pointer on the screen that allows you to select commands and manipulate drawing entities.

cutaway pictorial assembly drawing a drawing that shows a product as if a section had been cut out of it so that the detail inside can be seen. Similar to a broken-out section, except the purpose of this drawing is to facilitate assembly.

cutting plane the plane used to cut through an object to define a sectional view.

cylinder any tube-shaped object; dimension cylinders by specifying their length and diameter.

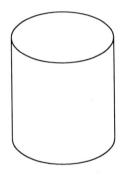

datum a point, line, or surface whose dimensions are assumed to be exact; other entities can be dimensioned from datums.

dedendum on a gear, the depth of the teeth from the pitch diameter to the root.

default value In AutoCAD, the value that is stored in the system and appears automatically for certain operations. Many of the default values in AutoCAD are controlled by system variables and can be configured by the user.

degrees a unit of measure for angles; there are 360 degrees in a full circle, 180 degrees in a half circle, 90 degrees in a quarter circle (right angle), and so on. See also *circle*.

design size usually the same as basic size, unless allowances are given.

developments the individual shapes that make up a pattern. See also *pattern*.

diagonal a line across opposite corners of a square or rectangle.

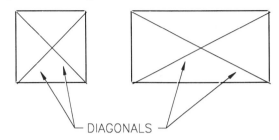

DIAGONALS

dialogue boxes graphical pop-up boxes that allow you to set defaults by checking boxes or entering text.

diameter the length of a line segment that begins at any point on a circle, goes through the center of the circle, and ends at a point on the circle directly opposite the first point. The diameter of a circle equals 2 times the radius.

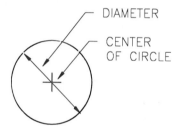

DIAMETER

CENTER OF CIRCLE

diametrical pitch the number of teeth per inch of pitch diameter on a gear.

digitizer an advanced pointing device that looks something like a mouse. It allows you to select commands or trace and digitize intricate shapes as you click at different places on a special digitizing drawing board.

dimension line   the line that specifies the distance from one end of a feature to another and gives its dimensions, or size, in units you specify. In most cases, the dimension line runs from extension line to extension line.

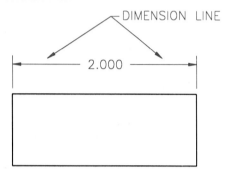

dimensioning variables   variables AutoCAD uses to control the appearance and structure of dimensions in a drawing.

dimensions   a numerical size description of an object.

displacement diagram   a graph of the maximum rise of a cam follower on the Y axis through a 360° rotation of the cam on the X axis.

dividers   a drawing instrument used in mechanical drawing to mark off equal distances or to divide lines, arcs, or circles. Dividers may also be used to transfer measurements or distances from one object on a drawing to another.

drafting   the use of carefully drawn and documented pictures to describe an object precisely so that it can be constructed.

drafting machine   an instrument used in mechanical drafting to replace T-squares and triangles for most purposes.

dwell   the period during which the follower on a cam neither rises or falls; dwell commonly occurs between different types of motion.

eccentric   off-center; a circle is eccentric if it does not share the same center point as another circle in a drawing.

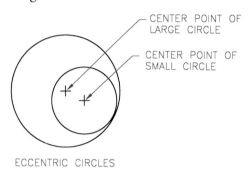

edit   to change the characteristics, shape, or size of an entity that already exists on the screen.

elbow   a short piece used to change the direction of a duct or pipe.

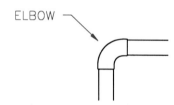

elevation   *1)* the level at which the bottom of a three-dimensional object begins. AutoCAD's default elevation is 0. *2)* in architecture, a side view of a structure that shows locations of features such as doors and windows.

ellipse   a circle as viewed from an angle to its primary plane. To draw an ellipse in AutoCAD, use the ELLIPSE command. An ellipse has both a major (longer) axis and a minor (shorter) axis.

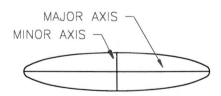

elliptical   shaped like an ellipse.

entity   a predefined drawing element that you can manipulate in drawings. AutoCAD's standard entities are points, lines, arcs, circles, text, dimensions, polylines, traces, solids, shapes, blocks, attributes, 3D polylines, 3D faces, 3D meshes, polyface meshes, and viewports.

equilateral triangle   a triangle in which all the sides are the same length. All of the angles in an equilateral triangle equal 60°.

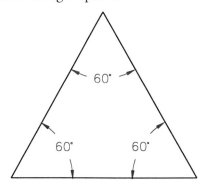

EQUILATERAL TRIANGLE: ALL
SIDES AND ANGLES ARE EQUAL

exploded pictorial assembly drawing   a drawing in which parts are not yet assembled, but are in their proper placement for assembly.

extension lines   the lines that extend from an object to the dimension line that describes its size. You can change the amount of space between the object and the extension lines using the DIMEXO dimensioning variable. You can control how far the extension lines extend beyond the dimension line using the DIMEXE dimensioning variable.

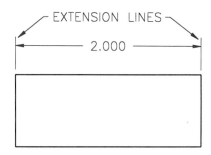

external thread   a thread on the outside surface of a cylinder or cone.

fastener   a part that holds other parts together.

ferrous metal   any metal that contains iron.

fillet   a small radius on an inside corner of an object. In AutoCAD, you can create a fillet using the FILLET command.

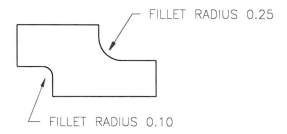

finish symbols   standard symbols that accompany finish information on a drawing. The three most common are the check, the "V," and the italicized *f*, although the *f* is now becoming obsolete.

flaws   surface defects in a material to be used for construction.

font   a typeface design that includes a complete set of characteristics for the typeface.

freeze   a method of telling AutoCAD to ignore one or more drawing layers. When a layer is frozen, the entities on that layer do not appear on the screen. See also *thaw*.

front auxiliary view   (also called *vertical auxiliary view*) an auxiliary view in which the "glass box" hinges to the front view.

frontal plane revolution   a revolution in which the axis of revolution (the axis around which the view is revolved) appears as a point in the front view.

full section   a section in which the cutting plane passes all the way through the object.

geometry   the study of the relationships of points, lines, angles, and figures in space.

graphic communication   communication that relies on pictures and printed material such as textbooks and magazines.

graphics screen   the screen in AutoCAD where drawing and editing occur. In Release 12 and later releases, this screen appears by default when you load AutoCAD.

grid   a drawing tool provided by AutoCAD, consisting of a set of evenly spaced dots on the screen. You can set the grid at any intervals using the GRID command. The F7 function key toggles the grid on and off.   The grid does not print; it is only a reference tool. The figure below shows the grid behind the AutoCAD crosshairs.

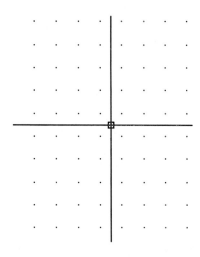

grips   small blue boxes that appear when you select an entity when no command has been entered. The grips allow you to manipulate the entity directly by stretching it, moving it, copying it, or changing its shape.

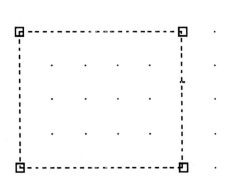

half section   a section in which the cutting plane passes halfway through the object and a quarter of the object is removed so that you can see inside.

harmonic motion   non-linear motion that provides for a smooth start and stop. The displacement diagram of harmonic motion is a sine curve.

hatch lines   the individual lines that make up a hatch pattern in AutoCAD.

hatching   See *crosshatching*.

hexagon   a closed six-sided figure. In a **regular hexagon**, all six sides are the same length. An **irregular hexagon** has six sides, but they are not all the same length. You can draw a regular hexagon in AutoCAD by entering the POLYGON command and specifying six sides.

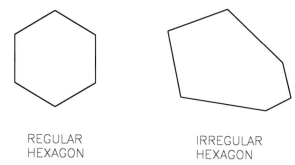

REGULAR
HEXAGON

IRREGULAR
HEXAGON

hidden line   a dashed line used to show object boundaries and features that would ordinarily be hidden by other features if you were looking at the actual object.

- - - - - - - - - - - - - -

HIDDEN LINE

hieroglyphics   an early Egyptian language in which modified pictures were used; developed from pictograms.

hydration   a chemical reaction between the water and lime in concrete that strengthens the concrete; the longer the curing time (reaction time), the stronger the concrete will be.

hypotenuse the side opposite the right (90°) angle in a right triangle.

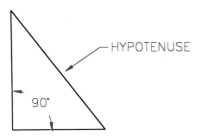

ideogram similar to a pictogram, but the picture and meaning are often more abstract; they convey ideas rather than specific objects.

inclined lines lines that are neither horizontal nor vertical.

INCLINED LINES

inscribed a polygon that fits entirely within a circle whose radius is equal to the distance from one vertex to the opposite vertex of the polygon.

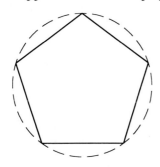

PENTAGON INSCRIBED
IN A CIRCLE

interference fit a fit in which minimum shaft size is larger than maximum hole size, so that interference will always occur when the two are mated.

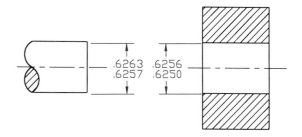

internal thread a thread on the inside surface of a cylinder or cone.

International Standards Organization See *ISO*.

intersection the point at which entities meet or cross.

INTERSECTIONS

irregular curves mechanical drafting instruments used to draw noncircular curves; also called French curves.

ISO the International Standards Organization, an organization that specifies standards common to many countries around the world.

isosceles triangle a triangle in which two sides are of equal length.

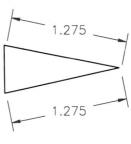

ISOCELES TRIANGLE

isometric grid in AutoCAD, a feature that allows you to set the grid along lines that make it easier to create an isometric drawing. The orientation of the crosshairs changes to follow the grid.

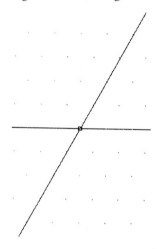

isometric sketch a sketch in which the object is drawn on axes that are 120 degrees apart.

justification the alignment of text on the screen. AutoCAD defaults to a left justification (all text lines are aligned on the left). To change the justification, use the Justify option of the TEXT or DTEXT command.

key a metal piece that locks two pieces together so that the pieces cannot move relative to each other.

lay the direction of the grain (of wood) or predominant surface pattern of a material.

layers the electronic equivalent of transparent overlays on a manual drawing. Layers help CAD drafters organize the drawing; for example, all the dimensions can be placed on one layer. Since a layer can be "frozen" so that its contents become invisible, the drawing can be plotted with or without the dimensions. To create a layer or to change layers, use the LAYER command or enter DDLMODES for a dialogue box. In three-dimensional drawings that contain more than one viewport, use the VPLAYER command to control each viewport separately.

layout the arrangement of views on a drawing. See also *sheet layout*.

leader a line that has an arrowhead at one end. Leaders are used to connect drawing notes and other specifications with the place on the drawing to which they refer.

LEADER

left-hand thread a thread in which the spiral grooves wind in a counterclockwise direction when viewed along the axis of the fastener.

lettering text used on a drafted document.

limit dimensioning a method of dimensioning that includes a range of acceptable values (tolerance).

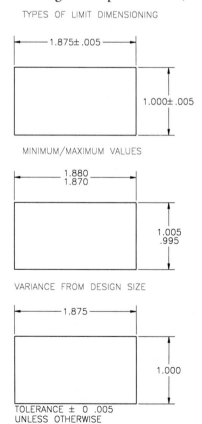

NOTE

limits the limits of an AutoCAD drawing are the drawing boundaries. AutoCAD is versatile; it allows you to set limits large enough to hold any complete drawing at full size. For example, if you were doing an architectural drawing, you could set the limits at 96 feet wide and 72 feet long. Within AutoCAD, the drawing is stored with these actual dimensions. To set the limits for an AutoCAD drawing, use the LIMITS command.

limits of size in limit dimensioning, the maximum and minimum tolerance values. Any value within the limits of size meets specifications.

major axis the longer of the two axes in an ellipse.

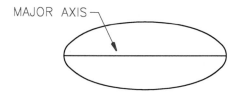

mechanical (manual) drafting drafting that is done with traditional instruments, without the aid of a computer.

midpoint the point in the exact middle of a line segment or arc.

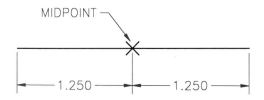

minor axis the shorter of the two axes in an ellipse.

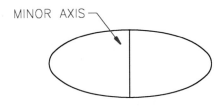

mirror image an image that is identical to the original image, but reflected as it would be in a mirror. In AutoCAD, you can create a mirror image using the MIRROR command.

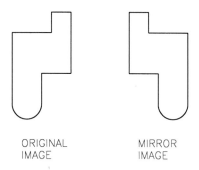

mirror line an imaginary line you specify in AutoCAD to create and place a mirrored copy of an object in a drawing. The copy is rotated around the mirror line to create the mirror image.

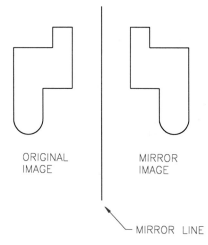

model space the standard drawing space in AutoCAD, in which most drawings are created.

modeling creating a drawing in three dimensions within AutoCAD. Since the objects in the drawing actually exist in three dimensions, you can change your viewpoint to see them from any side or angle.

mouse a pointing device commonly used with CAD programs.

multiview drawing   a two-dimensional representation of a three-dimensional object; the object can be described fully through the use of more than one view. The views most commonly found in a multiview drawing are the front view, top view, and right-side view. Additional views may be used if necessary. See also *view*.

multiview sketch   a sketch that contains more than one view of an object. See also *multiview sketch*.

nominal size   general identification of a size; nominal size is not always equal to the actual size. For example, a board that has a nominal size of 2" × 4" has an actual size of 1½" × 3½".

normal views   the front, top, and right-side views of an object.

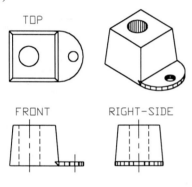

THE THREE NORMAL VIEWS

nut   a bolt retainer; the internal threads on the nut mate with the external threads on a bolt so that the nut can be used to tighten, hold, or loosen the bolt. See *bolt* for illustration.

object snap   a feature in AutoCAD that allows you to snap to specific places or points on entities. For example, the ENDpoint object snap allows you to snap the endpoints of a line. To use the object snap, enter the first three letters of the name of the object snap you want to use at the *To point:* prompt (END for ENDpoint, NOD for NODe, INT for INTersection, and so on). See also *running object snap*.

object snap, running   See *running object snap*.

oblique cone   a cone in which the vertical axis is not perpendicular to the base.

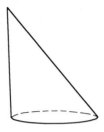

OBLIQUE CONE

oblique prism   a prism in which neither the top nor the bottom forms a 90° angle with the sides.

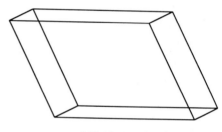

OBLIQUE PRISM

oblique sketch   a sketch in which the front of the object is undistorted, and the top and side of the object are drawn at an angle other than 90 degrees.

obliquing angle   the slope of individual text characters. This text characteristic can be set in AutoCAD using the STYLE command.

THE OBLIQUING ANGLE FOR THIS TEXT IS SET TO 15°

THE OBLIQUING ANGLE FOR THIS TEXT IS SET TO −15°

obtuse angle   an angle that measures more than 90°.

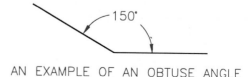

AN EXAMPLE OF AN OBTUSE ANGLE

octagon   an eight-sided polygon.

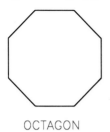

OCTAGON

offset   to create a line parallel to another line at a given distance; to create an arc or circle that is concentric with the original arc or circle at a given distance. Use the OFFSET command to offset lines, circles, and arcs.

offset section   a section in which the cutting plane is staggered at an angle of 90° to show two or more features that are not ordinarily aligned using just one section.

origin   the point at which the axes on the Cartesian coordinate system intersect (point 0,0,0). See *Cartesian coordinate system* for illustration.

Ortho   a drawing aid in AutoCAD that forces all lines to be exactly vertical or horizontal. The ORTHO command and the F8 function key toggle Ortho on and off.

orthographic projection   an object seen as a series of several single views that show each face of the object in its true size and shape.

overscore   a line that runs above a line of text. To overscore text, enter %%O where you want the overscore to begin, enter the text to be overscored, and enter %%O again to turn off the overscore feature. Note that in the %%O control characters, the O is the letter O of the alphabet, not a zero.

THIS TEXT IS OVERSCORED.

paper space   a drawing space in AutoCAD that allows you to lay out multiview drawings and plot all the views at the same time.

parallel lines   lines in which every point on one line is at an equal distance from its corresponding point on the other line. Parallel lines never intersect.

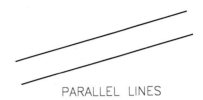

PARALLEL LINES

partial auxiliary view   an auxiliary view that has been simplified so that it does not include all of the features normal to an auxiliary view. Partial auxiliary views are used for complicated objects to make the view easier to understand.

pattern   a flat piece of material that can be cut, formed, shaped, or joined to create a finished product.

pentagon   a five-sided polygon.

PENTAGON

perpendicular lines   lines that intersect each other at a 90-degree angle.

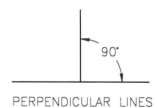

PERPENDICULAR LINES

perspective sketch   a sketch that shows an object as it appears to the eye rather than showing actual dimensions.

phantom line   a thin line composed of a repeating pattern of a long segment followed by two short segments; used to show lines on a drawing that are not present in the real object.

—— – – ––– – – –––– – – –––– – – ——

PHANTOM LINE

pictogram   pictures that are recognized within a society as having a specific meaning.

pictorial sketch   a sketch that has a "picture-like" quality.

piercing point   the intersection of a line with a straight or curved surface.

pitch diameter   the diameter of an imaginary circle formed by the tangent of mating gears.

pixels   short for "picture elements"—the tiny individual dots of light that form the image on a computer screen.

plan view   the top view of an object; AutoCAD's default view.

point   *1)* a numerical value that describes a unique location on your drawing or screen; for example, the point (4,6) is a location four units to the right of the origin and six units up from the X axis on the Cartesian coordinate system. If the Z coordinate is not given, the point is assumed to lie flat to the screen (4,6,0). *2)* An AutoCAD entity that can appear as a small dot, X, circle, square, or various combinations of these. You can snap to these AutoCAD points using the NODe object snap.

polar coordinates   coordinates that describe the distance and the angle of a line from the origin (absolute) or from the previous point (relative). To specify a set of polar coordinates, enter the line length first, then an angle symbol (<), and then the angle of the line. If the polar coordinates are relative, then the @ symbol should precede the line length.

polygon   a closed figure made of line segments. A **regular polygon** is one in which all the line segments have the same length.

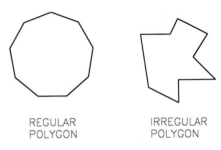

REGULAR POLYGON   IRREGULAR POLYGON

polyline   a connected series of line segments and arcs that AutoCAD treats as a single entity. Use the PLINE command to create a polyline, and use the PEDIT command to edit the polyline.

pressure angle   the angle of contact between the teeth of two mating gears.

primary auxiliary view   an auxiliary view that is projected from the vertical, horizontal, or profile plane of an object.

prism   any three-dimensional box-like shape.

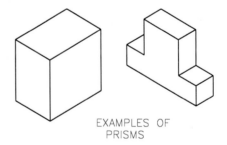

EXAMPLES OF PRISMS

proportions   the size of each individual part as compared to the size of all the other parts.

prototype drawing   a drawing that holds the default settings you will use in most or all of your drawings.

protractor   a drafting instrument that allows you to measure and construct accurate angles.

pull-down menus menus that run across the top of the AutoCAD screen. These menus are only visible when you place the cursor near the top of the screen. To make selections from these menus, use the pointing device.

quadrant a quarter of a circle; 90°.

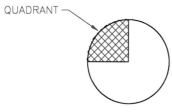

A CIRCLE CONTAINS FOUR
QUADRANTS, OR QUARTERS.

quadrilateral a polygon that has four sides.

QUADRILATERAL

rack a gear in which the teeth are laid out in a row instead of on a circular base; racks are usually used with pinion spur gears.

radians a unit of measure for angles.

radius the distance from the center of a circle to any point on the circle. The radius of a circle is half the diameter of the circle.

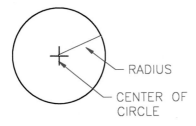

rectangle a special quadrilateral in which opposite sides have the same length and all angles between sides are right (90°) angles.

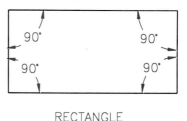

RECTANGLE

reference assembly drawing a drawing that shows how parts go together and how they work. Reference assembly drawings are not dimensioned, but each part is identified precisely so that the parts can be assembled more readily by someone who is not familiar with the finished product.

reference dimensions dimensions that are for general information or convenience only. They are not exact measurements and should not be used to construct or manufacture a part.

reference line a line that represents the reference plane in a view in which the reference plane is perpendicular to the line of sight.

reference plane a plane that is parallel to the incline on an object to be shown in an auxiliary view.

regenerate to cause AutoCAD to recalculate all the vectors that make up a drawing. Use the REGEN command to regenerate a drawing in AutoCAD.

region a closed two-dimensional area that results when you combine two or more two-dimensional objects into a single entity.

REGION

relative coordinates coordinates that define the distance from the current point to the next point. For example, if you begin a line at (2,3) and specify relative coordinates (5,2), the line goes from (2,3) to absolute point (7,5). Relative coordinates contain an @ symbol before the coordinate values.

removed section a detail drawing section that has been moved from its normal place on a drawing and placed in a more convenient location; because it is a detail drawing, it may be drawn to a larger scale than the rest of the drawing. If so, its scale must be clearly marked.

rendering the simulation of the effects of perspective, reflection, light, and shadow on the appearance of an object. To render an object, you must first load AutoCAD's Render module and enter the RENDER command. An elementary example of a rendered object is shown below.

revolution a view in which the object is revolved around an axis so that the true size of an inclined surface can be shown in the front, top, or side view.

revolved section a section that shows the cross section of a spoke, rib, airfoil, or other part of an object that must be described at a specific location.

rhomboid a quadrilateral that contains no right angles; all opposite sides are parallel.

RHOMBOID

rhombus a quadrilateral that contains no right angles but in which all four sides are of equal length.

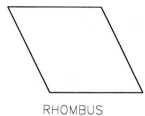

RHOMBUS

right angle an angle that measures exactly 90°.

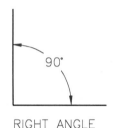

RIGHT ANGLE

right cone a cone in which the axis is perpendicular to the base.

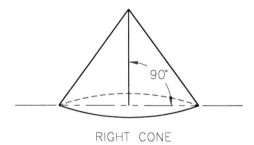

RIGHT CONE

right cylinder a cylinder in which both bases are perpendicular to the side.

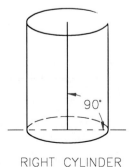

RIGHT CYLINDER

right prism   a prism that has two parallel bases perpendicular to three or more sides.

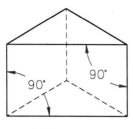

RIGHT PRISM

right rectangular pyramid   a pyramid that is built on a rectangular base and has a vertical axis that is perpendicular to the base.

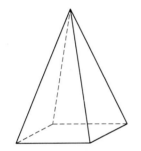

RIGHT RECTANGULAR PYRAMID

right triangle   a triangle that contains a right (90°) angle.

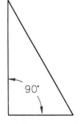

RIGHT TRIANGLE

right-hand thread   a thread in which the spiral grooves wind in a clockwise direction when viewed along the axis of the fastener.

right-side auxiliary view   (also called *profile auxiliary view*) an auxiliary view in which the "glass box" hinges to the right side.

rivet   a permanent fastener that fastens sheet metal parts and steel plates; rivets cannot be removed after they are installed without breaking the rivet.

roughness   the general condition of the surface after forming or machining.

round   a small radius of an outside corner of an object. In AutoCAD, you can create a round using the FILLET command.

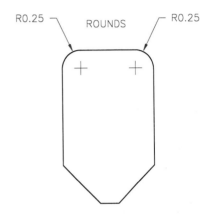

running object snap   one or more object snap modes set to run continuously as you work in AutoCAD. When you specify a running object snap, it affects every command you enter (and every point you pick) until you turn the object snap off again. To set a running object snap, use the OSNAP command.

scale   *1)* a rule that allows you to measure distances accurately on drawings. Some rules have more than one scale, so that if you have reduced or enlarged the drawing, you can still measure actual distances directly. *2)* to increase or decrease the size of a drawing to make it easier to see or plot. *3)* to increase or decrease the actual size of an object in an AutoCAD drawing. To scale an object in this manner, use the SCALE command.

scalene triangle   a triangle in which all the sides have different lengths.

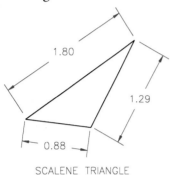

SCALENE TRIANGLE

screen menu   a menu of commands that appears on the right side of the AutoCAD screen. To enter the commands, you select them using the pointing device.

screw   an externally threaded fastener that can be tightened and released by using a screwdriver to torque its head.

secondary auxiliary view   an auxiliary view used to describe a surface that is inclined to all three primary planes; secondary auxiliary views are projected from primary auxiliary views.

section lines   lines that are drawn to indicate material that has been cut away from the object.

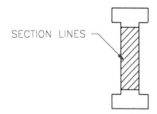

SECTION LINES

sectional view   a drawing view in which an imaginary cutting plane is passed through an object. The material on one side of the cutting plane is then removed so that you can see what is inside.

selection set   a set of entities you select to be operated on by a command. You can create a selection set using a window or a crossing window. If you want to select all the entities on the screen, respond "All" to the *Select objects:* prompt.

semi-circle   half of a circle; 180°.

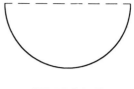

SEMICIRCLE

shading   a quick method of shadowing a model in AutoCAD to give it visual depth. You can shade objects in AutoCAD using the SHADE command.

sheet layout   in mechanical drafting, preparing the paper for drafting by fastening it to the drawing board and constructing the borders and/or title block.

side plane revolution   a revolution in which the axis of revolution (the axis around which the view is revolved) appears as a point in the side view.

site plan   a plan that shows a building as it will be situated on a site or lot of land.

sketching   drawing without the aid of mechanical devices such as rules and triangles; freehand drawing.

Snap   an invisible grid in AutoCAD that controls the cursor movement so that the cursor lands on exact points. For example, if you set the snap for 1 unit, the cursor would jump in 1-unit intervals, and you could not pick points that are not whole units, such as point (4,5.5). You can set the snap to the same intervals as the grid, or you can set each one differently. You can set the Snap intervals using the SNAP command. The F9 function key toggles Snap on and off.

spotface a very shallow cut meant to level a surface so that a bolt head, washer, or nut will seat properly on the surface.

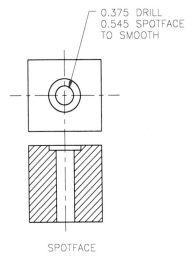

0.375 DRILL
0.545 SPOTFACE
TO SMOOTH

SPOTFACE

square a special quadrilateral in which all four sides are of equal length and all angles between sides are right (90°) angles.

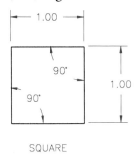

SQUARE

status line the line at the top of the AutoCAD screen that shows the current layer, the color of the current layer, and the status of toggles such as Snap and Ortho. The tracking figures are also located on this line.

stretchout line a line that represents the opened-up length of a pattern.

style a variation of a font. You can set the text style by using the STYLE command or the Style option of the TEXT or DTEXT command. See also *font*.

supplementary angles two angles that, when placed so that they share one common side (vector), measure 180°.

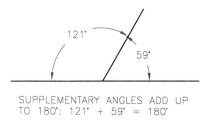

SUPPLEMENTARY ANGLES ADD UP
TO 180°: 121° + 59° = 180°

surface finish (also called *surface texture*) the roughness, waviness, lay, and flaws allowed in a finished product.

surface modeling creating three-dimensional surfaces that are contoured to a shape you specify.

symbol library a file (or group of files) that stores a group of blocks that you can insert into other drawings. The blocks in a symbol library are usually, but not always, related. You can create and use as many symbol libraries as you need to make your work easier.

T-square an instrument used in mechanical drafting to draw horizontal lines and to provide a base for the 45-degree and 30-60-degree triangles.

tangent touches a line, circle, or arc at exactly one point.

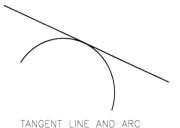

TANGENT LINE AND ARC

technical illustrator someone who prepares illustrations that must be technically correct, such as illustrations for science or medical textbooks.

text all of the letters, numbers, characters, and symbols used to supplement the information in a drawing.

thaw the opposite of freezing: when you thaw a frozen layer, AutoCAD no longer ignores the layer and its contents are displayed on the screen.

thickness another term for height. You can create three-dimensional objects by setting the THICKNESS variable to a value other than 0 and then using the basic drawing commands as you would to create a two-dimensional object. The prism shown here was created by setting THICKNESS to 3 and then using the LINE command to create a simple rectangle. To see the three-dimensional prism, use the VPOINT command to change the viewpoint.

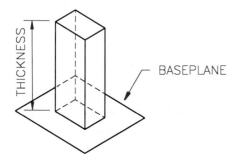

thread classes the allowances and tolerances that govern the fit between external and internal threads.

thread form the profile of a thread in cross section.

threads spiral grooves cut into the surface of a cylindrical or cone-shaped fastener to allow it to grip the material in which it is placed.

three-dimensional (3D) an object that has depth as well as width and height; also a drawing that shows or describes such an object.

tolerance in limit dimensioning, the maximum amount that a dimension can vary and still meet specifications. Tolerances are necessary because it is impossible to manufacture parts to exact sizes. In the example below, the dimension can vary by 0.0002 above the design size and 0.0004 below the design size.

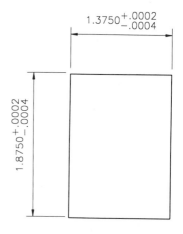

top auxiliary view (also called *horizontal auxiliary view*) an auxiliary view in which the "glass box" hinges to the top view.

top plane revolution a revolution in which the axis of revolution (the axis around which the view is revolved) appears as a point in the top view.

transition fit a fit in which there may be interference or clearance between a shaft and hole, depending on the exact size (within tolerance) of the shaft and hole.

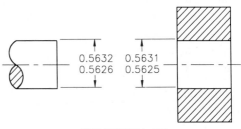

TRANSITIONAL FIT

transition piece a piece that connects ducts and pipes of different shapes and sizes. One end of the transition piece conforms to the end of one duct or pipe, and the other end conforms to a duct or pipe of different shape or dimensions.

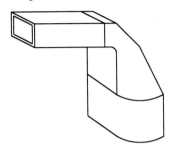

TRANSITION PIECE

trapezium a quadrilateral that contains no right angles and has no sides parallel.

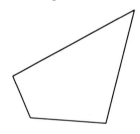

TRAPEZIUM

trapezoid a quadrilateral that contains no right angles; two opposite sides are parallel.

TRAPEZOID

triangle *1)* a special polygon that has only three sides. The terms "regular" and "irregular" do not apply to triangles. *2)* An instrument used in mechanical drafting that has three angles of known values. The triangles most commonly used in drafting are the 45-degree triangle and the 30-60-degree triangle.

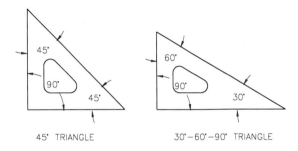

45° TRIANGLE   30°–60°–90° TRIANGLE

true perfectly straight.

truncated cylinder a cylinder in which one end has been cut at an angle other than 90°.

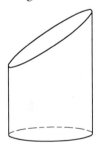

TRUNCATED CYLINDER

truncated prism a prism in which part of the prism is cut off at an angle.

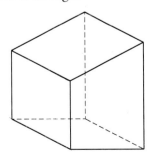

TRUNCATED PRISM

truncated right cone a right cone in which the top has been cut off at an angle to the axis.

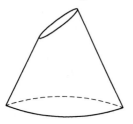

TRUNCATED CONE

truncated right pyramid a right pyramid in which the top has been cut off at an angle to its axis.

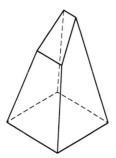

TRUNCATED PYRAMID

two-dimensional (2D) an object that has only width and height; also a drawing that shows or describes such an object.

UCS icon In AutoCAD, an icon that indicates the type of coordinate system in use at any given time. The icon appears in the lower left corner of the AutoCAD screen. See also *user coordinate system.*

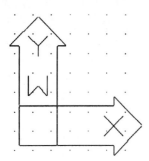

underscore a line that runs beneath text. To underscore text in AutoCAD, enter %%U where you want the underscore to begin, enter the text to be underscored, and then enter %%U again to turn off the underscore feature.

THIS TEXT IS UNDERSCORED.

unidirectional method of placing dimensions a placement method in which dimensions are always horizontal, regardless of the orientation of the dimension line. See also *aligned method of placing dimensions.*

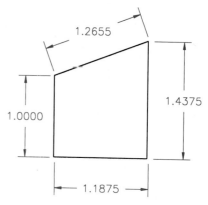

ALIGNED METHOD OF
PLACING DIMENSIONS

uniform motion straight-line motion in which the follower on a cam rises and falls at a constant speed.

uniformly accelerated and decelerated motion a very smooth form of motion in which the follower speeds up and slows down steadily or uniformly. The displacement diagram for this type of motion is a parabola.

unilateral tolerance in limit dimensioning, a tolerance that is allowed only on one side of the design size.

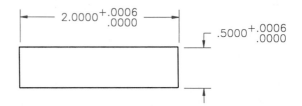

update to redisplay one or more dimensions that are already on the screen using the current variable settings.

user coordinate system (UCS) In AutoCAD, a coordinate system specified or defined by the user. The default coordinate system in AutoCAD is the world coordinate system. However, using the UCS command, the user can change one or more of the axes to customize the coordinate system for specific purposes. See also *world coordinate system.*

variables internal settings that control various features of a program. In AutoCAD, you can set many of the variables directly at the prompt line.

vertex (*plural: vertices*) in a polygon, the point at which two line segments intersect. The number of vertices a polygon has is equal to the number of sides it has.

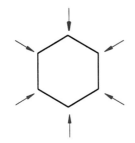

THE ARROWS POINT TO
THE VERTICES IN THIS
HEXAGON.

view a drawing that shows a single side of an object or shows the object from a single angle or side. The six basic views for drafting are top, front, right-side, left-side, bottom, and back. The top, front, and right-side views are used by convention to describe most three-dimensional objects.

viewports logical divisions on an AutoCAD screen that provide multiple viewing areas. For example, you can set up viewports that allow you to see simultaneously the front, right-side, and top views of a three-dimensional object as well as a pictorial view of the object. To activate viewports in AutoCAD, use VPORTS (in model space) or MVIEW (in paper space).

visible line a continuous linetype used to show basic object outlines and features.

VISIBLE (CONTINUOUS) LINE

visualization the ability to imagine what an object looks like.

waviness a measure of the flatness of a surface. A warped board, for example, has a high degree of waviness.

width factor the expansion or compression of individual text characters. This text characteristic can be set in AutoCAD using the STYLE command.

WIDTH FACTOR = 0.5

WIDTH FACTOR = 1.0

WIDTH FACTOR = 1.5

window a selection box created with the pointing device in AutoCAD. You pick opposite corners of the box, beginning with the upper (or lower) left side. All entities that lie entirely within the window become selected. However, entities that lie only partially inside the window are not selected. See *crossing window.*

wireframe model a model that consists only of lines that define its shape. Since it has no closed sides, you can see through a wireframe model.

world coordinate system AutoCAD's default coordinate system, which consists of the top right quadrant (Quadrant I) of the Cartesian coordinate system. See *Cartesian coordinate system.*

worm gear a gear that transmits motion between shafts that are at right angles to each other but do not intersect.

Appendix A
Reference Tables

Table 1

Metric and English Equivalents

Linear Measure

1 kilometer = 0.6214 mile.   1 mile = 1.609 kilometers.

1 meter = { 39.37 inches.   1 yard = 0.9144 meter.
3.2808 feet.   1 foot = 0.3048 meter.
1.0936 yards.   1 foot = 304.8 millimeters.

1 centimeter = 0.3937 inch.   1 inch = 2.54 centimeters.

1 millimeter = 0.03937 inch.   1 inch = 25.4 millimeters.

Square Measure

1 square kilometer = 0.3861 square mile = 247.1 acres.
1 hectare = 2.471 acres = 107,639 square feet.
1 are = 0.0247 acre = 1076.4 square feet.
1 square meter = 10.764 square feet = 1.196 square yards.
1 square centimeter = 0.155 square inch.
1 square millimeter = 0.00155 square inch.

1 square mile = 2.5899 square kilometers.
1 acre = 0.4047 hectare = 40.47 ares.
1 square yard = 0.836 square meter.
1 square foot = 0.0929 square meter = 929 square centimeters.
1 square inch = 6.452 square centimeters = 645.2 square millimeters.

Cubic Measure

1 cubic meter = 35.315 cubic feet = 1.308 cubic yards.
1 cubic meter = 264.2 U.S. gallons.
1 cubic centimeter = 0.061 cubic inch.
1 liter (cubic decimeter) = 0.0353 cubic foot = 61.023 cubic inches.
1 liter = 0.2642 U.S. gallon = 1.0567 U.S. quarts.

1 cubic yard = 0.7646 cubic meter.
1 cubic foot = 0.02832 cubic meter = 28.317 liters.
1 cubic inch = 16.38706 cubic centimeters.
1 U.S. gallon = 3.785 liters.
1 U.S. quart = 0.946 liter.

Weight

1 metric ton = 0.9842 ton (of 2240 pounds) = 2204.6 pounds.
1 kilogram = 2.2046 pounds = 35.274 ounces avoirdupois.
1 gram = 0.03215 ounce troy = 0.03527 ounce avoirdupois.
1 gram = 15.432 grains.

1 ton (of 2240 pounds) = 1.016 metric ton = 1016 kilograms.
1 pound = 0.4536 kilogram = 453.6 grams.
1 ounce avoirdupois = 28.35 grams.
1 ounce troy = 31.103 grams.
1 grain = 0.0648 gram.

1 kilogram per square millimeter = 1422.32 pounds per square inch.
1 kilogram per square centimeter = 14.223 pounds per square inch.
1 kilogram-meter = 7.233 foot-pounds.
1 pound per square inch = 0.0703 kilogram per square centimeter.
1 calorie (kilogram calorie) = 3.968 Btu (British thermal unit).

Table 2
Decimal Equivalents

4ths	8ths	16ths	32nds	64ths	To 4 Places	4ths	8ths	16ths	32nds	64ths	To 4 Places
				1/64	.0156					33/64	.5156
			1/32		.0312				17/32		.5312
				3/64	.0469					35/64	.5469
		1/16			.0625			9/16			.5625
				5/64	.0781					37/64	.5781
			3/32		.0938				19/32		.5938
				7/64	.1094					39/64	.6094
	1/8				.1250		5/8				.6250
				9/64	.1406					41/64	.6406
			5/32		.1562				21/32		.6562
				11/64	.1719					43/64	.6719
		3/16			.1875			11/16			.6875
				13/64	.2031					45/64	.7031
			7/32		.2188				23/32		.7188
				15/64	.2344					47/64	.7344
1/4					.2500	3/4					.7500
				17/64	.2656					49/64	.7656
			9/32		.2812				25/32		.7812
				19/64	.2969					51/64	.7969
		5/16			.3125			13/16			.8125
				21/64	.3281					53/64	.8281
			11/32		.3438				27/32		.8438
				23/64	.3594					55/64	.8594
	3/8				.3750		7/8				.8750
				25/64	.3906					57/64	.8906
			13/32		.4062				29/32		.9062
				27/64	.4219					59/64	.9219
		7/16			.4375			15/16			.9375
				29/64	.4531					61/64	.9531
			15/32		.4688				31/32		.9688
				31/64	.4844					63/64	.9844
					.5000						1.0000

Table 3

Millimeters to Decimal Inches

mm	in.	mm	in.	mm	in.	mm	in.	mm	in.
1 = 0.0394		21 = 0.8268		41 = 1.6142		61 = 2.4016		81 = 3.1890	
2 = 0.0787		22 = 0.8662		42 = 1.6536		62 = 2.4410		82 = 3.2284	
3 = 0.1181		23 = 0.9055		43 = 1.6929		63 = 2.4804		83 = 3.2678	
4 = 0.1575		24 = 0.9449		44 = 1.7323		64 = 2.5197		84 = 3.3071	
5 = 0.1969		25 = 0.9843		45 = 1.7717		65 = 2.5591		85 = 3.3465	
6 = 0.2362		26 = 1.0236		46 = 1.8111		66 = 2.5985		86 = 3.3859	
7 = 0.2756		27 = 1.0630		47 = 1.8504		67 = 2.6378		87 = 3.4253	
8 = 0.3150		28 = 1.1024		48 = 1.8898		68 = 2.6772		88 = 3.4646	
9 = 0.3543		29 = 1.1418		49 = 1.9292		69 = 2.7166		89 = 3.5040	
10 = 0.3937		30 = 1.1811		50 = 1.9685		70 = 2.7560		90 = 3.5434	
11 = 0.4331		31 = 1.2205		51 = 2.0079		71 = 2.7953		91 = 3.5827	
12 = 0.4724		32 = 1.2599		52 = 2.0473		72 = 2.8247		92 = 3.6221	
13 = 0.5118		33 = 1.2992		53 = 2.0867		73 = 2.8741		93 = 3.6615	
14 = 0.5512		34 = 1.3386		54 = 2.1260		74 = 2.9134		94 = 3.7009	
15 = 0.5906		35 = 1.3780		55 = 2.1654		75 = 2.9528		95 = 3.7402	
16 = 0.6299		36 = 1.4173		56 = 2.2048		76 = 2.9922		96 = 3.7796	
17 = 0.6693		37 = 1.4567		57 = 2.2441		77 = 3.0316		97 = 3.8190	
18 = 0.7087		38 = 1.4961		58 = 2.2835		78 = 3.0709		98 = 3.8583	
19 = 0.7480		39 = 1.5355		59 = 2.3229		79 = 3.1103		99 = 3.8977	
20 = 0.7874		40 = 1.5748		60 = 2.3622		80 = 3.1497		100 = 3.9371	

The American Society of Mechanical Engineers (ASME) publishes American National Standards Institute (ANSI) standards. To obtain a catalog of ASME publications, contact the American Society of Mechanical Engineers, 22 Law Drive, P.O. Box 2300, Fairfield, NJ 07007-2300, telephone 800-843-2763 or 201-887-1717, or contact ASME through CompuServe at 73302,1017.

Table 4 lists ANSI Y14 drafting standards, which provide guidelines for creating drawings.

Table 4

Y14 Drafting Standards

Y14.1-1980(R1987)	Drawing Sheet Size and Format
Y14.2M-1979(R1987)	Line Conventions and Lettering
Y14.3-1975(R1987)	Multi and Sectional View Drawings
Y14.4M-1989	Pictorial Drawing
Y14.5M-1982(R1988)	Dimensioning and Tolerancing
Y14.6-1978(1987)	Screw Thread Representation
Y14.7.1-1971(R1988)	Gear Drawing Standards — Part 1: For Spur, Helical, Double Helical, and Rack
Y14.7.2-1978(R1984)	Gear and Spline Drawing Standards — Part 2: Bevel and Hypoid Gears
Y14.8M-1989	Castings and Forgings
Y14.13M-1981(R1987)	Mechanical Spring Representation
Y14.15-1966(R1988)	Electrical and Electronics Diagrams (With 1971 and 1973 Supplements)
Y14.18M-1986	Optical Parts
Y14.24M-1989	Types and Applications of Engineering Drawings
Y14.26M-1989	Digital Representation for Communication of Product Definition Data
Y14.34M-1989	Parts Lists, Data Lists, and Index Lists
Y14.36-1978(R1987)	Surface Texture Symbols
Y14	Technical Report 4-1989: A Structural Language Format for Basic Shape Description

Tables 5 through 14 are extracted from the ANSI B series of standards. They provide the information required to make parts such as common screws, bolts, and nuts. Fig. 1 provides a graphic description of the parts described in the tables. The lettered dimensions in Fig. 1 refer to dimensions given in the tables.

Fig. 1

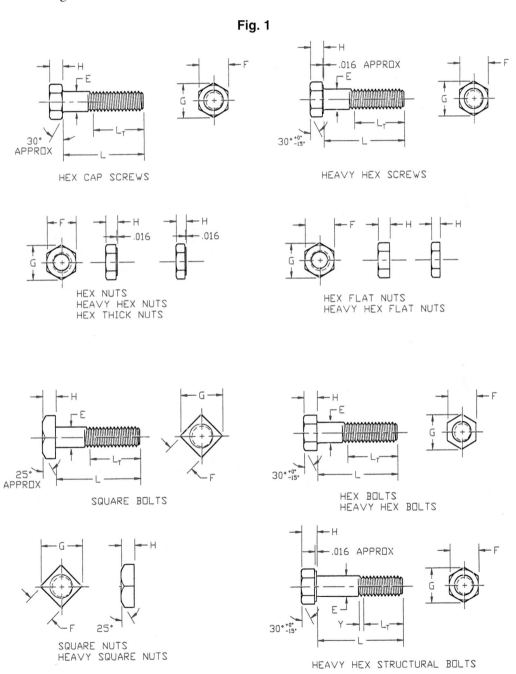

Extracted from ANSI B18.2.1-1981(R1992) *Square and Hex Bolts and Screws* and ANSI B18.2.2-1987(R1993) *Square and Hex Nuts*, with permission of the publisher, The American Society of Mechanical Engineers.

Table 5

Unified Inch Screw Threads, Fine Thread Series, UNF

Nominal Size[1]	Major Diameter[2] (D)	Threads/ in.	Pitch Diameter (E)	Minor Diameter External Threads (Ref.)	Minor Diameter Internal Threads
0 (0.060)	0.0600	80	0.0519	0.0451	0.0465
1 (0.073)*	0.0730	72	0.0640	0.0565	0.0580
2 (0.086)	0.0860	64	0.0759	0.0674	0.0691
3 (0.099)*	0.0990	56	0.0874	0.0778	0.0797
4 (0.112)	0.1120	48	0.0985	0.0871	0.0894
5 (0.125)	0.1250	44	0.1102	0.0979	0.1004
6 (0.138)	0.1380	40	0.1218	0.1082	0.1109
8 (0.164)	0.1640	36	0.1460	0.1309	0.1339
10 (0.190)	0.1900	32	0.1697	0.1528	0.1562
12 (0.216)*	0.2160	28	0.1928	0.1734	0.1773
¼	0.2500	28	0.2268	0.2074	0.2113
5/16	0.3125	24	0.2854	0.2629	0.2674
3/8	0.3750	24	0.3479	0.3254	0.3299
7/16	0.4375	20	0.4050	0.3780	0.3834
½	0.5000	20	0.4675	0.4405	0.4459
9/16	0.5625	18	0.5264	0.4964	0.5024
5/8	0.6250	18	0.5889	0.5589	0.5649
¾	0.7500	16	0.7094	0.6763	0.6823
7/8	0.8750	14	0.8286	0.7900	0.7977
1	1.0000	12	0.9459	0.9001	0.9098
1⅛	1.1250	12	1.0709	1.0258	1.0348
1¼	1.2500	12	1.1959	1.1508	1.1598
1⅜	1.3750	12	1.3209	1.2758	1.2848
1½	1.5000	12	1.4459	1.4008	1.4098

[1] Asterisks denote secondary sizes.
[2] All dimensions are in inches.

Extracted from ANSI B1.1-1989 *Unified Inch Screw Threads,* with permission of the publisher, The American Society of Mechanical Engineers.

Table 6

Unified Inch Screw Threads, Coarse Thread Series, UNC

Nominal Size[1]	Major Diameter[2] (D)	Threads/ in.	Pitch Diameter (E)	Minor Diameter, External (Ref.)	Minor Diameter, Internal
1 (0.073)*	0.0730	64	0.0629	0.0544	0.0561
2 (0.086)	0.0860	56	0.0744	0.0648	0.0667
3 (0.099)*	0.0990	48	0.0855	0.0741	0.0764
4 (0.112)	0.1120	40	0.0958	0.0822	0.0849
5 (0.125)	0.1250	40	0.1088	0.0952	0.0979
6 (0.138)	0.1380	32	0.1177	0.1008	0.1042
8 (0.164)	0.1640	32	0.1437	0.1268	0.1302
10 (0.190)	0.1900	24	0.1629	0.1404	0.1449
12 (0.216)*	0.2160	24	0.1889	0.1664	0.1709
¼	0.2500	20	0.2175	0.1905	0.1959
⁵⁄₁₆	0.3125	18	0.2764	0.2464	0.2524
⅜	0.3750	16	0.3344	0.3005	0.3073
⁷⁄₁₆	0.4375	14	0.3911	0.3525	0.3602
½	0.5000	13	0.4500	0.4084	0.4167
⁹⁄₁₆	0.5625	12	0.5084	0.4633	0.4723
⅝	0.6250	11	0.5660	0.5168	0.5266
¾	0.7500	10	0.6850	0.6309	0.6417
⅞	0.8750	9	0.8028	0.7427	0.7547
1	1.0000	8	0.9188	0.8512	0.8647
1⅛	1.1250	7	1.0322	0.9549	0.9704
1¼	1.2500	7	1.1572	1.0799	1.0954
1⅜	1.3750	6	1.2667	1.1766	1.1946
1½	1.5000	6	1.3917	1.3016	1.3196
1¾	1.7500	5	1.6201	1.5119	1.5335
2	2.0000	4½	1.8557	1.7353	1.7594
2¼	2.2500	4½	2.1057	1.9853	2.0094
2½	2.5000	4	2.3376	2.2023	2.2294
2¾	2.7500	4	2.5876	2.4523	2.4794
3	3.0000	4	2.8376	2.7023	2.7294
3¼	3.2500	4	3.0876	2.9523	2.9794
3½	3.5000	4	3.3376	3.2023	3.2294
3¾	3.7500	4	3.5876	3.4523	3.4794
4	4.0000	4	3.8376	3.7023	3.7294

[1] Asterisks denote secondary sizes.
[2] All dimensions are in inches.

Extracted from ANSI B1.1-1989 *Unified Inch Screw Threads,* with permission of the publisher, The American Society of Mechanical Engineers.

Table 7

Square Bolts[1]

Nominal Size[2]		Body Diameter, Maximum (E)	Width Across Flats (F)	Width Across Corners, Maximum (G)	Height (H)	Thread Length[3] (L$_T$)
¼	0.2500	0.260	⅜	0.530	¹¹⁄₆₄	0.750
⁵⁄₁₆	0.3125	0.324	½	0.707	¹³⁄₆₄	0.875
⅜	0.3750	0.388	⁹⁄₁₆	0.795	¼	1.000
⁷⁄₁₆	0.4375	0.452	⅝	0.884	¹⁹⁄₆₄	1.125
½	0.5000	0.515	¾	1.061	²¹⁄₆₄	1.250
⅝	0.6250	0.642	¹⁵⁄₁₆	1.326	²⁷⁄₆₄	1.500
¾	0.7500	0.768	1⅛	1.591	½	1.750
⅞	0.8750	0.895	1⁵⁄₁₆	1.856	¹⁹⁄₃₂	2.000
1	1.0000	1.022	1½	2.121	²¹⁄₃₂	2.250
1⅛	1.1250	1.149	1¹¹⁄₁₆	2.386	¾	2.500
1¼	1.2500	1.277	1⅞	2.652	²⁷⁄₃₂	2.750
1⅜	1.3750	1.404	2¹⁄₁₆	2.917	²⁹⁄₃₂	3.000
1½	1.5000	1.531	2¼	3.182	1	3.250

[1] Threads are Unified Coarse, Fine, or 8-thread series, Class 2A.
[2] All dimensions are in inches.
[3] Thread Length (L$_T$) is a reference dimension intended for calculation purposes only. Basic thread length equals twice the basic thread diameter plus 0.25 inches for nominal bolt lengths up to and including 6 inches, and twice the basic thread diameter plus 0.50 inch for nominal lengths over 6 inches.

Extracted from ANSI B18.2.1-1981(R1992) *Square and Hex Bolts and Screws,* with permission of the publisher, The American Society of Mechanical Engineers.

Table 8

Hex Bolts[1]

Nominal Size[2]		Body Diameter (E)	Width Across Flats (F)	Width Across Corners, Maximum (G)	Height (H)	Thread Length[3] (L$_T$)
¼	0.2500	0.260	⁷⁄₁₆	0.505	¹¹⁄₆₄	0.750
⁵⁄₁₆	0.3125	0.324	½	0.577	⁷⁄₃₂	0.875
⅜	0.3750	0.388	⁹⁄₁₆	0.650	¼	1.000
⁷⁄₁₆	0.4375	0.452	⅝	0.722	¹⁹⁄₆₄	1.125
½	0.5000	0.515	¾	0.866	¹¹⁄₃₂	1.250
⅝	0.6250	0.642	¹⁵⁄₁₆	1.083	²⁷⁄₆₄	1.500
¾	0.7500	0.768	1 ⅛	1.299	½	1.750
⅞	0.8750	0.895	1 ⁵⁄₁₆	1.516	³⁷⁄₆₄	2.000
1	1.0000	1.022	1 ½	1.732	⁴³⁄₆₄	2.250
1 ⅛	1.1250	1.149	1 ¹¹⁄₁₆	1.949	¾	2.500
1 ¼	1.2500	1.277	1 ⅞	2.165	²⁷⁄₃₂	2.750
1 ⅜	1.3750	1.404	2 ¹⁄₁₆	2.382	²⁹⁄₃₂	3.000
1 ½	1.5000	1.531	2 ¼	2.598	1	3.250
1 ¾	1.7500	1.785	2 ⅝	3.031	1 ⁵⁄₃₂	3.750
2	2.0000	2.039	3	3.464	1 ¹¹⁄₃₂	4.250
2 ¼	2.2500	2.305	3 ⅜	3.897	1 ½	4.750
2 ½	2.5000	2.559	3 ¾	4.330	1 ²¹⁄₃₂	5.250
2 ¾	2.7500	2.827	4 ⅛	4.763	1 ¹³⁄₁₆	5.750
3	3.0000	3.081	4 ½	5.196	2	6.250
3 ¼	3.2500	3.335	4 ⅞	5.629	2 ³⁄₁₆	6.750
3 ½	3.5000	3.589	5 ¼	6.062	2 ⁵⁄₁₆	7.250
3 ¾	3.7500	3.858	5 ⅝	6.495	2 ½	7.750
4	4.0000	4.111	6	6.928	2 ¹¹⁄₁₆	8.250

[1] Threads are Unified Coarse, Fine, or 8-thread series, Class 2A.

[2] All dimensions are in inches.

[3] Thread Length (L$_T$) is a reference dimension intended for calculation purposes only. Basic thread length equals twice the basic thread diameter plus 0.25 inches for nominal bolt lengths up to and including 6 inches, and twice the basic thread diameter plus 0.50 inch for nominal lengths over 6 inches.

Extracted from ANSI B18.2.1-1981(R1992) *Square and Hex Bolts and Screws,* with permission of the publisher, The American Society of Mechanical Engineers.

Table 9

Heavy Hex Bolts[1]

Nominal Size[2]		Body Diameter, Maximum (E)	Width Across Flats (F)	Width Across Corners, Maximum (G)	Height (H)	Thread Length[3] (L$_T$)
½	0.5000	0.515	⅞	1.010	¹¹⁄₃₂	1.250
⅝	0.6250	0.642	1¹⁄₁₆	1.227	²⁷⁄₆₄	1.500
¾	0.7500	0.768	1¼	1.443	½	1.750
⅞	0.8750	0.895	1⁷⁄₁₆	1.660	³⁷⁄₆₄	2.000
1	1.0000	1.022	1⅝	1.876	⁴³⁄₆₄	2.250
1⅛	1.1250	1.149	1¹³⁄₁₆	2.093	¾	2.500
1¼	1.2500	1.277	2	2.309	²⁷⁄₃₂	2.750
1⅜	1.3750	1.404	2³⁄₁₆	2.526	²⁹⁄₃₂	3.000
1½	1.5000	1.531	2⅜	2.742	1	3.250
1¾	1.7500	1.785	2¾	3.175	1⁵⁄₃₂	3.750
2	2.0000	2.039	3⅛	3.608	1¹¹⁄₃₂	4.250
2¼	2.2500	2.305	3½	4.041	1½	4.750
2½	2.5000	2.559	3⅞	4.474	1²¹⁄₃₂	5.250
2¾	2.7500	2.827	4¼	4.907	1¹³⁄₁₆	5.750
3	3.0000	3.081	4⅝	5.340	2	6.250

[1] Threads are Unified Coarse, Fine, or 8-thread series, Class 2A.

[2] All dimensions are in inches.

[3] Thread Length (L$_T$) is a reference dimension intended for calculation purposes only. Basic thread length equals twice the basic thread diameter plus 0.25 inches for nominal bolt lengths up to and including 6 inches, and twice the basic thread diameter plus 0.50 inch for nominal lengths over 6 inches.

Extracted from ANSI B18.2.1-1981(R1992) *Square and Hex Bolts and Screws,* with permission of the publisher, The American Society of Mechanical Engineers.

Table 10

Heavy Hex Structural Bolts[1]

Nominal Size[2]		Body Diameter, Maximum (E)	Width Across Flats, Maximum (F)	Width Across Corners, Maximum (G)	Height, Maximum (H)	Thread Length[3] (L_T)	Transition Thread (Y)
½	0.5000	0.515	0.875	1.010	0.323	1.00	0.19
⅝	0.6250	0.642	1.062	1.227	0.403	1.25	0.22
¾	0.7500	0.768	1.250	1.443	0.483	1.38	0.25
⅞	0.8750	0.895	1.438	1.660	0.563	1.50	0.28
1	1.0000	1.022	1.625	1.876	0.627	1.75	0.31
1⅛	1.1250	1.149	1.812	2.093	0.718	2.00	0.34
1¼	1.2500	1.277	2.000	2.309	0.813	2.00	0.38
1⅜	1.3750	1.404	2.188	2.526	0.878	2.25	0.44
1½	1.5000	1.531	2.375	2.742	0.974	2.25	0.44

[1] Threads are Unified Coarse, Fine, or 8-thread series, Class 2A.
[2] All dimensions are in inches.
[3] Thread Length (L_T) is a reference dimension intended for calculation purposes only.

Extracted from ANSI B18.2.1-1981(R1992) *Square and Hex Bolts and Screws,* with permission of the publisher, The American Society of Mechanical Engineers.

Table 11

Square Nuts[1]

Nominal Size[2]		Width Across Flats[1] (F)	Width Across Corners, Maximum (G)	Height (Thickness) (H)
¼	0.2500	⁷⁄₁₆	0.619	⁷⁄₃₂
⁵⁄₁₆	0.3125	⁹⁄₁₆	0.795	¹⁷⁄₆₄
⅜	0.3750	⅝	0.884	²¹⁄₆₄
⁷⁄₁₆	0.4375	¾	1.061	⅜
½	0.5000	¹³⁄₁₆	1.149	⁷⁄₁₆
⅝	0.6250	1	1.414	³⁵⁄₆₄
¾	0.7500	1⅛	1.591	²¹⁄₃₂
⅞	0.8750	1⁵⁄₁₆	1.856	⁴⁹⁄₆₄
1	1.0000	1½	2.121	⅞
1⅛	1.1250	1¹¹⁄₁₆	2.386	1
1¼	1.2500	1⅞	2.652	1³⁄₃₂
1⅜	1.3750	2¹⁄₁₆	2.917	1¹³⁄₆₄
1½	1.5000	2¼	3.182	1⁵⁄₁₆

[1] Threads are Unified Coarse series, Class 2B.
[2] All dimensions are in inches.

Extracted from ANSI B18.2.2-1987(R1993) *Square and Hex Nuts,* with permission of the publisher, The American Society of Mechanical Engineers.

Table 12

Heavy Square Nuts[1]

Nominal Size[2]		Width Across Flats (F)	Width Across Corners, Maximum (G)	Height (Thickness) (H)
¼	0.2500	½	0.707	¼
⁵⁄₁₆	0.3125	⁹⁄₁₆	0.795	⁵⁄₁₆
⅜	0.3750	¹¹⁄₁₆	0.973	⅜
⁷⁄₁₆	0.4375	¾	1.060	⁷⁄₁₆
½	0.5000	⅞	1.237	½
⅝	0.6250	1 ¹⁄₁₆	1.503	⅝
¾	0.7500	1 ¼	1.768	¾
⅞	0.8750	1 ⁷⁄₁₆	2.033	⅞
1	1.0000	1 ⅝	2.298	1
1⅛	1.1250	1 ¹³⁄₁₆	2.563	1⅛
1¼	1.2500	2	2.828	1¼
1⅜	1.3750	2 ³⁄₁₆	3.094	1⅜
1½	1.5000	2 ⅜	3.359	1½

[1] Threads are Unified Coarse series, Class 2B.
[2] All dimensions are in inches.

Extracted from ANSI B18.2.2-1987(R1993) *Square and Hex Nuts,* with permission of the publisher, The American Society of Mechanical Engineers.

Table 13

Hex Nuts[1]

Nominal Size[2]		Width Across Flats (F)	Width Across Corners, Maximum (G)	Height (Thickness), Nuts (H)
¼	0.2500	⁷⁄₁₆	0.505	⁷⁄₃₂
⁵⁄₁₆	0.3125	½	0.577	¹⁷⁄₆₄
⅜	0.3750	⁹⁄₁₆	0.650	²¹⁄₆₄
⁷⁄₁₆	0.4375	¹¹⁄₁₆	0.794	⅜
½	0.5000	¾	0.866	⁷⁄₁₆
⁹⁄₁₆	0.5625	⅞	1.010	³¹⁄₆₄
⅝	0.6250	¹⁵⁄₁₆	1.083	³⁵⁄₆₄
¾	0.7500	1 ⅛	1.299	⁴¹⁄₆₄
⅞	0.8750	1 ⁵⁄₁₆	1.516	¾
1	1.0000	1 ½	1.732	⁵⁵⁄₆₄
1 ⅛	1.1250	1 ¹¹⁄₁₆	1.949	³¹⁄₃₂
1 ¼	1.2500	1 ⅞	2.165	1 ¹⁄₁₆
1 ⅜	1.3750	2 ¹⁄₁₆	2.382	1 ¹¹⁄₆₄
1 ½	1.5000	2 ¼	2.598	1 ⁹⁄₃₂

[1] Threads are Unified Coarse series, Class 2B.
[2] All dimensions are in inches.

Extracted from ANSI B18.2.2-1987(R1993) *Square and Hex Nuts,* with permission of the publisher, The American Society of Mechanical Engineers.

Table 14

Heavy Hex Nuts[1]

Nominal Size[2]		Width Across Flats (F)	Width Across Corners, Maximum (G)	Height (Thickness), Nuts (H)
$1/4$	0.2500	$1/2$	0.577	$15/64$
$5/16$	0.3125	$9/16$	0.650	$19/64$
$3/8$	0.3750	$11/16$	0.794	$23/64$
$7/16$	0.4375	$3/4$	0.866	$27/64$
$1/2$	0.5000	$7/8$	1.010	$31/64$
$9/16$	0.5625	$15/16$	1.083	$35/64$
$5/8$	0.6250	$1\,1/16$	1.227	$39/64$
$3/4$	0.7500	$1\,1/4$	1.443	$47/64$
$7/8$	0.8750	$1\,7/16$	1.660	$53/64$
1	1.0000	$1\,5/8$	1.876	$63/64$
$1\,1/8$	1.1250	$1\,13/16$	2.093	$1\,7/64$
$1\,1/4$	1.2500	2	2.309	$1\,7/32$
$1\,3/8$	1.3750	$2\,3/16$	2.526	$1\,11/32$
$1\,1/2$	1.5000	$2\,3/8$	2.742	$1\,15/32$
$1\,5/8$	1.6250	$2\,9/16$	2.959	$1\,19/32$
$1\,3/4$	1.7500	$2\,3/4$	3.175	$1\,23/32$
$1\,7/8$	1.8750	$2\,15/16$	3.392	$1\,27/32$
2	2.0000	$3\,1/8$	3.608	$1\,31/32$
$2\,1/4$	2.2500	$3\,1/2$	4.041	$2\,13/64$
$2\,1/2$	2.5000	$3\,7/8$	4.474	$2\,29/64$
$2\,3/4$	2.7500	$4\,1/4$	4.907	$2\,45/64$
3	3.0000	$4\,5/8$	5.340	$2\,61/64$
$3\,1/4$	3.2500	5	5.774	$3\,3/16$
$3\,1/2$	3.5000	$5\,3/8$	6.207	$3\,7/16$
$3\,3/4$	3.7500	$5\,3/4$	6.640	$3\,11/16$
4	4.0000	$6\,1/8$	7.073	$3\,15/16$

[1] Threads are Unified Coarse, Fine, or 8-thread series, Class 2B. Unless otherwise specified, coarse thread series is furnished.
[2] All dimensions are in inches.

Extracted from ANSI B18.2.2-1987(R1993) *Square and Hex Nuts,* with permission of the publisher, The American Society of Mechanical Engineers.

Tables 15 through 17 provide information extracted from the ANSI B series of standards for machine screws. Fig. 2 provides a graphic description of the parts described in the tables. The lettered dimensions in Fig. 2 refer to dimensions given in the tables.

Fig. 2

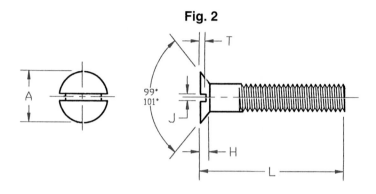

SLOTTED 100° FLAT COUNTERSUNK HEAD MACHINE SCREW

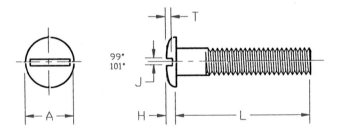

SLOTTED ROUND HEAD MACHINE SCREW

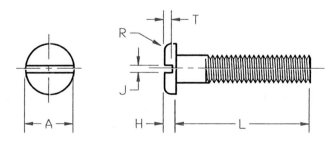

SLOTTED PAN HEAD MACHINE SCREW

Extracted from ANSI B18.6.3-1972(R1991) *Machine Screws and Machine Screw Nuts,* with permission of the publisher, The American Society of Mechanical Engineers.

Table 15

Slotted 100° Flat Countersunk Head Machine Screws

Nominal Size[1]		Head Diameter, Maximum (A)	Head Height (H)	Slot Width, Maximum (J)	Slot Depth, Maximum (T)
0000	0.0210	.043	.009	.008	.008
000	0.0340	.064	.014	.012	.011
00	0.0470	.093	.020	.017	.013
0	0.0600	.119	.026	.023	.013
1	0.0730	.146	.031	.026	.016
2	0.0860	.172	.037	.031	.019
3	0.0990	.199	.043	.035	.022
4	0.1120	.225	.049	.039	.024
6	0.1380	.279	.060	.048	.030
8	0.1640	.332	.072	.054	.036
10	0.1900	.385	.083	.060	.042
1/4	0.2500	.507	.110	.075	.055
5/16	0.3125	.635	.138	.084	.069
3/8	0.3750	.762	.165	.094	.083

[1] All dimensions are in inches.

Extracted from ANSI B18.6.3-1972(R1991) *Machine Screws and Machine Screw Nuts,* with permission of the publisher, The American Society of Mechanical Engineers.

Table 16

Slotted Round Head Machine Screws

Nominal Size[1] or Basic Screw Diameter		Head Diameter, Maximum (A)	Head Height, Maximum (H)	Slot Width, Maximum (J)	Slot Depth, Maximum (T)
0000	0.0210	.041	.022	.008	.017
000	0.0340	.062	.031	.012	.018
00	0.0470	.089	.045	.017	.026
0	0.0600	.113	.053	.023	.039
1	0.0730	.138	.061	.026	.044
2	0.0860	.162	.069	.031	.048
3	0.0990	.187	.078	.035	.053
4	0.1120	.211	.086	.039	.058
5	0.1250	.236	.095	.043	.063
6	0.1380	.260	.103	.048	.068
8	0.1640	.309	.120	.054	.077
10	0.1900	.359	.137	.060	.087
12	0.2160	.408	.153	.067	.096
1/4	0.2500	.472	.175	.075	.109
5/16	0.3125	.590	.216	.084	.132
3/8	0.3750	.708	.256	.094	.155
7/16	0.4375	.750	.328	.094	.196
1/2	0.5000	.813	.355	.106	.211
9/16	0.5625	.938	.410	.118	.242
5/8	0.6250	1.000	.438	.133	.258
3/4	0.7500	1.250	.547	.149	.320

[1] All dimensions are in inches.

Extracted from ANSI B18.6.3-1972(R1991) *Machine Screws and Machine Screw Nuts,* with permission of the publisher, The American Society of Mechanical Engineers.

Table 17

Slotted Pan Head Machine Screws

Nominal Size[1] or Basic Screw Diameter		Head Diameter, Maximum (A)	Head Height, Maximum (H)	Head Radius, Maximum (R)	Slot Width, Maximum (J)	Slot Depth, Maximum (T)
0000	0.0210	.042	.016	.007	.008	.008
000	0.0340	.066	.023	.010	.012	.012
00	0.0470	.090	.032	.015	.017	.016
0	0.0600	.116	.039	.020	.023	.022
1	0.0730	.142	.046	.025	.026	.027
2	0.0860	.167	.053	.035	.031	.031
3	0.0990	.193	.060	.037	.035	.036
4	0.1120	.219	.068	.042	.039	.040
5	0.1250	.245	.075	.044	.043	.045
6	0.1380	.270	.082	.046	.048	.050
8	0.1640	.322	.096	.052	.054	.058
10	0.1900	.373	.110	.061	.060	.068
12	0.2160	.425	.125	.078	.067	.077
1/4	0.2500	.492	.144	.087	.075	.087
5/16	0.3125	.615	.178	.099	.084	.106
3/8	0.3750	.740	.212	.143	.094	.124
7/16	0.4375	.863	.247	.153	.094	.142
1/2	0.5000	.987	.281	.175	.106	.161
9/16	0.5625	1.041	.315	.197	.118	.179
5/8	0.6250	1.172	.350	.219	.133	.197
3/4	0.7500	1.435	.419	.263	.149	.234

[1] All dimensions are in inches.

Extracted from ANSI B18.6.3-1972(R1991) *Machine Screws and Machine Screw Nuts,* with permission of the publisher, The American Society of Mechanical Engineers.

Table 18 provides information extracted from the ANSI B series of standards for cotter pins. Fig. 3 provides a graphic description of the parts described in the table. The lettered dimensions in Fig. 3 refer to dimensions given in Table 18.

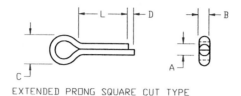

EXTENDED PRONG SQUARE CUT TYPE

Fig. 3

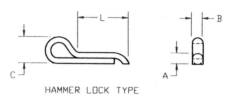

HAMMER LOCK TYPE

Extracted from ANSI B18.8.1-1972(R1983) *Clevis Pins and Cotter Pins,* with permission of the publisher, The American Society of Mechanical Engineers.

Table 18

Cotter Pins

Nominal Size[1]	Diameter, Maximum (B)	Head Diameter, Minimum (C)	Prong Length, Minimum (L)	Hole Size
1/32	.032	0.06	.01	.047
3/64	.048	0.09	.02	.062
1/16	.060	0.12	.03	.078
5/64	.076	0.16	.04	.094
3/32	.090	0.19	.04	.109
7/64	.104	0.22	.05	.125
1/8	.120	0.25	.06	.141
9/64	.134	0.28	.06	.156
5/32	.150	0.31	.07	.172
3/16	.176	0.38	.09	.203
7/32	.207	0.44	.10	.234
1/4	.225	0.50	.11	.266
5/16	.280	0.62	.14	.312
3/8	.335	0.75	.16	.375
7/16	.406	0.88	.20	.438
1/2	.473	1.00	.23	.500
5/8	.598	1.25	.30	.625
3/4	.723	1.50	.36	.750

[1] All dimensions are in inches.

Extracted from ANSI B18.8.1-1972(R1983) *Clevis Pins and Cotter Pins,* with permission of the publisher, The American Society of Mechanical Engineers.

Tables 19 and 20 provide information extracted from the ANSI B series of standards for small solid rivets. Fig. 4 provides a graphic description of the parts described in the tables. The lettered dimensions in Fig. 4 refer to dimensions given in the tables.

Fig. 4

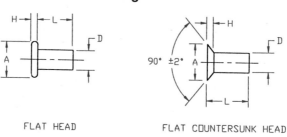

FLAT HEAD   FLAT COUNTERSUNK HEAD

Extracted from ANSI B18.1.1-1972(R1989) *Small Solid Rivets,* with permission of the publisher, The American Society of Mechanical Engineers.

Table 19

Small Solid Rivets: Flat Head

Nominal Shank Diameter[1] (D)	Head Diameter, Maximum (A)	Head Height, Maximum (H)
1/16	.140	.027
3/32	.200	.038
1/8	.260	.048
5/32	.323	.059
3/16	.387	.069
7/32	.453	.080
1/4	.515	.091
9/32	.579	.103
5/16	.641	.113
11/32	.705	.124
3/8	.769	.135
13/32	.834	.146
7/16	.896	.157

[1] All dimensions are in inches.

Extracted from ANSI B18.1.1-1972(R1989) *Small Solid Rivets,* with permission of the publisher, The American Society of Mechanical Engineers.

Table 20

Small Solid Rivets: Flat Countersunk Head

Nominal Shank Diameter[1] (D)	Head Diameter, Maximum (A)	Head Height, Maximum[2] (H)
$\frac{1}{16}$	.118	.027
$\frac{3}{32}$	.176	.040
$\frac{1}{8}$	.235	.053
$\frac{5}{32}$	.293	.066
$\frac{3}{16}$	.351	.079
$\frac{7}{32}$	.413	.094
$\frac{1}{4}$	.469	.106
$\frac{9}{32}$	.528	.119
$\frac{5}{16}$	.588	.133
$\frac{11}{32}$	.646	.146
$\frac{3}{8}$	.704	.159
$\frac{13}{32}$	.763	.172
$\frac{7}{16}$	.823	.186

[1] All dimensions are in inches.

[2] Given for reference only. Variations in head height are controlled by the head and shank diameters and the included angle of the head.

Extracted from ANSI B18.1.1-1972(R1989) *Small Solid Rivets*, with permission of the publisher, The American Society of Mechanical Engineers.

Appendix B
Release 10 Cross Reference

AutoCAD Release 10 (R10) and Release 12 (R12) differ in many respects. *AutoCAD Drafting* was written for Release 12. This appendix cross-references Release 12 with Release 10 chapter by chapter. When a chapter discusses an area in which R10 performs differently than R12, an R10 icon appears in the margin of the text page. Refer to this cross reference to find out more about the difference in each case.

R10 is different from R12 in many minor details. Examples of these differences include:

- R10 does not use dialogue boxes as frequently as does R12. When dialogue boxes are used in R10, they appear considerably different.
- The command structure is slightly different in R10 than it is in R12. Therefore, the wording and options of many of the prompts initiated by various AutoCAD commands differ in the two releases.
- Although the graphics screens of the releases look similar, the pull-down menus and screen menus contain different commands and command arrangements.

These and other detail differences are not included in this cross reference. The differences described below are confined to those that affect the command sequence, the logic of the drawing process, or the drawing itself. If a difference described in one chapter is encountered in a later chapter, it is not described or even pointed out again.

The following differences are given in general terms; specific details are provided only when necessary. If you have trouble understanding the differences, or if your computer does not respond as you expect it to, consult the reference manuals that came with your version of AutoCAD or ask your instructor for help.

Chapter 1

Page	Description
No differences	

Chapter 2

Page	Description
No differences	

Chapter 3

Page	Description
No differences	

Chapter 4

Page	Description
104	In R10, AutoCAD begins at the Main Menu. The first five options (0 through 4) deal with drawings directly. Options 5 through 8 are concerned with configuration and files. To reach the drawing screen, you must either begin a new drawing or edit an existing drawing.
112	The DDUNITS command does not exist in R10.
117	"2" and "X2" are not available in R10. Use the LTSCALE command for all linetype adjustments.
119	No dialogue box appears in R10. Instead, AutoCAD responds with *File to search \<AutoCAD\>:*.
120	To change a Layer Control dialogue box entry, you must pick each entity separately, not as an entire layer.
123	At the start of a new AutoCAD session, AutoCAD asks you to specify the name of your new drawing. There are two ways to save the drawing:

- Type SAVE to save and continue with the drawing session or to save the drawing with a new name.

- Type END to save the drawing with its current name and exit AutoCAD.

124 To begin a new drawing with the ACAD.DWG prototype drawing, select option 1, "Begin a NEW drawing," from the Main Menu. At the *Enter NAME of drawing:* prompt, type the name of your new drawing. To begin a new drawing with a custom prototype drawing, type the name of your new drawing followed by an equals sign (=) and the prototype name. Do not include spaces before or after the equals sign. For example:

Enter NAME of drawing:
BRACKET=START

This creates a new drawing named BRACKET patterned after a prototype named START.

125 To open an existing drawing from the Main Menu, type "2" and enter the name of the drawing including its path, if necessary.

Chapter 5

Page	Description
136	The Drawing Aids dialogue box does not exist in R10. Set Ortho, Snap, and Grid from the keyboard using the ORTHO, SNAP, and GRID commands.
152	You can select both a printer and a plotter from the Main Menu. Enter option 5, "Configure AutoCAD," and follow the prompts to select and configure your printer and/or plotter. Consult your instructor before changing any of the existing configurations. To get a hard copy of your drawing, select PRINT or PLOT, depending upon your device, from the FILE pull-down menu. Follow the prompts.

Chapter 6

Page	Description
170	The Running Object Snap dialogue box does not exist in R10. Set the running object snap from the keyboard using the OSNAP command.
179	The RECTANG command does not exist in R10.

Chapter 7

Page	Description
186	R10 allows you to change PDMODE and PDSIZE only by way of the SETVAR command.
193	ALL is not an acceptable response to the *Select objects:* prompt in R10. Use a selection window instead.
202	Grips do not exist in R10.

Chapter 8

Page	Description
230	Paper space and the commands associated with it do not exist in R10. You can create viewports in model space (the only space recognized by R10) using the VPORTS command.

Chapter 9

Page	Description
247	To change the current font in R10, activate the pull-down menus and pick Options and Fonts.... To select one of the 20 fonts and symbols, pick the box next to it.
256	The DDEDIT command does not exist in R10. You cannot edit text directly. You have to erase it and then re-enter it using the TEXT command.

Chapter 10

Page	Description
265	Dimensioning variables are different in R10. Consult the AutoCAD reference manuals that came with your software for a list of dimensioning variables and their meanings. You cannot change the value of a dimensioning variable by entering the variable name at the *Command:* prompt. You must enter the SETVAR or DIM command to change the value.
275	The Dimension Styles and Variables dialogue box does not exist in R10. Ticks replace arrows when the value of the variable that controls tick size (DIMTSZ) is not zero. Dots are not available except to be inserted as blocks.
304	The TEDIT command does not exist in R10.

Chapter 11

Page	Description
340	The BHATCH command is not available. To display the hatch patterns visually, pick the Select Hatch Pattern dialogue box from the pull-down menu HATCH command.

Chapter 12

Page	Description
No differences	

Chapter 13

Page	Description
397	You cannot add the DD prefix to attribute commands to get dialogue boxes in R10. However, you can edit attributes with the DDATTE command.

Chapter 14

Page	Description
No differences	

Chapter 15

Page	Description
No differences	

Chapter 16

Page	Description
No differences	

Chapter 17

Page	Description
558	The double line routine does not exist in R10. Use a polyline and offset it to obtain the proper wall thickness.

Chapter 18

Page	Description
574	Solid modeling is not available in R10.
581	Shading is not available in R10.
582	Rendering is not available in R10.
585	Region modeling is not available in R10.
586	The Advanced Modeling Extension (AME) software package is not an available option in R10.

Appendix C
Release 11 Cross Reference

AutoCAD Release 11 (R11) and Release 12 (R12) differ in many respects. *AutoCAD Drafting* was written for Release 12. This appendix cross-references Release 12 with Release 11 chapter by chapter. When a chapter discusses an area in which R11 performs differently than R12, an R11 icon appears in the margin of the text page. Refer to this cross reference to find out more about the difference in each case.

R11 is different from R12 in many minor details. Examples of these differences include:

- R11 dialogue boxes may appear different from those in R12.
- The command structure is slightly different in R11 than it is in R12. Therefore, the wording and options of many of the prompts initiated by various AutoCAD commands differ in the two releases.
- Although the graphics screens of the releases look similar, the pull-down menus and screen menus contain different commands and command arrangements.

These and other detail differences are not included in this cross reference. The differences described below are confined to those that affect the command sequence, the logic of the drawing process, or the drawing itself. If a difference described in one chapter is encountered in a later chapter, it is not described or even pointed out again.

The following differences are given in general terms; specific details are provided only when necessary. If you have trouble understanding the differences, or if your computer does not respond as you expect it to, consult the reference manuals that came with your version of AutoCAD or ask your instructor for help.

Chapter 1

Page	Description
No differences	

Chapter 2

Page	Description
No differences	

Chapter 3

Page	Description
No differences	

Chapter 4

Page	Description
104	In R11, AutoCAD begins at the Main Menu. The first five options (0 through 4) deal with drawings directly. Options 5 through 9 are concerned with configuration and files. To reach the drawing screen, you must either begin a new drawing or edit an existing drawing.
112	The DDUNITS command does not exist in R11.
120	To change a Layer Control dialogue box entry, you must pick each entity separately, not as an entire layer.
123	At the start of a new AutoCAD session, AutoCAD asks you to specify the name of your new drawing. There are two ways to save the drawing:

- Type SAVE to save and continue with the drawing session or to save the drawing with a new name.

- Type END to save the drawing with its current name and exit AutoCAD.

124 To begin a new drawing with the ACAD.DWG prototype drawing, select option 1, "Begin a NEW drawing," from the Main Menu. At the *Enter NAME of drawing:* prompt, type the name of your new drawing. To begin a new drawing with a custom prototype drawing, type the name of your new drawing followed by an equals sign (=) and the prototype name. Do not include spaces before or after the equals sign. For example:

Enter NAME of drawing:
BRACKET=START

This creates a new drawing named BRACKET patterned after a prototype named START.

125 To open an existing drawing from the Main Menu, type "2" and enter the name of the drawing including its path, if necessary.

Chapter 5

Page	Description

152 You can select both a printer and a plotter from the Main Menu. Enter option 5, "Configure AutoCAD," and follow the prompts to select and configure your printer and/or plotter. Consult your instructor before changing any of the existing configurations. To get a hard copy of your drawing, select PRINT or PLOT, depending upon your device, from the FILE pull-down menu. Follow the prompts.

Chapter 6

Page	Description

170 The Running Object Snap dialogue box does not exist in R11. Set the running object snap from the keyboard using the OSNAP command.

179 The RECTANG command does not exist in R11.

Chapter 7

Page	Description

193 ALL is not an acceptable response to the *Select objects:* prompt in R11. Use a selection window instead.

202 Grips do not exist in R11.

Chapter 8

Page	Description

No differences

Chapter 9

Page	Description

247 To change the current font in R11, activate the pull-down menus and pick Options, DTEXT OPTIONS, Text Font. To select one of the 20 fonts and symbols, pick the box next to it. You can also control text alignment, text height, and text rotation from the Options, DTEXT OPTIONS pull-down menu.

Chapter 10

Page	Description

275 The Dimension Styles and Variables dialogue box does not exist in R11. Ticks replace arrows when the value of the DIMTSZ variable is not zero. Dots can be inserted only as blocks.

Chapter 11

Page	Description

338 You can select the HATCH command using the keyboard, the pull-down menus, or the screen menu. Select Hatch Options on the Options pull-down menu to display the hatch pattern and style, as well as to set the hatch scale and hatch angle.

340 The BHATCH command does not exist in R11.

Chapter 12

Page   Description

No differences

Chapter 13

Page   Description

397   You cannot add the DD prefix to attri-
       bute commands to get dialogue boxes in
       R11. However, you can edit attributes
       with the DDATTE command.

Chapter 14

Page   Description

No differences

Chapter 15

Page   Description

No differences

Chapter 16

Page   Description

No differences

Chapter 17

Page   Description

558   DLINE is an AutoLISP routine in R11.
       Load it from the *Command:* prompt by
       typing:

       (load"dline")

       Type the command exactly as shown, includ-
       ing the parentheses and quotation marks.

Chapter 18

Page   Description

582   Rendering is not available in R11.

585   Region modeling is not available as a sub-
       set of the Advanced Modeling Extension
       (AME). The commands which make up
       region modeling in R12 are included in the
       optional AME software package for R11.

Index

Photo Credits

AimaFast: 487

Tony Aimer, Tony Aimer & Associates: 347

Artwork Courtesy of Alvin & Company, Inc.: 73(b), 74(l)

American Airlines: 413(b)

American Iron & Steel Institute: 528(c)

Anchor Engineering Services, Inc.: 158, 159

Arnold & Brown: 9, 348, 488

Autodesk Inc.: 127, 208, 380, 570, 575(r)

Roger B. Bean: 6, 8, 17(tr), 160, 162(tl), 408, 413(tl,tr)

Keith M. Berry: 21(t), 184

The Bettmann Archive: 17(tl), 19(t), 210(t)

Boston Gear: 514, 515

Michele Bousquet: 517

Ryan K. Brown: 432, 438, 439, 441, 447, 448, 451, 526, 550(t), 551, 552, 554, 555

The Ceco Corporation, Oakbrook Terrace, IL: 533

Cincinnati Milacron: 465

Circle Design: 17(c,b), 18(l,r), 28, 37, 39(b), 44(r), 47(b), 49, 50, 59, 67, 68, 69(t,b), 71, 72, 73(t), 74(r), 75, 77(t,b), 78, 84, 86, 111(t), 121, 219(b), 230, 266, 268, 403, 414, 415, 461

Control Data: 7

Design Works Inc.: 525(b)

DeVlieg Machine Co.: 320

Ann Garvin: 528(t,b), 530(c), 537(b), 538

Roy Hirshkowitz: 527, 582

Dennis Huette: 387, 393

Sung Su Hur: 391, 452, 462, 466, 467, 470, 478, 479, 481

The Image Bank Chicago/Harald Sund: 32, 34(br)

IMI Cash Valve, Inc.: 383, 384(t), 385, 386, 389, 394, 396

Charles P. Johnson: 412, 433, 434, 435, 436, 437(t), 443, 446(b), 449, 450

Keller-Williams Realty, Denton, TX; Meg Boliver: 537(b)

Ketiv Industries: 569

Koehler: 540

Logitech, Inc.: 105(l,br)

Yisong Luo: 384(b), 392, 463, 464, 473, 474, 475(b), 480, 482, 483, 485

Joe MacRae: 63

The Mayline Co.: 70(t)

METALSOFT INC.: 457

Mishima: 102

Monroe, Inc.: 379

Monticello/Thomas Jefferson Memorial Foundation: 19(b)

National Bureau of Standards: 535

Noah Herman Sons/House of the 1990s, Roger B. Bean: 537(t)

Brent Phelps: 5, 21(b), 25(t,b), 26, 34(bl), 85, 88, 211, 244, 262, 458, 460, 575(l)

Michelle Pillers, GVO, Inc.: 594

Cloyd Richards: 20, 34(t), 35, 52, 162(b), 212(l), 218, 219(t), 220, 225, 265, 297, 303, 508, 520

Jesse Robbins: 523(b), 545(b), 546, 548

SACON, Adelaide, S. Australia: 518

Ed Sargent: 101

Staedtler, Inc.: 64

Summagraphics: 105(c, tr), 130

Paul Swanberg: 31, 522, 524, 525(t), 541, 543, 544, 545(t), 549, 550(b), 553(t,b), 556(t), 557

Teledyne Post: 66

Thet-Win: 261

TJI® Joist: 530(b)

Dave Tosti-Lane, Cornish College of the Arts: 406, 407

William J. Townsend, Bushido Bicycles: 207

Photo Courtesy of TransCanada Pipelines: 319

UPI/Bettmann: 14, 16

View-Master®: 210(b)

Melanie Yamamoto: 129

We gratefully acknowledge the following individuals for their technical assistance: Benedict Wong, Ph.D., University of North Texas, Denton, TX; and Ryan Brown, Ed.D., Illinois State University, Normal, IL.